WITHDRAWN
UTSA Libraries

PHILOSOPHY OF MATHEMATICS
IN THE TWENTIETH CENTURY

PHILOSOPHY OF MATHEMATICS IN THE TWENTIETH CENTURY

SELECTED ESSAYS

Charles Parsons

HARVARD UNIVERSITY PRESS

Cambridge, Massachusetts

London, England

2014

Copyright © 2014 by the President and Fellows of Harvard College
All rights reserved
Printed in the United States of America

Library of Congress Cataloging-in-Publication Data

Parsons, Charles, 1933–
   [Essays. Selections]
   Philosophy of mathematics in the twentieth century : selected essays / Charles Parsons.
     pages cm
   Includes bibliographical references and index.
   ISBN 978-0-674-72806-6 (alk. paper)
   1. Mathematics—Philosophy. 2. Logic, Symbolic and mathematical. I Title.
QA9.2.P372   2014
510.1—dc23   2013030762

Library
University of Texas
at San Antonio

For Solomon Feferman, Wilfried Sieg, and William Tait

CONTENTS

The present volume is the second of two volumes of selected essays about other philosophers, the first being *From Kant to Husserl* (2012). The division is essentially chronological, although the present volume is also more focused on philosophy of mathematics. The writings discussed range in date from the first decade of the twentieth century into the first decade of the twenty-first. The arrangement is roughly chronological with respect to subject matter. I will not undertake here to sketch the content of individual essays. The Introduction that follows will undertake to do that at least to a certain extent, while placing them in the context of the broader developments of which the matters discussed in these essays are a part.

Although I have followed more of the development of the foundations of mathematics in the twentieth century than of the (vast) development of philosophy from Kant to Husserl, this volume, like its predecessor, consists of essays and leaves out much of the relevant history considered as a whole. Important figures whose contribution was mainly mathematical, such as Zermelo, Skolem, Herbrand, and many post-war figures, are left out because the essays concern philosophy of mathematics. I have also been selective in the parts of the philosophical history I have discussed. The most significant omission from the early part of the century is undoubtedly Bertrand Russell. I have also not undertaken to write about Wittgenstein, in part because so many other able philosophers, some of them close to me personally, have devoted a major part of their energy to understanding his work. Furthermore, Wittgenstein's philosophy of mathematics, whatever its intrinsic interest might be, stands apart from the main development. Another omission is that of the Vienna Circle and especially Rudolf Carnap. That omission is more serious because I have written about two critics of their views,

Kurt Gödel and W. V. Quine. It is not possible for me to rectify this at this late date, but in the Postscript to Essay 6 I do comment briefly on recent discussions of Gödel's arguments by writers whose interest centers on Carnap.

I have divided the essays into two parts. The individuals who are the principal subjects of Essays 1 through 7 belong to a remarkable sequence of philosophizing mathematicians.[1] This sequence begins around 1870 with Dedekind, Cantor, and their great opponent Leopold Kronecker.[2] This phenomenon was a product of the revolution in mathematics of the late nineteenth century and the foundational problems that it gave rise to, which shaped discussion of the foundations of mathematics down to at least 1930 and still influences it today. In many ways Kurt Gödel was the last of this sequence, although one could argue that Alonzo Church and A. M. Turing both belong to it.

Gödel looms large in this collection not only because of his stature but because at the end of 1986 I became one of the editors of his posthumous writings and thus became involved with documents concerning his thought that were unpublished at the time. Assessments of his standing as a philosopher vary quite widely. I try to present my own, unavoidably briefly, in Essay 4, a survey article written not long after the editors' work ended.

The second part is entitled "Contemporaries," because the careers of each of the subjects (W. V. Quine, Hao Wang, Hilary Putnam, and William Tait) have overlapped significantly with my own. I have known each fairly well personally.[3] It is admittedly a stretch to list Quine as a contemporary, because he was twenty-five years older. But he lived for nearly forty years after I ceased to be his student, and he was active as a philosopher for nearly all that time. Furthermore, in my writing about him (except in Essay 8) I have treated him more as an older con-

---

[1] As the opening sentence in Essay 6 states, I view that essay as primarily about Gödel.

[2] Others who belong in this company but do not figure prominently in these essays are Frege, Poincaré, and Tarski. Two essays on Frege were for chronological reasons placed in *From Kant to Husserl*. Whitehead is better described as a mathematician turned philosopher. Russell, while always a philosopher, also functioned effectively as a mathematician from 1900 or so until at least 1910.

[3] I never met or saw Gödel, but I did have two encounters with Bernays (in 1956 and 1972) and a little correspondence with him.

temporary than as a historical figure.[4] Quine and Wang were important teachers of mine, though as regards my dissertation, Quine was subordinate to Burton Dreben, and Wang offered only a little (but strategic) advice in correspondence. Hilary Putnam, whom I first met in 1959, was an active colleague from 1989 until his retirement in 2000. Quine responded to Essays 6 and 9 and reviewed *Mathematics in Philosophy.* Putnam has responded to Essay 11 and to another essay that could not be included here.[5] I have had exacting comments from Tait on what I have written about his work.

Thus my debts to the authors treated in Essays 8 through 12 are great and go back much further than to the time at which the essays were written. Concerning Quine and Wang (as well as Dreben), I will not repeat what I have written in earlier publications.[6] Putnam's writings and conversations had been a source of instruction and stimulus long before I became his colleague. In December 1965 I commented on an early version of "Mathematics without Foundations" in a symposium at a meeting of the Eastern Division of the American Philosophical Association. I date from that experience the idea of using modality to interpret mathematical existence. That idea does not play a role in any of these essays, but it is prominent in some earlier writings of mine. I first got to know Bill Tait at the legendary Summer Institute of Symbolic Logic at Cornell University in 1957, and he has been instructing me about the foundations of mathematics and its history ever since.

Needless to say, I have many other debts, some of which are mentioned in the places cited in note 6. On Brouwer, my principal debts are to Dirk van Dalen, my student Richard Tieszen, and especially Mark van Atten. Much earlier I had stimulating conversations with Johan de Iongh. Evidently the work of William Ewald, Michael Hallett, and Wilfried Sieg has instructed me about Hilbert, and I have learned about

---

[4] A happy consequence of his longevity was that he could be both. Essay 9 was written for a conference on Quine's philosophy at Washington University, St. Louis, in 1988. He was able to comment on speakers whose approach was that of a historian of analytical philosophy. It was as if a Kant scholar could question Kant!

[5] "Putnam on Realism and 'Empiricism'."

[6] See *Mathematics in Philosophy,* p. 11, *Mathematical Thought and Its Objects,* p. xvi, and Parsons and Link, *Hao Wang,* p. 3, n.2. I learned much about the history of logic and foundations at the "logic lunches" at Harvard in 1962–1964, where Dreben, Quine, and Wang were regular participants, and Jean van Heijenoort attended from time to time.

Hilbert's program and about Bernays from Sieg for many years. On Gödel, my principal debts have been to Wang and my fellow editors of Gödel's posthumous writings, especially John Dawson, Solomon Feferman, Warren Goldfarb, and Sieg. In addition I owe much to Tony Martin, Mark van Atten, Richard Tieszen, and especially Peter Koellner. As regards Quine, James Higginbotham, Isaac Levi, and especially Sidney Morgenbesser were helpful during my time at Columbia University. At Harvard one must close one's ears not to find instruction about Quine. Dreben instructed me even when I was not at Harvard. Quine himself, Putnam, Goldfarb, and my student Øystein Linnebo have all helped my understanding. Among others my greatest debt is to Dagfinn Føllesdal. On Putnam, I owe most to the man himself. The breadth of his philosophy far surpasses my ability to comment on it, but two conferences on his work, in Dublin in 2007 and at Brandeis and Harvard in 2011, organized by Maria Baghramian and Alan Berger respectively, have been very instructive. On Wang and Tait, again my principal debt is to the authors, but Montgomery Link on Wang and Koellner and my student Douglas Marshall on Tait deserve mention.

I am grateful to Lindsay Waters, editor at Harvard University Press, for his encouragement of my complex and rather slow project and to his assistant, Shanshan Wang, for her help and advice. As regards the arrangement of the collection, what essays of mine should be included, I am indebted to Waters and to anonymous referees. Thanks once again to John Donohue of Westchester Publishing Services and the copy editor, Ellen Lohman, for their careful work and attention to detail. Thanks also to Zeynep Soysal for preparing the index.

Essay 1 appears for the first time in this volume. The other essays have been reprinted unrevised except for bringing the references into the style that I prefer, but some notes have been added, always in square brackets. Several essays have been provided with Postscripts, where issues raised in later literature seemed to me to require comment.

Sol Feferman, Wilfried Sieg, and Bill Tait have been valued friends and discussion partners on the foundations of mathematics over a period ranging from thirty-five to fifty years. Each has helped to compensate for the limitations of my knowledge, especially on the mathematical side. This volume is dedicated to them.

PHILOSOPHY OF MATHEMATICS
IN THE TWENTIETH CENTURY

INTRODUCTION

The writings discussed in these essays range in date from the first decade of the twentieth century into the first decade of the twenty-first. In what follows I will make some brief remarks about the larger history of which the work discussed in these essays is a part and briefly indicate where the essays fit into that history.

Much thought about the foundations of mathematics in the twentieth century, especially before the Second World War, can reasonably be viewed as a rather direct continuation of the reflection on the revolution in mathematics of the second half of the nineteenth century. However, the beginning of the new century does mark a new turn. A public event that marks it is the famous lecture of David Hilbert on mathematical problems at the International Congress of Mathematicians in Paris in 1900. The first problem was the continuum problem: what is the power, or cardinality, of the real numbers? The second problem was to give a proof of the consistency of the theory of real numbers. Both problems have given rise to a lot of mathematical research and philosophical reflection. Neither can count as solved to this day.

Hilbert's publication on foundations began with his *Grundlagen der Geometrie* of 1899, and although we tend to identify his concern with consistency proofs with the proof-theoretic program of the 1920s and his response at that time to the radical challenge posed by intuitionism, it actually began in the earlier period, as is shown by his listing a proof of the consistency of the real numbers as his second problem. This concern might very roughly be said to grow out of the broadly structuralist approach to mathematics of which the *Grundlagen* is a prime exemplar.[1]

---

[1] The background of Hilbert's concern with foundations and the earlier development of his thinking are beautifully analyzed in Ewald, "Hilbert's Wide Program."

However, quite apart from Hilbert's motivations, the development of the foundations of mathematics in the ensuing decades was decisively influenced by two developments of the first decade. The first was the coming to light of the paradoxes of logic and set theory. Cantor had long been aware that he could not regard all "multiplicities" as sets, and contradictions resulting from the assumption of a set of all cardinals or ordinals were discussed in the late 1890s. Russell's paradox, discovered by him in 1901 and communicated to Frege in 1902, had a more definite impact, since Russell proved that Frege's logic is inconsistent, and this was noted in print in 1903, by Russell in *Principles of Mathematics* and by Frege in the appendix to the second volume of his *Grundgesetze der Arithmetik*. It spurred Russell to undertake the construction of a comprehensive logic that would "solve" paradoxes, not only those based on set-theoretic ideas but a much wider range that included what are now called the semantical paradoxes.

The second was the beginning of the constructivist tendency known as intuitionism. The criticism of some assumptions of set theory, particularly Zermelo's axiom of choice, by French mathematicians early in the decade was soon followed by a more radical constructivist critique by L. E. J. Brouwer, which extended to basic logic. He gave the name intuitionism not only to his views but also to the distinctive kind of mathematics that he developed.[2]

By the time of the First World War, Hilbert, Russell, and Brouwer could have been identified as the dominant figures in the foundations of mathematics, at least as the sources of the principal ideas. A principal new element in the 1920s was Hilbert's program of formalizing mathematical theories and proving consistency by his finitary method. Hilbert was the leader of an actual school pursuing this program. His junior collaborator Paul Bernays played an important role in the mathematical work and a role probably as great as Hilbert's in formulating the accompanying philosophy. Thus we came to the situation where logicism, "formalism," and intuitionism were seen as the principal and maybe the only options in foundations.[3] Logicism at the time meant

---

[2] Leopold Kronecker had developed constructivist views a generation earlier, but he did not carry out a critique of logic comparable to Brouwer's. Nonetheless his ideas posed a challenge to Hilbert and had a positive influence on Hilbert's conception of the finitary method.

[3] "Formalism" is a somewhat misleading label for Hilbert's views and program, for reasons that others have presented at length.

Whitehead and Russell's *Principia Mathematica* and attempts to defend or improve it. Probably the leading heir of that tradition at the time was F. P. Ramsey, whose views, like Russell's later views and those of the Vienna Circle, were greatly influenced by the Wittgenstein of the *Tractatus*. The main new development in intuitionism was Brouwer's development of intuitionistic analysis. Somewhat acrimonious debates, particularly between the Hilbert school and intuitionism, marked the period.

About the development of logicism there is little in these essays.[4] Although Hilbert's program and views are briefly discussed in Essays 1 and 3 and in connection with Gödel, I have not tried to match the scholarship on Hilbert of others, especially William Ewald, Michael Hallett, and Wilfried Sieg. Bernays figures more prominently, but with reference to this period only as an instance of the "Kantian legacy" discussed in Essay 1. However, he had a long career after the effective end of the Hilbert school.[5] Essay 3 concentrates on the development of his philosophical views in later years. That development shows a continuing responsiveness to developments in foundational research as well as a strong general philosophical interest, in particular a movement away from the Kantian views of his earlier thought.

One might expect more in these essays about intuitionism. I could have written more about Brouwer's philosophy, but it has been well treated in recent years by Mark van Atten. Something that would have added to the present collection would be a treatment of some of the debate involving Brouwer, Hilbert, Weyl, and others. I may yet write on that subject, but I cannot delay this collection in order to undertake it. One issue, that of impredicativity, is treated in Essay 2.

The 1930s are a new period, chiefly because of the coming to maturity of mathematical logic. Gödel's completeness and incompleteness theorems were a watershed, as many have observed and as is noted in Essay 4. The analysis of computability was another step occurring a little later, where Gödel had a role, but it was eclipsed by that of Church and especially Turing. The 1920s had seen the codification of the Zermelo-Fraenkel system (ZF) of set theory, but it was only gradually

---

[4] But cf. Potter, *Reason's Nearest Kin*, especially chs. 6–8.

[5] This was brought about by the Nazi destruction of the Göttingen mathematical institute as it had been built up by Felix Klein and Hilbert. That was motivated not by hostility to Hilbert's program but by the fact that several important figures in Göttingen mathematics were Jews.

that it came to eclipse other approaches, in particular those growing out of the tradition of *Principia*. Zermelo's paper of 1930, "Über Grenzzahlen und Mengenbereiche," gave a clear picture of what standard models of ZF would look like, but this was not widely appreciated until after the war. It was only post-war work that made clear the relation of ZF to the simple theory of types.

The late 1930s saw Gödel's work on the axiom of choice and the generalized continuum hypothesis. A contribution of that work to reflection on the foundations of mathematics generally, not often remarked on, is that it brought higher set theory more to the forefront of the consciousness of thinkers on the subject than before. There had been important mathematical work in set theory after Cantor and the early Zermelo, by Hausdorff and others in Germany, by the French figures who founded descriptive set theory, and then by others in Poland and Russia. But the arguments of the 1920s and 1930s centered on arithmetic and analysis and the role of set theory in the latter. Gödel was keenly aware quite early of the foundational significance of higher set theory, as is shown by his lecture of 1933, "The Present Situation in the Foundations of Mathematics," given when he was at an early stage of the work that led to his consistency proof and the underlying theory of constructible sets.[6] His two extended published philosophical papers, "Russell's Mathematical Logic" and "What Is Cantor's Continuum Problem?," are dominated by reflection on set theory. It is in these essays that he avows a platonist philosophy.

Essay 4 surveys Gödel's work generally. Although it devotes a lot of space to his philosophical work, I remark that his mathematical work, especially the incompleteness theorem, may have had a greater impact on philosophy than his properly philosophical writing. Essay 5, originally written before I became an editor, is the only introduction I wrote for the Gödel edition that I thought should be reprinted here.[7]

In Essay 7 I discuss the development of Gödel's philosophical thought and maintain that his mathematical work was a major factor in giving

---

[6] It is unfortunate that this lecture was not published at the time. It appears in *Collected Works* (hereafter CW), vol. 3, pp. 45–53, with an instructive introductory note by Solomon Feferman.

[7] Some of the others are quite brief, and many are dominated by documentary issues and by narrative. The other one with the most philosophical content is probably that to the correspondence with Gotthard Günther (CW IV 457–476), but the philosophy in question is Günther's, which I don't consider of sufficient interest.

him confidence in set theory and thus in defending a realist view of it. Gödel's more controversial ideas about mathematical intuition appear later, most clearly in the 1964 version of the continuum paper, but he surely viewed his conception of intuition as a natural development of his realistic views. Essay 6 discusses some of Gödel's arguments against logical empiricist views of mathematics and their parallel to arguments deployed by W. V. Quine.

The post-war history of the foundations of mathematics appears to some extent already in the essays on Gödel and is of course a major background for the remaining essays. A very significant development was the explosion of mathematical logic in the period from shortly after the war until the 1970s. All branches of scientific research saw substantial expansion during that period, but in the case of mathematical logic it really changed the character of the subject. Many of its leading figures and a lot of others wanted to see it as a branch of mathematics more or less like any other. The subject came to be understood as having four branches: proof theory (generally taken to include constructivism), set theory, model theory, and recursion (computability) theory. Of these the last two came to be largely dissociated from foundational concerns, and even in the first two the foundational concern often receded into the background. Paul Cohen's proof in 1963 of the independence of the continuum hypothesis brought into set theory the new method of forcing, which has proved to have very wide application; it led to a "take-off" of set theory and the proof of a vast array of independence results. But the study of large cardinal numbers had been developing at the same time, and both large cardinals and forcing are integral to set theory as now pursued. Set theory raises many interesting philosophical questions, about which something is said in Essay 7 on Gödel, Essay 10 on Wang, and Essay 12 on Tait. I have also commented on these questions in other, more systematically oriented writings.[8]

Gödel dominates the middle part of this collection, and Quine dominates the last part, although he appears in this collection only as a philosopher. How should we relate Quine to the main history of the foundations of mathematics in the twentieth century? His graduate work in philosophy was done in some haste, and at the time Harvard logic, such as it was, was in the shadow of *Principia Mathematica*. A lot of

---

[8] See *Mathematics in Philosophy*, part III, *Mathematical Thought and Its Objects*, ch. 4 and §55, and "Analyticity for Realists."

his early work was devoted to making ideas in *Principia* precise and straightforward. He came into contact with modern mathematical logic during his decisive year in central Europe, 1932–1933. Gödel's completeness and incompleteness theorems entered into his teaching, and Tarski's theory of truth played an important role in his later philosophical thought. He also learned of natural deduction and embodied an original version of it in his classic textbook *Methods of Logic.* He did not absorb deeply the ideas of either intuitionism or the Hilbert school. There is no doubt that Quine's writing on technical logic as well as his philosophical writing reflects a sophisticated understanding of the developments of the 1930s and many later ones. But if one looks at his logical writings, including his most interesting contribution, the set theory of "New Foundations," he must be regarded as a peripheral figure in the development of mathematical logic. On the other hand, his persistent efforts toward precision and clarity had great importance for the development of logic in *philosophy*, even though philosophical logic after 1960, with its emphasis on modal and related logics, took a direction that owed little to him and was uncongenial to him.

Quine's place in philosophy of logic and mathematics is another matter. His contribution to the philosophy of logic is enduring though contested. In my view, his most important contribution to the philosophy of mathematics was to develop the most defensible empiricist view of mathematical knowledge, which has been a model for many other thinkers, and which avoids what I persist in considering a desperate expedient, the view that mathematical statements are in some essential way empty, akin to tautologies. Other contributions of Quine are important but not of the same magnitude: That is true of his emphasis on ontology and clarification of questions surrounding it, although they have had a great influence on my own work. I find Quine's views about the significance of the necessity of mathematics for science too complicated to discuss in this introduction, but they are of course one of the pillars of his empiricist epistemology of mathematics.

Essay 8 draws heavily on Quine's early writings, from a time when his philosophical work was close to logic. His early thought on meaning and a view about predication that he continued to hold until the end of his career are more central than questions of philosophy of mathematics. But relevant to the latter subject are his views on second-order logic and his engagement in the late 1930s and 1940s with nominalism about mathematics.[9] Essay 9 concerns his somewhat neglected work *The Roots of Reference,* with emphasis (not exclusive) on logical issues. Essay 11, on Putnam, deals with an issue originally framed by Quine.

A significant development whose beginning is not easy to trace is the gradual differentiation of philosophy of mathematics from logic, especially from mathematical logic as a research subject. The technical topics that have come to be classified as "philosophical logic" are very often not much connected with the foundations of mathematics, so that the growth of those areas has also reinforced the differentiation. These developments have been manifest in my own work and career. For example, in the "job market" in my youth I was represented as a logician. Today my reputation is probably largely as a philosopher of mathematics. This development is a special case of the increase in specialization in almost all areas of inquiry. But noticeable in the English-speaking world is an accompanying change in the philosophy of mathematics itself.

The period from Cantor to Gödel gave rise to a remarkable sequence of philosophizing mathematicians. They varied in the amount of philosophical training they had and the extent to which they undertook to develop philosophical views. It is from that group that the subjects of the essays in the first half of this volume were chosen. The remaining subjects (Quine, Wang, Putnam, and Tait) all made significant contributions to mathematical logic. Much work in the philosophy of mathematics both earlier in the twentieth century and in more recent years has mathematical logic as an essential part of its background, whether or not the authors involved were active in research on the mathematical side. That is true of much of my own writing and of that of many others who are known principally as philosophers of mathematics.

However, another genre of philosophy of mathematics has gradually grown up, which does not have an organic connection with mathematical logic. The work of Imre Lakatos made this tendency salient. As in other parts of the philosophy of science, the practitioners have paid a lot of attention to history, in this case of mathematics more generally rather than of what we think of as foundations. It has brought

---

[9] Essay 7 of *Mathematics in Philosophy* gives a fuller discussion of other issues concerning Quine's philosophy of mathematics.

into salience for philosophers something that the logic-based tradition has somewhat neglected: the application of mathematics in the physical sciences. It is a deficiency of my own work that I have not explored this matter in any depth. It may be that the type of work less connected to logic will come to dominate the philosophy of mathematics.[10] But the continuing challenge of questions about higher set theory is likely to prevent the more logical tradition from disappearing anytime soon.

Something should be said here about the relation of the history that I have sketched here to the history of analytical philosophy, which has been studied in recent years with considerable intensity and scholarly depth. I have never considered myself such a historian, even though my teacher Burton Dreben was one of the founders of the subject. The principal canonical figures are Frege, Russell, Moore, Wittgenstein, Carnap, and Quine. Of these only Quine is the subject of essays in this collection, and except in Essay 8 I treat him more as an older contemporary than as a historical figure. I have followed much of the work on this history with interest and have been on some relevant dissertation committees. Although all of these figures except Moore were to some degree actors in the history of the foundations of mathematics, only Frege and Russell played a central role in that history. The central development of analytical philosophy was much influenced by the latter history at least through the 1930s, first of all through the birth and early development of modern logic. Of the main tendencies in the foundations of mathematics, we think of logicism as most central to the development of analytical philosophy. This is largely true, but it should not be forgotten that Carnap was greatly influenced by Hilbert's metamathematics, and Wittgenstein in his later period was engaged to some degree with intuitionism and had significant personal interaction with Turing.[11]

---

[10] At the end of the last paragraph back I wrote "in the English-speaking world" because there are other traditions of philosophy of mathematics. I had in mind in particular a French tradition which goes back a hundred years or more, in which the history of mathematics is prominent but logic plays a much smaller role. This tradition is still little known here.

[11] Thanks to Peter Koellner and Dagfinn Føllesdal for comments on an earlier version of this Introduction.

PART I

SOME MATHEMATICIANS
AS PHILOSOPHERS

1

THE KANTIAN LEGACY IN TWENTIETH-CENTURY

FOUNDATIONS OF MATHEMATICS

Since my title refers to "twentieth-century" foundations of mathematics, you may think I haven't quite got the news that we live in the twenty-first century. The fact is, however, that what I could find to talk about are some tendencies in the thought about the foundations of mathematics of the *twentieth* century, indeed more prominent in its first half than in its second. I will try to say at the end what relevance these tendencies might still have today.

There is a picture of the foundations of mathematics with which many English-speaking philosophers grew up in the period between roughly 1920 and the 1950s. To some extent this picture was formed in reaction to Kantian views. A prominent part of the picture was that Kant's philosophy of mathematics was shown to be inadequate by developments in mathematics and physics, of which the most decisive are the development of non-Euclidean geometry and the theory of relativity, which applied non-Euclidean geometry to physics. Stated in this way, this is a claim with which one could hardly argue, but the picture goes on to hold that essential parts of Kant's apparatus, especially the idea of pure or a priori intuition and of space and time as forms of intuition, no longer have any relevance to mathematics. This, according to the picture, has been shown by developments in logic and mathematics, first various aspects of the late nineteenth-century revolution in mathematics, but especially by the logicist construction of mathematics in Whitehead and Russell's *Principia Mathematica*. (A present-day commentator on logicism might pay greater attention to Frege, but when the picture I am describing was ascendant, Frege tended to be viewed as a precursor of Russell and not much read.[1])

One could do much to fill in the details of this picture, but what is striking to anyone with more than a passing knowledge of the foundations of mathematics in the twentieth century is what it leaves out. Systematically, it leaves out the fact that aspects of *Principia* were contested from the beginning, especially the use of the axiom of reducibility and the axiom of infinity. Historically, it leaves out the two principal rivals of Russellian logicism as general points of view, the intuitionism of L. E. J. Brouwer (1881–1966), and what was called "formalism," in effect the point of view underlying the foundational program pursued in the 1920s by David Hilbert (1862–1943). One does not have to look far into these views to see that they did not agree that the ghost of Kant had been quite laid to rest.

It is another question how Kantian their views actually were. I propose to examine this question with reference to Brouwer, Hilbert, and Hilbert's junior collaborator Paul Bernays (1888–1977), who after the effective end of the Hilbert school in the 1930s had a long career during which he wrote quite a number of philosophical essays. Since Brouwer and Hilbert were the major Continental European figures in foundations for a fairly long period, it is natural to include them in a discussion of this kind. Bernays is chosen for another reason: He was philosophically the best trained of the three; it was probably partly for that reason that Hilbert chose him as his assistant in 1917. And starting as a disciple of the neo-Kantian Leonard Nelson, Bernays eventually came to a position rather removed from that of Kant, Nelson, or the more mainstream neo-Kantians. But there is still what might be described as a Kantian residue. I have written elsewhere on Bernays's later views and will concentrate in this essay on other elements of his thought.[2]

---

[1] There were of course exceptions, earlier, in addition to Russell himself, notably Wittgenstein and Carnap. In part the situation was due to the fact that before 1950 his writings were far from readily available. In the English-speaking world, this began to change with the publication of J. L. Austin's translation of *Grundlagen der Arithmetik* (1950) and Peter Geach and Max Black's *Translations from the Philosophical Writings of Gottlob Frege* (1952). Current German editions were not available until the 1960s.

[2] See Essay 3 of this volume.

1. Neo-Kantianism

Given the importance of neo-Kantianism in German philosophy from 1870 into the 1920s, one might expect the "Kantian legacy" in foundations during this period to have been mediated in an important way by neo-Kantianism. That is true only in a very limited way. The principal neo-Kantians were on the sidelines of the developments that interest me. Paul Natorp and Ernst Cassirer did write about the philosophy of mathematics.[3] What is most interesting about Cassirer's writing is the fact that in discussing contemporary developments, he began by concentrating on Russell's *Principles of Mathematics* and contemporary writings of Louis Couturat. Cassirer argues that this early logicism is largely compatible with his own point of view and thus (in particular against Couturat) with some Kantian theses. Influenced by the general developments undermining Kant's philosophy of geometry, he makes no attempt to defend the idea of space as an a priori form of intuition. Cassirer is also largely persuaded that the work of mathematicians leading up to and including Russell's had succeeded in eliminating geometrical intuition even from the theory of real numbers.[4] He makes the more general and striking remark:

> It is important and characteristic that the immanent further construction of the Kantian theory has led by itself to the same result that is demanded more and more distinctly by the progress of science. Like "logistic", modern critical logic has also progressed beyond Kant's theory of "pure sensibility."[5]

Similarly, Natorp remarks:

> The post-Kantian philosophy that took its point of departure from him, as well as the present-day neo-Kantian tendency, has more and more taken exception to the dualism of intuition and pure thought and finally broken decisively with it.[6]

Unlike many mathematicians commenting on Kant, Cassirer was fully aware that according to Kant the synthetic activity of the

---

[3] See Cassirer, "Kant und die moderne Mathematik" and *Substanzbegriff und Funktionsbegriff;* Natorp, *Die logischen Grundlagen der exakten Wissenschaften.*

[4] See esp. "Kant und die moderne Mathematik," pp. 12–14.

[5] Ibid., p. 31, my translation.

[6] Natorp, *Die logischen Grundlagen,* p. 2.

understanding has a fundamental role in mathematical cognition, although he tends to interpret it at the expense of the forms of intuition.[7] Concerning time and its role in arithmetic, he says that for Kant "it can only be a matter of the 'transcendental' conceptual determination of time, according to which it appears as the type of an *ordered sequence*."[8] In other words, it is the abstract structure embodied by time that matters for mathematics. Cassirer grants to "intellectual synthesis" much more power in mathematical cognition than Kant does, although he agrees with Kant that to be genuine cognition mathematics must be applicable to objects given in space and time, in particular in physics. Cassirer does not consider at this time that this view could be challenged by higher set theory.[9]

In *Substanzbegriff und Funktionsbegriff* Cassirer expresses views on the philosophy of arithmetic that are strongly influenced by Dedekind's *Was sind und was sollen die Zahlen?* There Cassirer takes a position that could be called structuralist, as Jeremy Heis argues at some length.[10] Some remarks are close to the slogans of recent structuralism, such as the following:

> What is here expressed [by Dedekind] is that there is a system of ideal objects whose whole content is exhausted in their mutual relations. The "essence" of the numbers is completely expressed in their positions.[11]

---

[7] Thus he remarks that the passages of the *Critique of Pure Reason* that emphasize sensibility as an independent principle of cognition are close in content to pre-critical writings (in particular the Inaugural Dissertation), whereas parts like the Transcendental Deduction "which contain the really new and original result of the critique of reason, let the functions of understanding appear as *prior conditions* of 'sensibility'" (op. cit., p. 35).

[8] Ibid., p. 34 n.1 (on p. 35). He goes on to say that "it is not the *concrete* form of the intuition of time that is the *ground* of the concept of number, rather that the purely logical concepts of sequence and order are already implicitly contained and embodied in that concept." He does not attribute this view to Kant. However, such remarks lead one to think that he would have found congenial the "logical" interpretation of the role of intuition in mathematics developed by E. W. Beth, Jaakko Hintikka, and Michael Friedman.

[9] Some comments on Cantor's transfinite numbers are made in *Substanzbegriff und Funktionsbegriff*, pp. 80–87, trans. pp. 62–67, but no connection is made with this issue.

[10] Heis, "'Critical Philosophy Begins at the Very Point Where Logistic Leaves Off'."

[11] *Substanzbegriff und Funktionsbegriff*, p. 51, trans. p. 39.

His main claim could be expressed by saying that the structure of the natural numbers is prior to other facts about them, and that in particular their application as cardinals can be derived. This is, I think, the main objection that he has to Frege's and Russell's treatment of number. Given certain assumptions, of course, the structure can be recovered from their definitions, but Cassirer thinks that that gets the conceptual order wrong.[12] Curiously, even in this slightly later work Cassirer does not take note either of the problem that the paradoxes pose for their treatment of number or of the theory that Russell develops in the face of them. Dedekind's use of set theory apparently does not bother him; he seems prepared to accept what Dedekind uses as belonging to logic.

Cassirer does not seem concerned to work out the structuralist view as a general ontology of mathematical objects or to deal with the problems it faces; a criticism he makes of Frege and Russell, that in their definition of the cardinal number 1 the number one is presupposed, does not seem to me to be in accord with structuralism.[13]

Cassirer is an interesting commentator on developments in foundations just before and after the turn of the century. But I don't know of evidence he or others of the main neo-Kantian schools had a substantial direct influence on Hilbert, Bernays, or Brouwer.[14] However, neo-Kantianism did have a more direct influence on Hilbert and Bernays through Leonard Nelson (1882–1927). Nelson owed his position in Göttingen in large part

---

[12] Ibid., pp. 62–63.

[13] Ibid., pp. 65–66, trans. p. 50. Essentially the same criticism is made by Natorp in *Die logischen Grundlagen,* pp. 114–115. It is of interest that in "Kant und die moderne Mathematik" Cassirer seems unaware of Frege, while in *Substanzbegriff* his discussion is focused more on Frege than on Russell. Natorp also discusses Frege at length. Natorp seems to know only the *Grundlagen;* Cassirer cites at one point *Grundgesetze,* vol. 2. In my opinion Cassirer's discussion is much superior.

[14] Bernays's philosophical writings are sparse in references to other philosophers. But he must have had some familiarity with the Marburg school. They are known for "historicizing" the Kantian conception of the a priori. In an essay written when he was beginning to move away from Nelson's position, Bernays characterizes the "theory of a priori knowledge" in an almost opposite way, of items of a priori knowledge (presumably synthetic) as fixed for all time. See "Grundsätzliche Betrachtungen zur Erkenntnistheorie" and my "Paul Bernays's Later Philosophy of Mathematics" (Essay 3 in this volume), §3. As a follower of Nelson he may have been quite consciously opposed to the Marburg school, but when he questioned Nelson he did not go in their direction, although the Gonsethian concept of the *préalable,* which he substituted for that of the a priori, is more "historical" than he understood Kant's conception to be.

to Hilbert's patronage, and he was Bernays's principal teacher in philosophy. He saw his mission in life as reviving and developing the Kantian philosophy of Jakob Friedrich Fries (1773–1843). In the philosophy of mathematics he wrote mainly about geometry, and here his position was a much more orthodox Kantianism, even at the end of his life.[15] There is no attempt in his writing to minimize or interpret away pure intuition as a decisive factor in geometrical cognition. We shall see how this position left its traces in the thought of Bernays, and possibly also in that of Hilbert, although neither could ignore the obvious reasons why a Kantian philosophy of geometry would clash with contemporary physics.

2. Brouwer

Intuitionism begins with L. E. J. Brouwer's dissertation of 1907, *Over de grondslagen der wiskunde.*[16] This is a most unusual mathematical dissertation, because a lot of its content is philosophical. I will not comment on it directly. There is one element of the above picture that Brouwer fits perfectly, the rejection of the idea that a pure intuition of *space* has any relevance to the foundations of geometry or of mathematics generally. Although the point is remarked on in the dissertation,[17] it is expressed most clearly in a lecture of 1909 on the nature of geometry.[18] In the first part he traces a series of steps that undermined and eventually made untenable Kant's philosophy of geometry, in the con-

---

[15] See in particular his lecture of 1927, "Kritische Philosophie und mathematische Axiomatik." Much information concerning Nelson, his school, and his position in Göttingen is to be found in Volker Peckhaus, *Hilbertprogramm und kritische Philosophie*, chs. 5–6. In my review of *Beiträge zur Philosophie der Logik und Mathematik*, I questioned Nelson's understanding of Hilbert's proof-theoretic program of the 1920s. One might discern such questioning already in the discussion remarks of Bernays and Richard Courant reported (by Nelson) in the published version of the above-mentioned lecture and in the prefatory note by Wilhelm Ackermann to the reprint in *Beiträge*.

[16] The writings of Brouwer in English, French, and German cited below are all reprinted in *Collected Works*, vol. 1, showing the original pagination; they are cited in the original. This volume is cited as CW. Vol. 2, containing Brouwer's topological papers, will not concern us. Writings in Dutch occur in CW only in English translation, and sometimes excerpted.

[17] Most explicitly *Over de grondslagen*, p. 121, CW, p. 71, which summarizes the second chapter.

[18] *Het wezen der meetkunde.* English translation in CW. This lecture did not figure in international foundational debates, no doubt largely because Brouwer published

text of characterizations of geometric concepts by invariance under groups of transformations. The mathematical foundations of what one might loosely call an arithmetization of pure geometry existed already with Descartes's coordinatization of geometry.

> Consequently it is natural to consider the geometry determined by a transformation group in a Cartesian space as the only exact Euclidean geometry and to see its approximative realization in the world of experience as a physical phenomenon, which in principle has the same character as, for instance, Boyle's law.[19]

The development of projective geometry was already a step toward alternatives to Euclidean geometry; in particular the group of projective transformations contains the classical non-Euclidean groups. On purely mathematical grounds Brouwer sees no reason for not giving "equal rights" to Euclidean and non-Euclidean geometry (p. 9, trans. p. 114), but, alluding to the special theory of relativity, he holds that the view that Euclidean geometry holds "physical apriority" is also untenable. He also dismisses the view that projective geometry is a priori, while experience determines the curvature and number of dimensions of space.[20] Here he notes the change from space and time to space-time and the centrality of the group of Lorentz transformations. That some weaker priority thesis might hold about space-time he does not consider.

This view of geometry forms the background of the negative part of Brouwer's most quoted statement about Kant, from his inaugural lecture of 1912:

> However weak the position of intuitionism seemed to be after this period of mathematical development, it has recovered by abandoning Kant's apriority of space but adhering the more resolutely to the apriority of time.[21]

By "intuitionism" Brouwer evidently meant at this time something much broader than his own view; it included Kant. He does not offer much of a characterization, but in the same lecture he says that the

---

it only in Dutch. It did not appear in English, French, or German until the publication of CW in 1975.

[19] Ibid., p. 7, trans. pp. 112–113.

[20] Brouwer attributes this view to Russell's *Essay on the Foundations of Geometry* (1897), which he criticizes on a number of grounds in the dissertation.

[21] *Intuitionisme en formalisme*, p. 11, trans. p. 85.

intuitionist answers the question "where mathematical exactness does exist" with "in the human intellect." He has in mind largely French mathematicians, of whom the most explicit about philosophical matters was Poincaré. All were critical of various aspects of the abstract and set-theoretical turn that mathematics had taken.

We will turn shortly to the more positive part of Brouwer's statement. But something should be said about Brouwer's philosophical background. It is likely that he was largely self-taught in philosophy. He did have some contact with the dominating figure in Dutch philosophy at the time, the idealist G. J. P. Bolland.[22] Concerning Kant, he wrote to his thesis supervisor D. J. Korteweg that he had "read all of the *Critique of Pure Reason* and studied many parts . . . repeatedly and seriously."[23] Brouwer's biographer Dirk van Dalen considers it probable that he attended the lectures of a philosopher teaching at Amsterdam, C. Bellaar-Spruyt, whom van Dalen describes as "a firm defender of Kant."[24]

Brouwer's conception of intuition has a certain affinity to Kant's of pure intuition and may well be of Kantian inspiration. But it occurs in the context of a general philosophy that was in important respects anti-Kantian. The general ideas of his philosophy came to him very early; they are present in the dissertation and even to some extent in a privately printed booklet of 1905, *Leven, kunst, en mystiek*. We will rely, however, on mature presentations, particularly his lecture in Vienna in 1928, its expanded Dutch version of 1933, and his lecture to the Tenth International Congress of Philosophy in Amsterdam in 1948.[25]

Brouwer begins these expositions from a quite un-Kantian premise. He seems to believe there is a state of consciousness that is prior to any in which there is a distinction of self and world or an engagement with the outer world. What he calls "mathematical attention," from which mathematical thought originates, is an act of *will* taken for self-

[22] Van Dalen, *Mystic, Geometer, and Intuitionist*: vol. 1, §2.4.
[23] Brouwer to Korteweg, November 5, 1906, trans. in van Dalen, op. cit., p. 94.
[24] Ibid., p. 106. In January 1965 Arend Heyting arranged for me to visit Brouwer at his home, and I asked Brouwer about the possible influence on him of Kant. In his reply, he went off on a tangent and did not answer my question.
[25] "Mathematik, Wissenschaft, und Sprache," hereafter MWS; "Willen, weten, spreken," hereafter WWS; "Consciousness, Philosophy, and Mathematics," hereafter CPM.

preservation (MWS 153, WWS 177[26]). Evidently Brouwer thinks that engagement of the self with the world is in some way optional:

> Mathematical attention is not a necessity but a phenomenon of life subject to the free will; everyone can find this out for himself by internal experience: Every human being can at will either dream away time-awareness and the separation between the self and the world of perception or call into being the world of perception [and] the condensation of separate things.[27]

In CPM, Brouwer describes a state similar to that described in the second sentence as "consciousness in its deepest home" (p. 1235). However, since mathematics arises only when the stance of mathematical attention has been taken and one has exited from the "deepest home," mathematical thinking and mystical experience are quite separate, as Mark van Atten emphasizes.[28] Here he may no longer think of mathematical attention as resulting from an act of will, although he is clear that the further step of causal attention is a "free-will phenomenon" (ibid.). Brouwer does not elaborate on his conception of will in any of these essays, but it will have to be very different from any even remotely Kantian conception. This is particularly evident in the view of MWS and WWS, since mathematical attention depends on the will but is prior to any conception he could have of reason. But I think the same holds for CPM, because for Kant any exercise of will presupposes causality.

Brouwer distinguishes temporal orientation or awareness[29] from causal orientation or attention, a distinction that in its working out recalls Kant's between mathematical and dynamical categories.[30] In causal attention one identifies in imagination certain series of phenomena with one another. This is the origin of the perception of a world of

---

[26] I am assuming that the German term "mathematische Betrachtung" and the Dutch term "mathematische beschouwing" are meant to be equivalent. I follow van Stigt in translating both as "mathematical attention." (He is also the translator of MWS in Mancosu's collection.)

[27] WWS 178, trans. 418–419.

[28] *On Brouwer,* p. 9.

[29] *Zeitliche Einstellung* in MWS, *tijdsgewaarwording* in WWS. I doubt that these were intended to be equivalent. In later English writings he speaks of the "perception of a move of time" ("Historical Background," p. 141; "Points and Spaces," p. 2), evidently closer to the Dutch formulation.

[30] As remarked in my "Brouwer, Luitzen Egbertus Jan."

objects and especially of the use of means: one produces a phenomenon that will be followed in a certain repeatable series by a desired phenomenon that cannot be directly produced. Brouwer holds that causal sequences exist only as correlates of a stance of the human will; in spite of a naturalistic element introduced by his view that this stance is in the service of self-preservation and more efficient gratification, his outlook is idealist. But there seems to be more of Schopenhauer than of Kant in it.

There are some delicate questions about the relation of Brouwer's later to his earlier formulations. First of all, the term "a priori" does not occur in these expositions. However, Brouwer's second Vienna lecture (following on MWS) indicates that he had not given up the view of 1912 that the basic temporal intuition is a priori. The lecture begins with a historical overview (such as one finds also in later texts), mentioning in the beginning the view, which he attributes to Kant and Schopenhauer, that the continuum as well as the rest of mathematics "is considered as pure intuition a priori . . . [and] as present exactly and unambiguously."[31] More tellingly, later he says that the intuitionistic construction of the continuum implies that "the previously mentioned view of the continuum as an intuition a priori, after Kant and Schopenhauer, can in light of intuitionism be maintained in essentials."[32] But then the term is virtually dropped from Brouwer's post-war writings.[33] I hesitate to infer from that that Brouwer gave up the view of his original intuition as a priori, but it does seem that for some reason the issue was no longer central for him.

A second question is what to make of the naturalistic element expressed by the claim in MWS and WWS that mathematical attention is a phenomenon of the will undertaken for self-preservation. In CPM he no longer asserts this, but as noted above he maintains such a view about causal attention. Even the earlier formulation may be less naturalistic than it appears, since the will may be "transcendental" and perhaps must be in view of Brouwer's generally subjective idealist

---

[31] *Die Struktur des Kontinuums*, p. 1, my translation.

[32] Ibid., p. 6, my translation. Thanks to Mark van Atten for pointing out this passage.

[33] The term does occur on pp. 1–2 of "Points and Spaces." The context is a historical overview. However, it does not occur in the similar overview in "Historical Background, Principles, and Methods of Intuitionism." (These observations are due to van Atten.)

stance.[34] But given how intimately mathematical attention and causal attention are connected in MWS and WWS, the apparent change of view in CPM is puzzling. One might conjecture that the later Brouwer wished to represent mathematics as less connected with the "fall from grace" inherent in the subject's engagement with the outer world.

The primary element of mathematical attention is what Brouwer calls the perception of a move of time. Here is a rather standard formulation: ". . . the falling apart of a life moment into two qualitatively distinct things, of which the one withdraws before the other and nevertheless is held on to by memory" (WWS 177; cf. MWS 153). The temporal structure of experience allows distinguishing experiences from each other and from the self. Similar formulations are found in many of Brouwer's writings going back to the dissertation.[35]

It is this temporal awareness that is important for mathematics, and here Brouwer introduces what could be described in Kantian terms as a form of intuition. What he calls mathematical abstraction consists in divesting the two-ity of its objectual content. One arrives at a "common substratum of all two-ities," which forms the "original intuition of mathematics" (MWS 154; cf. WWS 179). At this level at least, the kind of division that a move of time involves is iterable; that is (at least the present part of) an experience can itself be perceived as a new move of time, so that a "two-ity" becomes a three-ity, and so on. It should be noted that Brouwer distinguishes intuitive time from scientific time; in particular the former does not have a metric.[36]

---

[34] Connected with these issues is Michael Dummett's description of Brouwer's philosophy as "psychologistic through and through" (Critical Notice, p. 609), which I somewhat carelessly endorsed in "Intuition in Constructive Mathematics," p. 213 n.2. Van Atten gives documentary evidence that Brouwer rejected psychologism (*On Brouwer*, pp. 74–76, also *Brouwer Meets Husserl*, p. 21).

On the issues in this paragraph, see also *Brouwer Meets Husserl*, note 31 (flag on p. 21, text on pp. 132–133). Van Atten views Brouwer's subject as a transcendental subject analogous to Husserl's transcendental ego. But it appears that only in WWS and later does Brouwer explicitly idealize his subject's capacities. In correspondence, van Atten remarks that he did so implicitly much earlier, for example in regarding the natural numbers as constructible from the basic intuition. It doesn't follow that Brouwer saw the matter that way; he may not have worked out this implication of his own position.

[35] *Over de grondslagen*, p. 8 (CW 17); CPM 1235; "Historical Background," p. 141; "Points and Spaces," p. 2.

[36] *Het wezen der meetkunde*, p. 15 (CW 116); cf. *Over de grondslagen*, p. 99 (CW 61).

Brouwer often talks of the "self-unfolding" of this intuition. The iteration of the division of past and future would be an instance, yielding the structure of the natural numbers. In fact, Brouwer holds an extreme view, that all of mathematics is constructed from the basic intuition. In later writings he explains this briefly in his descriptions of the first and second acts of intuitionism.[37] By "intuitionism" he there means his own view and his own reconstruction of mathematics; the acts are stages in his own development. Before turning to them, I will make a general observation.

In Marburg neo-Kantianism the synthetic activity of the understanding takes over the role that for Kant himself was given to the forms of intuition. Brouwer's claim that all of mathematics is constructed from a basic intuition raises the question whether he does not allow the faculty of intuition, in some sense a priori, to absorb what would normally be taken as the functions of thought. That would be a departure from Kant in the opposite direction. I don't think that is what Brouwer does, but he is less articulate about the role of what for Kant would be the understanding. To see this, let us consider the "acts of intuitionism."

The "first act" consists not only in uncovering the intuition that is the basis of mathematics but also in "separating mathematics from mathematical language." Brouwer describes mathematics as "language-less." It is hard for an analytical philosopher to take that seriously. But it has an implication that drives his view of logic, where possibly his most enduring philosophical contribution is to be found. According to Brouwer, logic is completely derivative from mathematics. He has a kind of linguistic view of logic, although it is very different from that associated with logical positivism.

The second act introduces the concepts that are the key to Brouwer's development of intuitionistic analysis, first of all that of free choice sequence, introduced with Brouwer's concept of set *(Menge)*, later called in English "spread," and the concept of species, in effect class. It was with these concepts that Brouwer developed his remarkable theory of the continuum, where the sequences that give rise to real num-

---

[37] "Historical Background," pp. 140–143; "Points and Spaces," pp. 2–3; *Cambridge Lectures on Intuitionism*, pp. 4–9. The first presentation in these terms may have been in Brouwer's Berlin lectures of 1927; see *Intuitionismus*, pp. 21–23.

bers on a Cantorian treatment are generated by a succession of free choices that may be restricted (so that lawlike sequences are included).

Brouwer clearly thought in the 1920s that these new concepts still belonged to the self-unfolding of the basic intuition; in the Berlin lectures he describes the second act as "the recognition of the self-unfolding of the original mathematical intuition to the construction of sets."[38] And in the only slightly later MWS he says:

> . . . the self-unfolding [of the original intuition] introduces the infinite as reality of thought and moreover (in a way not gone into here) yields the totality of natural numbers as well as that of the real numbers and finally the whole of pure mathematics. (154–155)

Brouwer appears still to hold this view later.[39] The clearest expression of this is in a very short essay, "Richtlijnen der intuitionistische wiskunde" ("Guidelines of Intuitionistic Mathematics") of 1947. Brouwer seems never to have given up the view of his dissertation that the original intuition gives rise to the conception not only of discrete structures such as the natural numbers but of the continuum.

What is not articulate in Brouwer's philosophy is anything corresponding to Kant's theory of concepts and judgments. For some purposes what he has to say about language and communication will fill that gap, but not for the most essential purpose of saying what the "self-unfolding" of the intuition amounts to and why it yields some results and not others.[40] There is a kind of intuitive obviousness about two phenomenological starting points of Brouwer's view, the iterability of the process of separating past from present, and (essential for the continuum) seeing it as possible to insert a point *between* two places in an order generated in this way. But that there is a conceptual grasp of what iteration of these processes gives rise to, so that for example we can reason by mathematical induction, is something additional. And

---

[38] *Intuitionismus,* p. 23.

[39] In "Historical Background," p. 142, just before introducing the second act, Brouwer mentions the fear "that infinite sequences generated by the intuitionist self-unfolding of the basic intuition would have to be fundamental sequences," i.e., sequences generated according to a law. Presumably the second act is to dispel this fear.

[40] A similar lack seems to be pointed to by van Atten when he argues that Brouwer's philosophy lacks the capacity to reflect on itself. See his *Brouwer Meets Husserl,* §5.5.

something playing the role of a concept is necessary to conceive even a simple lawlike sequence like that of the even numbers. I have no reason to think that Brouwer would have denied that. He holds that all of mathematics is *constructed* from the basic intuition. This construction is an exercise of what Kant would call the active faculties of the mind, and Brouwer clearly thinks of it as an activity. But he says little about what this activity is, and he does not distinguish between the activity of construction, such as generating a sequence of the type of the natural numbers, and insight into what has been generated, such as the principle of induction.

One of Brouwer's great achievements was his criticism of classical logic and in particular of the principle of excluded middle. This was carried out with great thoroughness, so that starting with the idea that the principle is equivalent to the solvability of every mathematical problem, he was able to show in a large number of cases in classical mathematics that the law of excluded middle (LEM) is essential to their proofs. On his own grounds we have no reason to think these results true, and in many cases he could argue even that they are false, to be sure using assumptions about free choice sequences.

Brouwer's conception of mathematics as consisting of construction in intuition amounted to an epistemic view of truth in mathematics and led him implicitly to interpret logical connectives in that sense. As regards the LEM, we are in a position to assert A v ¬A only if we have at least a procedure that would allow us to assert one or the other. In earlier writings this view is implicit but underlies his logical arguments, but later he quite explicitly says that there are no non-experienced truths. The form the view has come to take is the so-called BHK (Brouwer-Heyting-Kolmogorov) interpretation of the logical connectives in terms of constructions or proofs. This was articulated by Brouwer's disciple Arend Heyting at the same time as his formalization of intuitionistic logic. Heyting refers to Husserl's intention-fulfillment theory of meaning. The connection with Kant becomes even more tenuous, although some have found an epistemic view of truth underlying some arguments in the *Critique,* for example the solutions to the Mathematical Antinomies. Brouwer's critique of LEM can be divorced from any controversial philosophy by observing that he discovered that the use of LEM in mathematical proofs can violate a methodological principle according to which all proofs should be constructive. If put in that way, it is noncontroversial. But of course the mainstream of mathematics has preferred accepting varying degrees of nonconstructivity to giving up classical logic.

The absence of a theory of concepts and judgments becomes visible in Brouwer's writings in the comments he makes about his opponents. In the 1912 lecture he opposed intuitionism to "formalism." That term naturally suggests Hilbert's views and program of the 1920s. In its time Brouwer criticized that,[41] but Hilbert's views had been an object of his criticism already in the dissertation. The answer Brouwer attributes in 1912 to the formalist as to where "mathematical exactness does exist" is "on paper."[42] Brouwer is not explicit here about who counts as a formalist, although he seems to have in mind at least Hilbert (both the *Foundations of Geometry* and his first step toward proof theory, the 1904 Heidelberg lecture). Much later he places in an "old formalist school" Dedekind, Cantor, Peano, Hilbert, Russell, Zermelo, and Couturat.[43] What I think distinguishes them from himself and those predecessors he called intuitionists or pre-intuitionists is adherence to the axiomatic method (without the demand for some kind of intuitive foundation for axioms, or at least an intuitive foundation he could accept) and taking logic to have some kind of autonomy. The quip about "on paper" and other remarks implausibly attributing to these writers formalism in a narrow sense result from projecting onto them Brouwer's own linguistic conception of logic. He isn't able to see any other ground for the validity of logic or for the kind of reasoning involved in set-theoretic mathematics. I don't know of any mention of Frege in Brouwer's published writings. How much he read any of Frege's writings I don't know, but I think he would have been quite deaf to Frege's rationalism.[44]

Is Brouwer's own mathematical practice, especially the theory of the continuum, better grounded in intuition? If we ask this question in

---

[41] "Intuitionistische Betrachtungen über den Formalismus."

[42] *Intuitionisme en formalisme*, p. 7, trans. p. 83.

[43] "Historical Background," p. 139.

[44] John Kuiper has discovered that Frege's *Grundlagen* and vol. 1 of *Grundgesetze* were in the library of the University of Amsterdam while Brouwer was a student. See Kuiper, "Ideas and Explorations," p. 223 n.11. Thanks to Mark van Atten for pointing this out to me. It is a priori likely that Brouwer would have seen Frege's two series of short essays "Über die Grundlagen der Geometrie," since they were published in the *Jahresbericht der deutschen Mathematiker-Vereinigung*. In fact, van Atten has found references to the second article in the first of these series in notebooks of Brouwer from the years 1905–1907.

an epistemological spirit, it is hard to see how the answer can be affirmative. It may be intuitively evident that the "falling apart of a life moment" yields a structure that is the same as before and so can "fall apart" in the same way, so that the process is iterable. But the conception of natural number (as a structure, leaving aside the question what *objects* the numbers may be[45]) involves having a concept of the product of this iteration, so that mathematical induction holds, and furthermore *operations* can be iterated, yielding primitive recursive functions. Are all these things intuitively evident? The question was discussed in the Hilbert school, and I have discussed it elsewhere.[46] There is something like a consensus, derived from analyses of Gödel and Kreisel in the 1950s, that because of its use of a general concept of construction or proof and full intuitionistic logic, intuitionistic mathematics goes beyond what is intuitively evident, even allowing for serious vagueness in that term.

This could be offered as a criticism of Brouwer, but I don't think he would have been moved by it. Although he claimed certainty for intuitionistic mathematics, I don't think the most fundamental role of intuition for Brouwer was epistemological. Intuition is first of all a medium of construction; its "self-unfolding" consists of mathematical constructions building on previous ones. Intuition gives a certain *metaphysical* status to mathematics. In one place he calls intuitionistic mathematics "inner architecture."[47] What drives Brouwer is a certain vision of the place of mathematics in human life; epistemology does not play at all the role for him that it plays for Bernays and even Hilbert.

3. Hilbert

David Hilbert was a native of Königsberg and absorbed a certain amount of Kant from his early environment.[48] But he was much less philosophical than the others I am discussing, in particular Brouwer. Philosophy does not seem to have played as significant a role in his studies as for Bernays, probably Brouwer, or some other philosophiz-

---

[45] Although Brouwer can hardly be described as a structuralist, I think he was not much moved by this question.

[46] See *Mathematical Thought and Its Objects*, ch. 7, esp. §44.

[47] CPM 1249. Adding the qualification "intuitonistic" may be an implicit concession to other mathematical practices that Brouwer often does not make.

[48] Cf. Reid, *Hilbert*, p. 3.

ing mathematicians such as Hermann Weyl and Kurt Gödel. Still, he chose as epigraph for his *Grundlagen der Geometrie* of 1899 a famous quotation from Kant's *Critique of Pure Reason*: "Thus all human cognition begins with intuitions, proceeds to concepts, and ends in ideas" (A702/B730). His idea was probably that in geometry intuition plays a role that is superseded (or reduced to being merely heuristic) by conceptual developments, particularly the rigorous axiomatization that Hilbert brought to completion in this work.

I will be comparatively brief about Hilbert, since he was self-consciously less a philosopher than many other figures in foundations; his program of the 1920s specifically aimed to give a *mathematical* solution to problems of foundations. I will also concentrate on that period. But it should be remembered that Hilbert's work in foundations really begins with his *Foundations of Geometry*, and although he did not publish in foundations between 1905 and 1918, he continued to lecture on the subject. It is now possible for scholars to see his work in the foundations of mathematics as a unity.[49] I will not try to do this but concentrate on the conception of finitism that provided the method for his proof-theoretic investigations of the 1920s. It is another case where a larger conception that is not particularly Kantian introduces a conception of intuition somewhat related to Kant's and probably at least indirectly inspired by him.

A constant of Hilbert's view of mathematics is the importance he gives to the axiomatic method. In the *Foundations of Geometry* it is developed in what we would call a structuralist way, so that the geometrical primitives do not have any content beyond what the axioms give them. This does not imply that geometry is just a "game with symbols," as Brouwer's identification of Hilbert as a formalist even at this stage might suggest. Hilbert did give great importance to the fact that the geometrical primitives can have different interpretations and exploited this possibility for the ingenious model-theoretic independence proofs that the book contains. But that still leaves the logical scaffolding as meaningful taken at face value. Hilbert might be regarded as a "formalist" in another sense, if we regard the logic as belonging to form as opposed to material content.

---

[49] That is undertaken in William Ewald, "Hilbert's Wide Program." However, his essay concentrates on earlier writings.

A consequence of this outlook is that there are not independent grounds on which geometrical axioms can be said to be *true*. This was one of the issues in the famous Frege-Hilbert controversy. In the geometrical work, Hilbert constructed models to show the independence of axioms, and he could also construct an analytical model to show the consistency of the axiomatization as a whole. Not at the time clearly distinguishing consistency from the existence of a model, he took the view (also held by others) that consistency is sufficient for mathematical existence.

The consistency problem that was attacked in the work of the 1920s already arose at this point. Hilbert gave an axiomatization of the theory of real numbers, and his model construction effectively showed the consistency of his geometry relative to that of the theory of real numbers. But what model construction would prove the consistency of the theory of real numbers? It was not obvious. Hilbert listed proving the consistency of analysis among the twenty-three problems in his famous address to the International Congress of Mathematicians in Paris in 1900. In his address to the next congress, in Heidelberg in 1904, he introduced the idea of a syntactic consistency proof, in which the theory whose consistency was to be proved was formalized and it was to be proved that no proof of a contradiction could be constructed. But the ideas were still in a very rudimentary state, and Hilbert did not return to the problem seriously until the beginning of his collaboration with Bernays near the end of World War I.

However, Poincaré and Brouwer pointed out an apparent circularity in the procedure that could be seen even at this stage. The theory whose consistency was to be proved included arithmetic, in particular mathematical induction. But the proof would itself proceed by induction, crudely, a proof of one line (i.e., an axiom) would not yield a contradiction, and if that is true of any proof of up to n lines, it will not be true of a proof of $n + 1$ lines. So it seems that induction is being used to prove that induction is consistent.

When Hilbert inaugurated his proof-theoretic program in the early 1920s, he had the idea of a solution to this problem. He and his collaborators had, partly by the study of Frege and Russell and partly by important further logical work beginning after Bernays joined him in the fall of 1917, formed the program of proving the consistency of formal systems in which logic and arithmetic or analysis would be formalized together (as Frege had envisaged), but without a lot of the logical complications of previous systems. The axiom system whose consistency is at issue is treated in complete abstraction from any meaning the expressions might have, as if the formulae and proofs were configurations of meaningless signs. Unfortunately this encouraged the interpretation of Hilbert as a formalist in the narrow sense of Brouwer's quip of 1912.

But the most important development from a philosophical point of view was the articulation of the finitary method. It was observed that formalization made the statement of consistency of the same character as elementary generalizations in arithmetic: that one cannot by certain rules operating on finite configurations (the rules for constructing proofs) arrive at two formulae in a simple relation, namely one being the negation of the other. The methods of the proof were then to be the most elementary and intuitive methods of a basically arithmetical character. As Hilbert and Bernays characterized them, they were of such an elementary character that, in particular, intuitionists could have no reason to object. The finitary method could use the law of excluded middle, but only because of restrictions on the use of quantification. Within the context of this method, Hilbert did not really disagree with Brouwer's critique.

So far, "intuitive" has been used only in a rough sense, not particularly philosophical. But Hilbert clearly believed that sign configurations of the sort that proof theory is about are objects of an intuition that is of an especially fundamental character. In his most famous lecture of the 1920s, "Über das Unendliche," he writes:

> As a condition for the use of logical inferences and the performance of logical operations, something must be already given to our faculty of representation, certain extralogical concrete objects that are intuitively present as immediate experience prior to all thought. If logical inference is to be reliable, it must be possible to survey these objects completely in all their parts, and the fact that they occur, that they differ from one another, and that they follow one another, or are concatenated, is immediately given intuitively, together with the objects, as something that neither can be reduced to anything else nor requires reduction.[50]

---

[50] "Über das Unendliche," p. 171, trans. p. 376.

Essentially the same statement is made in several other papers beginning in 1922. Just before the above passage Hilbert invokes Kant, who taught that "mathematics has a content that is secured independent of all logic." This might be read as a questionable interpretation of Kant, as denying that logic (or even concepts) have a role in mathematics.

Hilbert did not mean "concrete" in opposition to "abstract," as those terms are used in contemporary English-language discussions. It is rather clear from both this passage and the mathematical discussions that he understands the "concrete" signs as types and so abstract by contemporary usage. Rather, Hilbert means something like "perceptual."[51]

Although Hilbert does not use the noun *Anschauung* in this passage, it is clear from his writings and those of Bernays from the time of their collaboration that in their view cognition of signs and formal expressions rests on intuition. In contrast to Brouwer's view, the intuition involved seems to be as much spatial as temporal. Gödel remarked that Hilbert's intuition is Kant's space-time intuition, only restricted to finite configurations of discrete objects.[52]

Something that distinguishes Hilbert and Bernays's view from that of Brouwer (and probably makes it more Kantian) is that finitary proof is supposed to yield intuitive *knowledge*. It is Bernays who brings this out. For example he writes in a paper of 1922:

> According to this viewpoint we will examine whether it is possible to ground those transcendent assumptions [of classical mathematics] in such a way that only *primitive intuitive cognitions* come into play.[53]

To be sure, one might read Brouwer as holding that all mathematical knowledge is intuitive knowledge. I think the sense in which that could be true has to be quite different from what Bernays has in mind, in view of the modes of construction that intuitionist mathematics allows.

In a 1930 paper that was his most significant philosophical statement to date, Bernays describes the finitary standpoint as the standpoint of intuitive evidence *(Standpunkt der anschaulichen Evidenz)*.[54]

---

[51] For further discussion see *Mathematical Thought and Its Objects*, pp. 251–252.

[52] "On an Extension of Finitary Mathematics Which Has Not Yet Been Used," note b, p. 272.

[53] "Über Hilberts Gedanken zur Grundlegung der Arithmetik," p. 11, trans. p. 216.

[54] "Die Philosophie der Mathematik und die Hilbertsche Beweistheorie," *Abhandlungen*, p. 40, trans. p. 250.

In Hilbert's exposition this point of view involved restrictions on the use of quantification, and it has generally been thought that the method is best captured by a free-variable formalism, in which generality is expressed by free variables and there are no existential statements per se; existence of an object satisfying a condition can be asserted only by explicitly giving an instance. The best known such formalism is primitive recursive arithmetic (PRA); it allows introduction of function symbols by primitive recursion and inference by induction.[55] Hilbert and Bernays seem to have held that proofs in PRA convey intuitive knowledge. Bernays gives in the 1930 paper an argument that the introduction of exponentiation is compatible with that thesis, and in *Grundlagen der Mathematik* he recasts and generalizes the argument to cover primitive recursion in general.[56] Elsewhere I criticize these arguments.[57] But it may well be that the concept of intuitive knowledge cannot be made precise enough to determine whether the conclusion is true.

4. Bernays

Paul Bernays was a student in Berlin and then in Göttingen from about 1906 to 1912, when he received a doctorate in mathematics. He studied philosophy seriously and in Göttingen came into the circle of Leonard Nelson. His philosophical training was evidently a reason why Hilbert chose him to work with him on foundations. He was in Göttingen from 1917 but, being Jewish, was removed from his position after the Nazi takeover in 1933. He had inherited Swiss citizenship from his father and moved to Zürich in 1934. He did not hold a professorial position there until 1945 and was never an *Ordinarius* in Germany or Switzerland, although he was a mathematical logician of some eminence and well versed in other parts of mathematics.

---

[55] In his classic paper "Finitism," William Tait gives an analysis of finitism that does not invoke the concepts of intuition and intuitive knowledge discussed here, and concludes that PRA captures the extent of finitary mathematics. The question is still debated whether PRA captures what the Hilbert school intended to include. But the debate concerns whether they intended to allow going beyond PRA.

[56] Ibid., pp. 38–39; Hilbert and Bernays, *Grundlagen der Mathematik*, vol. 1, pp. 25–27.

[57] *Mathematical Thought and Its Objects*, §44.

During much of his life he wrote philosophical essays. Earlier ones were written when he was an active member of Hilbert's school.[58] In matters not directly related to the Hilbert program, he seems to have continued until about 1930 to be a disciple of Nelson, and he maintained a kind of loyalty to Nelson for the rest of his life, participating, for example, in the editing of Nelson's collected works. But over the course of the 1930s he moved away from Kantian views in general. Since I have written elsewhere about this evolution and the philosophy of mathematics that resulted, I will not go into those matters here.[59]

It is difficult to be clear about Bernays's view of arithmetic. One might even think that he has two parallel views, one of them about "intuitive arithmetic" conducted according to the finitary method, and one more structuralist but with some rationalist aspects.

About the indications of the first, the term "intuitive" *(anschaulich)* occurs frequently in his writings, and the term "intuition" *(Anschauung)* occurs a little less frequently but is not rare. But the underlying concept is elusive. I have not found any place where Bernays gives what might count as an "official" explanation of the concept of intuition. The discussions of intuitive arithmetic can be read as discussions of what intuition does but are not prefaced by an explanation of what intuition is. Bernays could certainly take the term to be familiar, but it is also hard to believe that he was unaware of how variously it had been used, even in discussions of mathematics. In a rather polemical review article of 1928, he complains that his author, Richard Strohal, does not undertake a "closer discussion" of the concept of intuition, but Bernays does not take the opportunity to offer such a discussion of his own.[60]

Bernays clearly often uses "intuition" in the informal, probably not very precise way in which it is often used by mathematicians, and doubtless not only by mathematicians. But there is one area in which it signifies more definite boundaries and a conception that is at least in a very general sense Kantian. That is when he discusses the finitary method that was to be the method of the proof theory of the Hilbert

---

[58] I leave out of consideration some very early publications in Nelson's *Abhandlungen der Freis'schen Schule.*

[59] See Essay 3 in this volume. Also relevant is my introduction to "Sur le platonisme" in Bernays, *Essays on the Philosophy of Mathematics.* This book will be cited as *Essays.*

[60] "Die Grundbegriffe der reinen Geometrie," p. 197.

school. Here, the most extensive discussions date from the time when he was an active member of the school: the above-mentioned 1930 paper, "Die Philosophie der Mathematik und die Hilbertsche Beweistheorie," especially part I, §3 and part II, §4, and the exposition of the method in Hilbert and Bernays, *Grundlagen der Mathematik,* volume 1, §2. A key statement in the latter text is the following:

> In number theory we have an initial object and a process of continuing. We must fix both intuitively in a definite way. The particular way of fixing is nonessential; only the choice once made must be maintained for the whole theory. We chose as initial thing the numeral 1 and as process of continuing the attaching of 1. (pp. 20–21)[61]

The symbol 1 and strings of 1 are referred to as signs *(Zeichen)* and as figures *(Figuren).* "Intuitively" seems to be meant in the same sense as in the passage from Hilbert quoted above.

After a discussion of what is included in finitary mathematics, Bernays writes:

> Our treatment of the basics of number theory and algebra above was meant to demonstrate, in its application and implementation, direct contentual inference, which takes place in thought experiments on intuitively presented objects and is free of axiomatic assumptions. (p. 32)

This remark expresses a well-known feature of the finitary method as Hilbert and Bernays conceived it, that it is not axiomatic.[62] For the practice of classical mathematics, Hilbert held something very close to what William Tait calls the axiomatic conception, but the finitary method was supposed to be based on direct intuitive evidences and not to involve axioms. That may be a reason why it has been so difficult to settle what were the limits of the finitary method as the Hilbert school conceived it, and why Bernays himself was uncertain about what methods introduced after Gödel's incompleteness theorem were proper extensions of finitary arithmetic.

In the 1930 paper, Bernays puts some of the issues about finitism in a broader context and makes remarks about arithmetic outside any

---

[61] Translations from this work are my own.
[62] In addition to the above passage, see p. 20.

explication of finitism. One conclusion he draws is that the classical foundation of analysis goes beyond the intuitive. But before arriving at that conclusion he discusses arithmetic and what is intuitive about it. He seems not especially concerned to separate finitary from classical arithmetic.

Bernays clearly maintains that logic and mathematics, arithmetic in particular, are entangled with each other in such a way that neither can be reduced to the other, contrary not only to logicism but to the opposed view of Brouwer.[63] Some of his arguments in 1930 rest on well-known features of modern formal logic, but there are also more obscure arguments concerning cardinality, directed at the Frege-Russell analysis. Underlying the number 2, Bernays suggests, is the idea of "a thing and yet another thing," one might say a structure consisting of objects a, b, such that a ≠ b, with no relations. That is an instance of what he calls "formal abstraction," singling out the structural aspect of a state of affairs involving objects. The Frege-Russell definition adds additional "logical clothing" to this idea.[64] Bernays wishes to claim that formal abstraction is more general than the abstraction involved in logic. But what is more essential is that the generality it involves is in a different direction from the generality of logic.[65]

But he insists that the basic idea does not yet give the number 2 as an object. It might have been natural to turn at this point to the concept of set. However, for the most elementary formal objects Bernays turns to numbers conceived ordinally, not in terms of an assumed ordering but in terms of the process that generates the sequence of natural numbers. It is in this way that he makes contact with the ideas expressed in his exposition of finitism.

---

[63] In my view this is an important thesis about logic, for which Bernays deserves credit. I explore it in "Some Consequences of the Entanglement of Logic and Mathematics," briefly crediting it to Bernays. (See note 6 of that paper.) I don't think Bernays states it prominently again, but it underlies some remarks in "Betrachtungen zu Ludwig Wittgensteins *Bemerkungen über die Grundlagen der Mathematik*" and in "Bemerkungen zur Philosophie der Mathematik."

[64] An idea reminiscent of what is said of this definition in Wittgenstein's *Remarks on the Foundations of Mathematics*. Cf. Bernays's comments in "Betrachtungen zu Ludwig Wittgensteins *Bemerkungen*,"p. 135. Essays by Bernays reprinted in *Abhandlungen* are cited in that pagination, others in the original pagination. Both are given in the margin of *Essays* for those essays reprinted there.

[65] On this issue see the introduction by Wilfried Sieg and W. W. Tait to "Die Philosophie der Mathematik" in *Essays*.

An ordinal number in itself is also not determined as an object; it is merely a place marker. We can, however, standardize it as an object, by choosing as place markers the simplest structures deriving from the form of succession. Corresponding to the two possibilities of beginning the sequence of numbers with 1 or with 0, two kinds of standardization *(Normierung)* can be considered. The first is based on a sort of things and a form of adjoining a thing; the objects are figures which begin and end with a thing of the sort under consideration, and each thing, which is not yet the end of the figure, is followed by an adjoined thing of that sort. In the second kind of standardization we have an initial thing and a process; the objects are then the initial thing itself and in addition the figures that are obtained by beginning with the initial thing and applying the process one or more times. (p. 31)[66]

In the first case the "figures" are what we might call strings, that is the procedure of the passage in Hilbert and Bernays noted above. The difference with the second is rather subtle, but he takes it to be expressed by a different kind of sequence of numerals, $0, 0', 0'', \ldots$.

Bernays goes on to say that to have the ordinals "free from all inessential features" we have to take as objects "in each case the bare schema of the respective features obtained by repetition"; however, this requires "a high degree of abstraction." "However, we are free to represent these purely formal objects by concrete objects ('number signs' or 'numerals')" (p. 32), again apparently arriving at one of the two "standardizations" he has referred to.

It seems to me possible that the whole discussion of formal abstraction and its application in criticizing logicism originated with reflection on discussions of the same issues in Marburg neo-Kantianism. As noted above, Cassirer and Natorp maintained that in the Frege-Russell definitions of number, the number 1 is presupposed. Bernays may have wished to work out what truth there may be behind this claim, although it must have been obvious to him that the definitions were not circular in any direct sense.[67]

The question arises how Bernays conceives of our knowledge of two very basic facts about these numbers, central to any modern

---

[66] Translation from *Essays*, modifying that of *From Brouwer to Hilbert*, p. 244.
[67] Cassirer comes closest to Bernays's formulation in *Substanzbegriff und Funktionsbegriff*, p. 66, trans. p. 50.

axiomatization of arithmetic. First, that either of the processes of continuation can at any point be continued one more step. Second, that either procedure gives rise to a corresponding induction principle. Can anything be said about why we should accept them?

Bernays seems quite unworried about the first point, not just about the continuability of the processes but about whether anything specific needs to be said about the basis of our knowledge of it. In the concrete case he, in keeping with the point of view of Hilbertian finitism, no doubt thought it intuitively evident, and probably (as Gödel thought) the intuition involved has a spatial character. But whether he thought it evident on the more abstract level, he does not say. A much later remark suggests that he might have come to that view eventually: "We are conscious of the freedom we have to advance from one position arrived at in the process of counting to the next one."[68] Relevant to this question is the Kantian view that there is no cognition of objects without intuition. In Bernays's context of 1930, that would imply that the "mere schema" of the figures generated by iteration of one or the other process of continuation is known to be nonempty only because in the case of the concrete representatives this is known intuitively. Bernays seems to have been inclined to that view; it would explain his rapid desertion of the more abstract level. The later remark just quoted is not strictly incompatible with this view, but Bernays's general movement away from Kantian views suggests that by then he no longer held it.

About mathematical induction, Bernays is probably clearest where he has discussed the essential introduction of the infinite into mathematics, which occurs in the step from elementary arithmetic to analysis. Analysis requires infinite sets or functions with infinite domains. In §1 of part II of the 1930 paper, Bernays argues that the existence requirements of analysis cannot be made out intuitively. He then returns to the subject of formal abstraction, which he views as essentially concerned with finite configurations, so that finiteness goes without saying. The "intuitive-structural" introduction of numbers fits only finite numbers. This is the basis for our knowledge of mathematical induction and of recursive definition.[69]

---

[68] "Die Mathematik als ein zugleich Vertrautes und Unbekanntes," p. 111.

[69] Bernays uses the phrase *finite Rekursion* without saying whether what is included goes beyond primitive recursion.

Bernays seems to admit that something conceptual is involved even in the finitist case he is focusing on. That is suggested by the rather obscure remark that follows:

Drawing on this representation of the finite of course goes beyond the intuitive evidence that is necessarily involved in logical reasoning. It corresponds rather to the standpoint from which one *reflects* already on the general characteristics of intuitive objects.[70]

Bernays's view that primitive recursion is admissible in intuitive mathematics, argued for in the 1930 paper and rather differently in Hilbert and Bernays, is called into question in a well-known remark a few years later, where he questions whether it is intuitively evident that there is an Arabic numeral for the number $67^{257^{729}}$.[71] More generally, in a postscript to the 1930 paper for the reprint in *Abhandlungen,* he concedes that ". . . the sharp distinction between the intuitive and the non-intuitive, which was employed in the treatment of the problem of the infinite, apparently cannot be drawn so strictly" (p. 61). But notably, among positive remarks scarcely in need of revision Bernays mentions "exhibiting the mathematical element in logic and emphasizing elementary arithmetical evidence."[72]

A discussion of Bernays's relation to Kant would be incomplete without some treatment of his views on geometry and the role of intuition in it. I have to limit myself to a few remarks. Geometrical intuition is mentioned quite often in his writings, and I think it is clear that it is supposed to have a substantial role. But for the most part this role belongs to what Tait calls "dialectic,"[73] that is motivating considerations for axioms that, once the axioms have been set up, play no logically necessary role in the proofs. Intuition could not play the role it

---

[70] P. 40, emphasis in the text. By "the intuitive evidence involved in logical reasoning," Bernays means to allude to part of his thesis of the entanglement of logic with mathematics but also more distantly to Hilbert's claim of the fundamental character to all reasoning of the apprehension of signs.

[71] "Sur le platonisme," p. 61.

[72] Ibid., translation modified.

[73] It should be said that this use of the term is quite different from that in Bernays's own later writings, which is derived from Ferdinand Gonseth. See Essay 3 in this volume, p. 73 n.13.

apparently plays for Kant and in Bernays's own time for Nelson, as yielding Euclidean geometry as the correct geometry.

Though it is still in the sphere of dialectic, a distinctive feature of Bernays's view is that the theory of the continuum, as developed for example by Cantor and Dedekind, is a rigorous formulation of an originally geometrical idea. "Analysis is concerned with making geometrical ideas conceptually precise."[74] In many places Bernays says that the classical theory of the continuum does not achieve a complete arithmetization, and in this place and elsewhere he rejects the demand for it. (He sees Weyl's *Das Kontinuum* and Brouwer's theory as aiming at that.[75]) The reason why a complete arithmetization is not achieved is the obvious one that the classical construction uses set-theoretic ideas. In one place Bernays refers to the classical theory as "a compromise that has succeeded very well."[76] It seems to have been Bernays's view that it is a constraint on theories of the continuum that they capture a geometrical intuition, even if, in the resulting theory, explicitly geometrical premises are absent.

5. Concluding Remarks

One might see the views we have been discussing as addressing the well-known difficulties of *Principia*. The issue about reducibility is one about the strength of the logic that is assumed there, and that has been a continuing issue in the foundations of mathematics addressed by both Brouwer and Hilbert. But there is nothing especially Kantian about the solutions offered. The notions of intuition introduced by Brouwer and Hilbert might be seen as a Kantian solution to the second problem, that of the axiom of infinity. That problem is certainly still with us; it is for example central to much of the discussion of Crispin Wright's neo-

---

[74] *Abhandlungen,* Vorwort, p. viii, my translation.

[75] This theme is prominent in what is probably his last philosophical essay, "Bemerkungen zu Lorenzens Stellungnahme in der Philosophie der Mathematik." But the interpretation of Brouwer as striving for complete arithmetization of analysis is surely not correct, since according to Brouwer, the intuition of time is itself the medium of construction of the continuum. Time is not given as a discrete sequence of "moments"; rather, one can always single out a point between any two, and continue this process. Cf. Brouwer's remark that the second act of intuitionism is just a special case of the first act (*Cambridge Lectures,* p. 93).

[76] "Bemerkungen zur Philosophie der Mathematik," p. 173, my translation.

Fregeanism. In some of my own writings I have explored the Kantian solution further.[77] It has to compete with a roughly empiricist solution, epitomized by Quine, and a rationalist solution, epitomized by Gödel, as well as various forms of formalism and skepticism. But there is no time to go into these matters now.[78]

---

[77] See now *Mathematical Thought and Its Objects,* chs. 5–7.

[78] The first version of this essay was presented at a conference, Universality and Normativity from a Kantian Perspective, University of California, Riverside, February 25, 2006. I am much indebted to Pierre Keller for the invitation and to the audience for comments. A version closer to the present one was presented to a conference at the University of Nancy 2 on November 3, 2010. I am also indebted to that audience. The debt of the section on Brouwer to Mark van Atten should be evident; he has commented on more than one version. The section on Bernays was not in the Riverside version. I had such a section but as I recall did not read it; it was closer in content to material in Essay 3 in this volume.

2

REALISM AND THE DEBATE ON
IMPREDICATIVITY, 1917-1944

It is fair to say that the acceptability of impredicative definitions and reasoning in mathematics is not now, and hasn't been in recent years, a matter of major controversy. Solomon Feferman may regret that state of affairs, even though his own work contributed a great deal to bringing it about. I see the work of Feferman and Kurt Schütte on the analysis of predicative provability in the 1960s as bringing to closure one aspect of the discussion of predicativity that began with Poincaré's protests against "non-predicative definitions" in the first decade of the twentieth century and with Russell's making a "vicious circle principle" a major principle by which constructions in logic should be assessed. Although in the 1950s Paul Lorenzen and Hao Wang had undertaken to reconstruct mathematics in such a way that impredicativity would be avoided, insistence on this (to which even Wang did not subscribe) was very much a minority view, and Feferman in particular sought principally to analyze what predicativity is, with the understanding that some aspects of this enterprise would require impredicative methods. The picture has changed since then by work to which he has also contributed, which has brought to light how much of classical analysis in particular can be done by methods that are logically very weak, in particular predicative.

The pioneer of this latter effort, as Feferman has analyzed in detail, was Hermann Weyl.[1] Weyl also brought about the most dramatic episode in the early history by claiming that there is a "vicious circle" of

---

[1] See *Das Kontinuum* and Feferman, "Weyl Vindicated." Page references to works cited in an original and an English translation are to the original, with the page reference to the translation in brackets. I have, however, not necessarily followed the translations cited.

the kind pointed to by Poincaré and Russell in some basic reasonings in analysis. Curiously, his promising beginning for a predicative reconstruction of analysis was not pursued further at the time either by him or by others.[2] It may seem that Weyl's charge of a vicious circle found few adherents, but this appearance may be misleading because intuitionist analysis, which was just then being developed seriously by Brouwer, was not thought to be subject to the same difficulty. For some time thereafter, however, workers in foundations who accepted classical mathematics thought it necessary to reply to Weyl.

How did we get from the situation in, say, the time from 1906 to 1925, when impredicative methods seemed questionable to many of the best minds, to a later situation in which one could examine in a calm way the scope and limits of predicative methods, and in particular it was accepted that the tools of this examination might themselves be impredicative? Obviously, that would be a longer story than could be told in one lecture. Furthermore, much of it has already been well told by others. What I would like to do is briefly to sketch Weyl's criticism of analysis as he found it, which presents with exemplary clarity the assumptions behind the rejection of impredicative methods, and then say something about reactions to it by some leading figures: Hilbert, Bernays, Ramsey, Carnap, and Gödel. There is a conventional wisdom, to which I myself have subscribed in some published remarks, that the defense of impredicativity in classical mathematics[3] rests on a realist or platonist conception. Such a view is fostered by Gödel's famous discussion of Russell's vicious circle principle. Still, I want to argue that this conventional wisdom is to some degree oversimplified, both as a story about the history and as a substantive view. I don't think it even entirely does justice to Gödel. The history will end with Gödel's 1944 paper on Russell. Thus I will not deal with Feferman's work or with its very interesting background in the logical work of the 1950s; fortunately he has given a clear and informative presentation of that background himself.[4]

---

[2] In the case of Weyl himself, this was in the first instance due to his temporary conversion to intuitionism; see Weyl, "Über die neue Grundlagenkrise" and van Dalen, "Hermann Weyl's Intuitionistic Mathematics."

[3] As opposed to constructive, in particular intuitionist.

[4] "Systems of Predicative Analysis," part I.

The debates about impredicativity, and the later work on the analysis of predicativity, both centered on analysis. Above and below that level there are issues left over, where there may still be room for controversy. One is the impredicativity involved in defining sets by quantification over the universe of sets. The other is the question whether there is a sense in which the concept of natural number is impredicative. I will be able to make only a brief remark about the first of these issues. To my regret, the second will have to be left for another occasion.[5]

1. Weyl's Charge of a Vicious Circle in Analysis

It is well known that Weyl doubted on grounds of impredicativity the theorem that every bounded set of real numbers has a least upper bound. What is less well known is the precise way in which Weyl developed this objection, why he thought it *was* an objection. Poincaré's and Russell's objections to impredicativity were much more theoretical, and in the case of Poincaré bound up with his objections to logicism and in the case of Russell insulated from actual mathematics by his adopting the axiom of reducibility.[6]

All three thinkers held views close to the view of Frege that what are called sets are extensions of concepts.[7] Weyl begins the exposition in *Das Kontinuum* with a discussion of judgments and predication. The particular formulations are visibly influenced by Husserl, but nothing derived from Husserl is essential except that Weyl puts things in terms of properties, relations, and judgments rather than beginning with language as most contemporary logicians would. However, he considers judgments formed from atomic ones by operations of first-order logic. Concerning existential judgments, however, he thinks of the range of the quantifier as being a category of objects (close to the range of significance of a propositional function for Russell), although apparently they might be further restricted; existential judgments presuppose that the range of the quantifier is "a closed system of definite objects existing in themselves" (p. 4 [8]). In §2 Weyl describes modes of

---

[5] A particular reason for regret is that on this matter Feferman has taken issue with published views of mine; see Feferman and Hellman.

[6] I will not discuss Russell's attempt, in the second edition of *Principia*, to dispense with the axiom of reducibility. But see Potter, *Reason's Nearest Kin*, which also goes more deeply into Ramsey's position than I do here.

[7] Weyl mentions this view of Frege with approval (*Das Kontinuum*, p. 35 [47]).

construction of judgments that correspond to operations of first-order logic; such judgments with empty places he calls derived properties and relations.

This development is the background of what Weyl says about the concept of set. Although finite sets can be given by enumerating their elements, an infinite set can be given only as the extension of a property. Since one rule for forming properties is the substitution of an object in an argument place of a relation, the means of giving a finite set is itself reducible to characterizing it by a property.[8] Weyl calls the "mathematical process" the construction, beginning with a domain of objects and certain basic properties and relations, of the domain of sets and relations in extension, essentially those that are first-order definable with parameters. He states quite clearly that sets are extensions of properties so constructed.[9] He rejects in advance what Bernays later called a "quasi-combinatorial" conception:

> The representation of an infinite set as a "collection" brought together by infinitely many individual arbitrary acts of choice, collected, and then overviewed as a whole by consciousness, is nonsensical. (p. 15 [23])[10]

He opposes his own conception to the "totally vague concept of function, which has become canonical in analysis since Dirichlet" and to the similar concept of set.

So far we might think that the only sets that Weyl admits are those first-order definable from some basic relations, including perhaps relations or functions introduced by recursion. However, he is willing to view the sets obtained in this way as themselves a domain of objects and to continue the process, so that by iterating it one obtains what for us would be a model of a ramified higher-order theory. In his positive construction of analysis, as is well known, he does not use this possibility,

---

[8] In the case of judgments with more than one open argument place, Weyl talks of "many-dimensional sets," what we would call relations (in extension). He also uses the term "functional connection," which is somewhat confusing because the relations involved need not be functions in the usual sense. But it is by way of such relations that he introduces functions, which I neglect in the text.

[9] If the domain is the natural numbers, some form of definition by induction, of which he admits he is not yet giving a precise formulation, is also available (p. 18 [26]). Such a formulation is given in §7; cf. note 11 below.

[10] Cf. also "Über die neue Grundlagenkrise," pp. 49–50 [94].

so that what he develops is what we would call arithmetic analysis.[11] But he observes that if one construes real numbers as Dedekind cuts, if one has given a bounded set of real numbers of order 1, then the definition of the least upper bound as the union of the lower sections of the corresponding cuts is of order 2.[12]

In *Das Kontinuum* Weyl proceeds on the basis of a conception of sets as extensions of predicates that evidently must be antecedently understood, and although he does not make very clear why the usual definitions, such as the unramified version of the definition of the least upper bound, should involve a vicious circle, it is easy enough to see the conflict between them and the underlying understanding of the concept of set. What is less clear is why Weyl rejects the alternative. He expands on the matter in "Der *circulus vitiosus.*" If a concept B is extensionally definite *(umfangsdefinit)*, then for a clearly understood property E of the objects falling under B the question whether there is an object satisfying B with the property E has a definite sense, and the law of excluded middle applies.

On the basis of what he calls the "intuition of iteration," we are convinced that the concept of natural number is extensionally definite. Weyl offers the general concepts of object and property as non-extensionally definite concepts, and then claims that one can prove this even for "property of natural numbers." This is based on the intuition that if we have an extensionally definite concept of "κ-property" of properties of natural numbers, and A is a property of properties of natural numbers, then the property E_A that a number x has if and only if there is a κ-property E satisfying A such that x has E is "according to its sense different from any κ-property." The point is maybe clearer if put linguistically: Suppose $A(F)$ is a second-level predicate, whose arguments are filled by predicates of numbers, and $K(F)$ is also a second-level predicate that is assumed to be extensionally definite. Then '$\exists F[K(F) \land A(F) \land Fx]$' has a definite sense, but this sense must be different from that of any predicate expressing a κ-property, i.e., satisfy-

---

[11] Feferman shows ("Weyl Vindicated," pp. 264–265) that Weyl's recursion principle creates a difficulty for this. He is able to reconstruct Weyl's construction of the elements of analysis so as to get around it.

[12] This example has continued to be a problem for later work in predicative analysis allowing much more powerful methods. Thanks to Stephen Simpson and Solomon Feferman for setting me straight on this point.

ing **K**.[13] Thus if we assume that properties are individuated by the senses of predicates, it follows that any extensionally definite property of properties of numbers cannot be satisfied by all such properties of numbers.

Weyl concludes that the property "property of rational numbers" is not extensionally definite, and therefore the attempt to define the least upper bound of a set of real numbers as the union, which on his view of sets involves a quantification over properties of rationals, is meaningless.

Recall that the κ-properties are assumed to be an extensionally definite domain. As Weyl notes, it does not follow from the above observations that this domain is not closed under existential quantification in the weaker sense that in the above case, the property E_A is *coextensive* with a κ-property; indeed such closure would follow from a translation into this situation of Russell's axiom of reducibility. About this he says:

> It is, however, at the outset extraordinarily improbable that it is possible to set up an extensionally definite concept of "κ-property" in an exact way, so that every property E_A to be defined by the above schema from the *totality* of κ-properties is coextensive with a κ-property. In any case we have *not the shadow of a proof* of such a possibility. (p. 87 [112])

It seems from Weyl's examples that properties are given by predicates constructed by logical operations from the fundamental relations of given structures, in particular the natural numbers. On that assumption, it certainly is improbable that there is such an extensionally definite domain of properties that is closed under existential quantification as Weyl envisages here. Weyl's view on this point was close to that of a number of researchers of the time in considering the axiom of reducibility; it was widely thought that the existence of a property or propositional function of lower order coextensive with a given one of higher order is not a logical truth and is not evident.

One might, however, consider a reply suggested by Russell's often cited remark that the axiom of reducibility accomplishes what is accomplished by assuming the existence of classes. If we have among our objects arbitrary *sets* of natural numbers, and allow ∈ as a fundamental relation, then by Weyl's own principles for a given set *a* the predicate

---

[13] **K** need not be extensional. Weyl clearly does not think that a property coextensive with a κ-property is necessarily a κ-property.

$x \in a$ will be admissible. Then if we assume the second-level predicates involved extensional, and assume the usual comprehension schema for sets of natural numbers, then for Weyl's E_A we can obtain a coextensive property by replacing quantification over properties by quantification over sets. So the assumption that the property "a is a set of natural numbers" is extensionally definite would accomplish the end. Obviously Weyl would regard this assumption as hopelessly question-begging, and indeed on the conception of sets as extensions of predicates, there is no reason to think it true. But it is still well to keep that reply in mind.

2. Hilbert and Bernays

A prime concern of Hilbert's foundational work was always to defend the practice of classical mathematics, so that one would expect him to respond to Weyl's critique of the elements of analysis. In fact, Hilbert and Bernays both commented on Weyl's views at the beginning of their publication on foundations in the 1920s. Since it derives from an earlier lecture than Hilbert's, I will comment first on Bernays, "Über Hilberts Gedanken zur Grundlegung der Arithmetik." He interprets Weyl as attempting to build up mathematics in a constructive way and as observing that in analysis and set theory "in the attempt to carry out consistently the replacement of existential axioms by methods of construction, one falls into logical circles at every step" (p. 13 [217–218]. This leads Weyl to limit the allowed inferences so that such circles are avoided.

Evidently Bernays saw Weyl's constructivism as consisting in his insistence that sets are the extensions of predicates constructed step by step, although he does not always distinguish the Weyl of 1918–1919 from the intuitionism of "Über die neue Grundlagenkrise." Both positions, however, require giving up some of existing analysis. And Bernays does not see that either approach leads to a superior theory, granted that existing analysis lacks a relevant kind of intuitive evidence. "Concerning the question of pure mathematics, what matters is only whether the usual, axiomatically characterized mathematical continuum is itself a possible, that is, a consistent, structure."[14] The question of consistency leads him into a discussion of Hilbert's views.

---

[14] "Über Hilberts Gedanken," p. 14 [218].

Hilbert also attributes Weyl's criticism to a constructive standpoint. He says that the circle claimed by Weyl does not arise under two alternative standpoints commonly taken by mathematicians.

> First one says something like: A real number is a partition of the rational numbers that possesses the Dedekind cut property; there the concept of a partition of the rational numbers is sharp in its content and exactly limited in its extension.[15]

Hilbert says the objection to this standpoint is that the concept of set has led to paradoxes. He replies that the fact that the general concept of set may lead to paradoxes in no way implies that the concept of set of natural numbers does so, and the known paradoxes offer no evidence of it. "On the contrary, all our mathematical experiences speak for the correctness and consistency of this concept" (ibid.).

The second standpoint is one alluded to briefly by Bernays: the axiomatic. Then the real numbers are simply a structure and are not defined by Dedekind cuts; instead one has the Archimedean axiom and the so-called completeness axiom. Hilbert says that "the standpoint presented is logically completely free of objections and it only remains undecided, whether a system of the required kind is thinkable, that is whether the axioms don't perhaps lead to a contradiction" (ibid., p. 159 [200]. Then Hilbert remarks that real analysis has been pursued more thoroughly than almost any domain of mathematics, with complete agreement in the results. It is thus reasonable to accept the axioms on which the theory rests, although the problem remains to prove their consistency.

Paolo Mancosu has called attention to a manuscript in Bernays's hand in Hilbert's papers that discusses Weyl's criticism.[16] This manuscript contains the basic points of Hilbert's response; in places the latter seems to be following it almost verbatim. Both of the standpoints Hilbert mentions are set forth in similar terms in the manuscript, and in common with both published papers it attributes to Weyl a constructive standpoint, opposed to the axiomatic one: "He would like

---

[15] "Neubegründung," p. 158 [199].
[16] *From Brouwer to Hilbert*, p. 75. I am indebted to Mancosu for providing me with a copy of this text. He thinks it likely that this was the text for a lecture Bernays gave in Zürich in 1917, thus responding to *Das Kontinuum* before its publication.

pure mathematics to create for itself the objects with which it deals and not put forth statements about an unknown system of things."

In Hilbert's lectures *Probleme der mathematischen Logik* (SS 1920) some of the same issues come up.[17] Hilbert views both Russell and Weyl as seeking to reduce the concept of set to predication. He agrees with Weyl that a vicious circle arises if one attempts to define a predicate by quantification over all predicates of certain objects:

> Namely we must ask what "there is a predicate P" should mean. In axiomatic set theory "there is" always relates to the underlying domain B. In logic we can, to be sure, think of predicates as collected to a domain, but this domain of predicates cannot in this case be considered as something given at the outset, but the predicates must be constructed by logical operations, and the domain of predicates is determined only afterward by the rules of logical construction.
>
> From this it is clear, that in the rules of logical construction of predicates reference to the domain of predicates cannot be allowed. For otherwise a *circulus vitiosus* would arise. (p. 31)

Russell's axiom of reducibility represents abandoning a constructive standpoint in favor of an axiomatic one.

I have gone on at some length describing these early replies of Hilbert and Bernays to Weyl. Now I wish to pose the question what it tells us about the conventional wisdom with which I began. It may seem supported by the fact that Weyl's position is described as constructive; constructivism vs. platonism or realism is a quite standard opposition in the philosophy of mathematics. On the other hand, we do not think of the Hilbert of the 1920s as a platonist, and there seems to be basic agreement in the responses of Bernays and Hilbert.

The second of these contrary impressions is supported by the fact that what Weyl's constructive standpoint is primarily contrasted with is an axiomatic standpoint; in fact Hilbert's second alternative to Weyl's position could well be described as structuralist. The relation of such views to platonism or realism has been much discussed in our own day, but there is no evidence that Hilbert saw the issue in those terms. The

---

[17] I am indebted to Wilfried Sieg for pointing this out and providing copies. [The lectures cited in the text are now published in Ewald and Sieg. The passage translated occurs on p. 362.]

axiomatic standpoint allows him to *assume* that the real numbers are, in Weyl's terms, an extensionally definite domain. The question how the facts about them relate to our knowledge, always the central concern of realism, is not raised. What matters is that the conception of such a domain, as described by the axioms, should be a coherent one, in particular that the resulting theory should be consistent.

It is true that Weyl holds, and Hilbert and Bernays deny, that sets are in some sense constructions. But one must exercise some care about this point. Weyl sees the positive integers as constructed by beginning with 1 and iterating the operation of passing to the successor. We can view the numbers as constructions because we obtain in this way *all* natural numbers; that is of course the idea that the principle of induction cashes in. Evidently for this reason, Weyl regards the natural numbers as extensionally definite and is ready to apply the law of excluded middle to statements involving quantification over them.

Weyl describes a similar way of constructing sets, by constructing properties and relations by the logical operations, reducing them by not distinguishing coextensive ones, and iterating what he calls the "mathematical process". But the resulting concept of set of natural numbers is what Michael Dummett calls an "indefinitely extensible" concept. In this case this has a pretty definite meaning: Our procedure of construction can always be continued, but we never have reason to believe that we have captured all sets of integers.[18] We don't have a conception of a procedure of construction that gives the conviction that it ultimately yields all sets of integers.

There may be reasons for doubting how fundamental the distinction of these two cases is; even Weyl himself came to doubt it when, under the influence of Brouwer, he questioned treating the natural numbers as extensionally definite. I should stress that the difference is not one of predicativity. Suppose one grants the thesis expressed by Dummett, myself, and Edward Nelson, that the concept of natural number is impredicative. I don't think that would have moved the Weyl of 1917–1919. The "intuition of iteration" yields the extensional definiteness of

---

[18] Putting Weyl aside, even if we introduced from outside arbitrary ordinals, so that the process would cease to yield new sets once we got beyond the countable ordinals, the conception we would get would be that of the constructible sets of integers.

the natural numbers, so that predicates formed by quantification over the natural numbers have a definite sense.

However, what is more relevant to my concern is the question of the extent to which Weyl's view reflects a form of anti-realism, and if so whether that is what separates him from Hilbert and Bernays. One might infer sympathy for such views from the evident influence of Husserl and the suggestion that Fichte could offer illumination on questions of the philosophy of logic,[19] and perhaps also from Weyl's conversion to intuitionism shortly after the time that concerns us. But the difference that makes the difference between them concerns the relation of sets or functions to constructions of a particular kind, logical or linguistic.

There is no doubt that later, in "Sur le platonisme dans les mathématiques," Bernays did link the defense of impredicativity to platonism. An important feature of Bernays's discussion is that he distinguishes platonistic assumptions with respect to different domains of mathematics, beginning with what he calls the assumption of the totality of integers. In its consequences, this is close to what Weyl calls regarding the integers as extensionally definite. The next step would be to make a similar assumption about sets or sequences of integers; Bernays asserts that acceptance of impredicative definitions involving quantification over real numbers depends on an assumption of this kind (p. 55 [260]). Eliminating the second platonistic assumption would give rise to a program of a more thorough arithmetization of analysis than the usual one. "One emphasizes that an infinite sequence or a decimal fraction can be given only by an arithmetical law, and one regards the continuum as a set of elements defined by such laws" (p. 57 [261]). The reference is to the arithmetic analysis undertaken in *Das Kontinuum*.

But what does Bernays mean by "platonism"? Before he introduces the term, he remarks that objections have been made to modes of reasoning in analysis and set theory:

[According to these modes of reasoning] the objects of a theory are viewed as elements of a totality such that one can reason as follows: For each property expressible using the notions of the theory, it is [an] objectively determinate [fact] whether there is or

---

[19] "*Das Kontinuum*," p. 2; cf. "Der *circulus vitiosus*," p. 86.

there is not an element of the totality which possesses this property. (pp. 52–53 [258])

He then contrasts Euclid's language of construction with the existential character of the axioms in Hilbert's axiomatization of geometry. "This example shows already that the tendency of which we are speaking consists[20] in viewing the objects as cut off from all links with the reflecting subject" (p. 53 [259]).

However, in part because platonism so characterized is restricted in that platonistic assumptions are introduced stepwise, he says that platonism "does not claim to be more than, so to speak, an ideal projection of a domain of thought" (p. 56 [261]). He contrasts this with "absolute platonism" which he characterizes as "conceptual realism, postulating the existence of a world of ideal objects containing all the objects and relations of mathematics." This view, he thinks, is refuted by the paradoxes.

Bernays does not explain what he means by an "ideal projection of a domain of thought". However, it is clear from much of what he says that the difference between the platonism he defends and alternatives is a methodological issue. Platonistic assumptions are a certain way of setting up theories, and whether or not it is appropriate depends on the nature of the investigation. He contrasts an intuitive point of view (of which the paradigm is Hilbert's version of finitism) with platonism and considers the former more appropriate to number theory, even though the platonistic assumption of the totality of numbers is acceptable and necessary for analysis. Bernays's linking platonism with the acceptance of cases of the law of excluded middle is a step in the direction of Dummett's later logical characterizations of realism. But so far as the Bernays of 1935 is a realist with regard to mathematics, his realism is a methodological stance. In particular, I don't think it commits him to the view that on any question that might be posed in the language of a platonistic theory, there must be an objectively determinate answer. Some years ago Charles Chihara discussed a hypothetical view he called "mythological Platonism." Bernays's view of 1935 might well be called "methodological platonism".

---

[20] The French is "consiste"; the later German version has "dahin geht," which might be rendered as "amounts to [viewing]" (*Abhandlungen,* p. 63). It's not clear that this difference makes a difference.

3. Ramsey

The idea that the acceptance of impredicativity requires a platonistic point of view may have originally derived its currency from a remark of Rudolf Carnap about F. P. Ramsey's attempt to reconstruct the system of *Principia Mathematica* in a way that would make the axiom of reducibility unnecessary. Carnap interprets Ramsey as assuming "that the totality of properties already exists before their characterization by definition" ("Die logizistische Grundlegung," p. 102 [50]). He goes on to say, "Such a conception, I believe, is not far removed from belief in a platonic realm of ideas which exist in themselves, independently of *if* and *how* finite human beings are able to think them."

What was the view that prompted this comment? "The Foundations of Mathematics," the paper to which Carnap refers, is a wide-ranging discussion of problems in foundations, attempting to reconstruct the position of *Principia* with the help of ideas from Wittgenstein's *Tractatus*.[21] Probably its most enduring contribution concerns the paradoxes: Ramsey worked out the distinction of mathematical and semantical paradoxes, pointed out that the simple theory of types was sufficient to circumvent the set-theoretical paradoxes, and showed how concepts of a different nature are required to generate the other paradoxes. But he tried to give a more fundamental defense of abandoning the position of the first edition of *Principia,* with the ramified theory and the axiom of reducibility. It is there that his defense of impredicativity comes in.

The guiding idea is that of Wittgenstein's view of logic (not his view of mathematics), that the "propositions" of logic are tautologies. Ramsey undertook to develop this view as a view of mathematics. It would follow that the propositions of mathematics "say nothing" and lack sense in the proper sense of that word, contrary to Carnap's reading and our conventional wisdom. Another Wittgensteinian idea, which Ramsey sees as the key to overcoming the difficulties of *Principia,* is that every proposition is a truth-function of elementary propositions. The decisive move is to admit such functions without regard to whether the number of arguments is finite:

---

[21] It is remarkable that this paper was submitted for publication in 1925, when Ramsey was 22 years old.

A *predicative function* of individuals is one which is any truth-function of arguments which, whether finite or infinite in number, are all either atomic functions of individuals or propositions. . . . Admitting an infinite number involves that we do not define the range of functions as those which could be constructed in a certain way, but determine them by a description of their meanings. (p. 39)

A little later, contrasting '$(\varphi).\varphi a$' with an ordinary finite truth-function, he remarks that the former cannot be elementarily expressed "owing to our inability to write propositions of infinite length, which is logically a mere accident" (p. 41).

Ramsey finds even this generous concept of predicative function not quite adequate, because of his view inherited from Wittgenstein that identities are not genuine propositions. So that at the level of individuals he introduces "functions in extension," which are hardly distinguishable from functions in the set-theoretic sense from individuals to propositions.[22]

Carnap's characterization of Ramsey's position seems to me on the whole just, and this case supports the conventional wisdom, even though Ramsey might have thought that the idea that logical truths are tautologies minimized the significance of his realistic commitment. But not long afterward, he saw difficulties about the axiom of infinity,[23] and before the end of his life changed his position more radically.

4. Carnap

Carnap's critical discussion of Ramsey's view prefaces a proposal of his own, which is quite different from those we have considered so far.[24] Carnap asks how we verify a second-order universal statement. If, in impredicative logic, we had to run through all cases for this verification, we would indeed encounter a circle. Rather, we can verify it by logical derivation, where the universally quantified second-order variable will in general play the role of a parameter for an arbitrary property.

---

[22] Ramsey argues that a corresponding move is not needed at higher levels.
[23] "Mathematical Logic," pp. 79–81.
[24] A similar suggestion is made in Langford, "On Inductive Relations."

If we reject the belief that it is necessary to run through individual cases and rather make it clear to ourselves that the complete verification of a statement about an arbitrary property means nothing more than its logical (more exactly, tautological) validity for an arbitrary property, we will come to the conclusion that impredicative definitions are logically admissible. (1931, p. 104 [51])

Carnap's suggestion is very sketchy, but it makes no noticeable appeal to any form of realism. Its focus on verification, in fact, makes it adaptable to intuitionism. It is hard to see that it says more than that we are able to obtain conclusions about arbitrary predicates by reasoning schematically, and that impredicativity is no obstacle to the verification of generalizations about them by such reasoning. The intuitionist will naturally ask whether Carnap's suggestion gives us any reason to believe that the law of excluded middle will hold where second-order quantifiers are involved or even tells us what meaning to assign to the negation of a second-order generalization.

Carnap returns to the matter in §44 of *Logical Syntax* but hardly gives more detail. However, the view of that work offers, rather than an answer to the questions just raised, a reason for brushing them off. Since the choice of predicative or impredicative logic is part of the setting up of a logical framework, and questions of meaning can only be answered in the context of such a framework, obviously the question of the meaning of negations is an improper question or is answered by giving the rules of inference involving negation. Since, by the principle of tolerance, "in logic there are no morals," neither the admission of impredicativity nor the use of classical logic with them can be rejected on the grounds of philosophical argument. The matter could be settled in a given context by working out the consequences of the different choices and deciding whether a given choice suits the purpose at hand. But on this specific issue, he doesn't do anything to elaborate the consequences.

Carnap's proposal is given a more direct reading by Gödel, who reads it and Langford's as interpreting "all" "as meaning analyticity or necessity or demonstrability."[25] The interpretation as necessity fits better Langford's proposal than Carnap's, since modalities are not

---

[25] "Russell's Mathematical Logic," p. 127.

mentioned in either of the Carnapian texts. Reflecting on the view of *Logical Syntax*, Carnap might well have said that $\forall F\, A(F)$ is true by virtue of the analyticity of $A(F)$, but not, I think, that $\forall F\, A(F)$ *means that $A(F)$* is analytic. The reading in terms of demonstrability is let in because in 1931 Carnap is not explicit about what he means by logical validity. Indeed, if it is put in terms of a semantic definition of higher-order consequence (not that clear to anyone in 1931), then it is not so clear how it differs from what Gödel rightly sees Carnap as seeking to avoid, namely the reading of the quantifier as infinite conjunction.[26]

5. Gödel: Platonism and the Vicious Circle Principle

Gödel already addresses the question of impredicativity in his Cambridge lecture of 1933,"[27] where he mentions it as one of the weak spots in justifying set theory. He has already said that in the application of excluded middle, the law is applied "just as if in some objective realm of ideas this question was settled quite independently of any human knowledge" (p. 49). In turning to the problem of predicativity, he says that the notion of class of integers is "essentially the same" as that of property of integers, and in giving an example of an impredicative definition formulates it as involving quantification over all properties (presumably of a given type).

> Again, as in the case of the law of excluded middle, this process of definition presupposes that the totality of all properties exists somehow independently of our knowledge and our definitions, and that our definitions merely serve to pick out certain of these previously existing properties. (p. 50)

Gödel is prompted by these considerations to conclude that "our axioms, if interpreted as meaningful statements, necessarily presuppose a

---

[26] [It has been brought out recently, just about the time when I completed this essay, that Carnap had studied higher-order logic more deeply than I or most others were aware. See Awodey and Carus, "Carnap, Completeness, and Categoricity," on the investigation he was engaged in at the end of the 1920s, and Reck, "Carnap and Modern Logic," for a lucid presentation of Carnap's work in and with logic. It is clear that one of the difficulties of Carnap's early work was the lack of a clear distinction between syntactic and semantic consequence.]

[27] "The Present Situation." Page references are to *Collected Works*, vol. III.

kind of Platonism, which cannot satisfy any critical mind and which does not even produce the conviction that they are consistent" (ibid.). It is well known that Gödel over the course of the next ten years did become satisfied.

It's not surprising that in the discussion of these issues in 1944, Gödel in effect defends the platonism about which he had made the above critical remark in 1933. He distinguishes three versions of the vicious circle principle, of which he says that only the first, "No totality can contain members definable only in terms of this totality," rules out impredicative definition.[28] Gödel observes, in agreement with both Weyl and the defenders of impredicativity, that classical mathematics does not satisfy the principle in this form; he considers this an objection to the principle rather than to classical mathematics; he concludes that the system of the first edition of *Principia* (with reducibility) also does not satisfy it if "'definable' means 'definable within the system'" (ibid.).

Gödel mentions the Carnap suggestion, but his main direct claim against the principle is that

> the vicious circle principle in its first form applies only if the entities involved are constructed by ourselves. In this case there must clearly exist a definition (namely the description of the construction) which does not refer to a totality to which the object defined belongs, because the construction of a thing can certainly not be based on a totality of things to which the construction itself belongs. (p. 127)

A little later he describes the standpoint from which the principle applies as constructivistic or nominalistic. This would, I think, include the definitionism of Poincaré and Weyl as well as what he sees as Russell's own tendency to try to eliminate reference to classes and concepts by paraphrase. The term "nominalistic" is probably inserted so as to include reduction of sets or classes to linguistic entities, i.e., predicates.

At this point Gödel makes a strong affirmation of realism concerning classes and concepts, making one of his most quoted statements:

---

[28] "Russell's Mathematical Logic," *Collected Works,* vol. II, p. 127. (References to this paper will be by page number in this volume.)

It seems to me that the assumption of such objects is quite as legitimate as the assumption of physical bodies and there is quite as much reason to believe in their existence. They are in the same sense necessary to obtain a satisfactory system of mathematics as physical bodies are necessary for a satisfactory system of our sense perceptions. (p. 128)

However, Gödel has clearly made some distinction, not made in the 1933 lecture, between independence of our constructions and definitions and independence of our knowledge. It's the former independence that is decisive for the defense of impredicativity. It seems possible to be quite realistic about what is constructible in the relevant sense and still reject impredicativity on the basis of this conception of set. Gödel does not quite say this, but I think it is implicit in the characterization of the constructivistic or nominalistic standpoint as one toward the *objects* of logic and mathematics. Intuitionism, by contrast, though it is often presented in that way, is first of all a standpoint about mathematical *truth* and its relation to proof.[29]

A distinctive feature of Gödel's realism is that it is not just realism about sets but also about "concepts". Although I can't defend this claim here, I don't think Gödel has as good an argument for the claim that the assumption of concepts on his conception of them is "necessary for a satisfactory system of mathematics" as he has for the corresponding more straightforward claim about sets. Moreover, his position on this point was a source of trouble for him; he seems never to have arrived at a formal theory of concepts that satisfied him, even very late in life he was troubled by what he called the "intensional paradoxes," and issues about concepts troubled him in a directly philosophical way. It appears from both "The Present Situation" and the Russell paper that he does not distinguish the issues about impredicativity with reference to properties (and in the terms of the latter, concepts) from those with reference to sets as clearly as most of us would today, even though he saw very clearly that the iterative conception was satisfactory as an

---

[29] This is not to say that intuitionistic mathematics does not differ in its ontology from classical; for example, as a referee suggests, Brouwer's argument for his Bar Theorem and the Brouwer-Heyting-Kolmogorov interpretation of the logical constants make proofs fundamental mathematical objects as they are not in standard formulations of classical theories.

account of the universe of sets, while nothing corresponding satisfied him as an account of concepts.

Although it goes beyond our historical limit, I will make one further remark about Gödel. Gödel clearly paid attention to the fact that ZF licenses what one might call "super-impredicative" definitions, specifications of sets by conditions involving quantification over the universe of sets. He refers to this explicitly in a justification of the axiom of replacement published by Wang.[30] Parallel to defenses of impredicativity at lower levels would be a defense of super-impredicativity by maintaining or assuming that the universe of sets is extensionally definite. Many of those who reflect on set theory reject this defense and hold that if that were right then the universe would be a set, which standard set theory denies. I think Gödel accepted it. It may have been connected in his mind with his realism about concepts; I don't now see clearly what the connection would be. But that he held this view is indicated by the following tantalizing remark in Wang's reconstruction of his conversations with Gödel:

8.3.4 To say that the universe of all sets is unfinishable does not mean objective undeterminedness, but merely a subjective inability to finish it.[31]

6. Conclusions

Feferman describes the position of Poincaré and of *Das Kontinuum*, and by implication of other opponents of impredicativity, as "definitionism":[32] sets are definable sets. What we have surveyed in §§2–5 are criticisms of definitionism that were made after Weyl dramatized the effect of this view on mathematical practice. The common denominator of the defenses of impredicativity might be called "anti-definitionism". This observation brings out the fact that what Poincaré and Weyl advanced, and the writers we have surveyed criticized, was a view about the *objects* of analysis. What makes this view constructive, as Weyl and others already made clear, is that it offers a way of taking

[30] *From Mathematics to Philosophy*, p. 186. Cf. also *A Logical Journey*, p. 259.
[31] *A Logical Journey*, p. 260.
[32] "Weyl Vindicated," p. 254.

sets of integers, and other sets that arise in mathematics, as constructed in a step-by-step way.

Some philosophers of mathematics have distinguished "object platonism" from "truth Platonism." Where the defense of impredicativity does rest on a form of platonism, the view presented might be described as a form of object platonism, asserting the independence of sets and real numbers from definitions and constructions. However, defenses of impredicativity along these lines are available from a broadly intuitionist point of view, so that it is not evident that the rejection of "definitionism" needs to commit us to any form of realism. Therefore the term "object platonism" may only be appropriate in the classical context.

"Truth platonism" would hold that there is an objective truth of the matter about the domains of mathematics at issue that is independent of our knowledge, perhaps even of our possibilities of knowledge, if the latter notion is not hopelessly idealized. Of course the existence of such objective truth implies something about the existence of the objects talked about, since among the truths will be some of the form 'there exist (sets, numbers) such that . . .'. There might be such truth about objects that are in some sense "constructions of our own," since it might be determined quite independently of our knowledge what we can construct, or what can be constructed by quite specific means such as first-order operations beginning with the basic relations of some given structure.[33] Thus it seems possible to hold truth platonism while denying object platonism in the specific form at issue here.

To what extent did the defenders of impredicativity we have surveyed assume truth platonism in this sense? Ramsey and Gödel certainly did; Carnap clearly did not. Gödel's actual argument against the first form of the vicious circle principle, however, does not in any obvious way turn on it. In §2 I argued that even Bernays's methodological platonism of 1935 falls short of truth platonism, if the latter view goes beyond admitting the law of excluded middle for the relevant domain of objects.

---

[33] The distinction of object- and truth-platonism made here is essentially that of my "A Distinction," where the former is called objects platonism. Others (in particular Isaacson, "Mathematical Intuition") have made such a distinction, not necessarily coinciding with the present one.

One way of distinguishing Bernays's platonism as a methodological stance adopted in specific domains of mathematics and platonism as a more metaphysical view would be to make the distinction between an internal and external perspective on a given theory. If we ask whether the continuum hypothesis must be true or false, of course there is a sense in which the answer is trivially affirmative: In set theory we assume CH ∨ ¬CH, but if CH, then CH is true, and if ¬CH, CH is false; therefore it is either true or false. So long as we are using the language of classical set theory and taking it at face value, there is no room for refusing to affirm that CH is either true or false. But it does not follow that, if one reflects on set theory and considers what grounds there might be for the truth or falsity of CH, one is forced to assume that such grounds for one or the other conclusion must exist. One relevant sense of "grounds" would be epistemic, possible reasons that an agent might have for affirming or denying CH. A view that one might call "holistic verificationism" would accept the law of the excluded middle, and thus the "internal" affirmation that CH is either true or false, but in the external stance would maintain that there is no sense to be made of the idea that CH is true independently of any possibility an epistemic agent might have of coming to know it. Such a view has had its defenders,[34] and it no doubt also has its difficulties. But there is no reason to think it requires a conception of set like the one underlying Weyl's arguments.

In the discussion of my lecture at the Reflections conference, Solomon Feferman asked whether the history I recounted shows that there is a nonplatonist defense of impredicativity. I assume he had in mind the context of classical mathematics. The history certainly offers suggestions in that direction, but it also shows that the answer depends on what one counts as "platonism." Holistic verificationism is a view that would be an example of what Feferman was asking for. Much recent work on the philosophy of mathematics seeks a middle ground between

---

[34] As a general view of truth, it is certainly implicit in the writings of W. V. Quine. Remarks expressing this tendency occur in Tait, "Truth and Proof" and in lectures by him I have heard, but he might reject the view as here stated because of skepticism about ideal possibility applied to epistemic concepts.

[Tait's view is made clearer in *The Provenance*, for example in the introduction. Although much of what is said in the text applies to it, the label "holistic verificationism" does not fit it. I attempt to do justice to Tait's view in §2 of Essay 12 in this volume.]

such a view and more full-blown truth platonism, what one might call transcendental realism. One might well shift Feferman's question and ask whether contemporary work, whatever its relation to the history, yields the kind of defense of impredicativity he is asking for. That is a larger question than can be attacked on the present occasion.[35]

[35] An earlier version of this essay was presented to the symposium Reflections, honoring Solomon Feferman on his 70th birthday, at Stanford University December 13, 1998. It was an honor and a pleasure to participate in this event honoring Sol, a valued friend whose leadership in the foundations of mathematics, both intellectual and institutional, has benefited all of us. In earlier years, when I worked in proof theory, I learned much from his writings, and he was a valuable source of information and advice for my own work. Since then, circumstances brought us closer, and we were collaborators on the long-running task of editing Gödel's works, where Sol's leadership and wide-ranging grasp of logic have been essential.

Section 6 of the essay is an inadequate response to a question asked by Sol after the lecture, which I continue to ponder. I am also indebted to Paolo Mancosu, Wilfried Sieg, W. W. Tait, and an anonymous referee.

In his review of the collection in which this essay appeared, David Charles McCarty makes the following comment about the essay:

> Parsons wishes to assay the prospects for defending impredicative definitions, assumptions, and principles in mathematics. This seems an odd project, there being little call to defend our hugely successful and thoroughly impredicative mathematics from the charges Russell and Poincaré brought against it. Their charges were neither cogently stated nor proven. They were so packed with fallacies, e.g. illicit congress between use and mention, that the worries they once fostered now seem more than mildly neurotic. Parsons looks hard and long to see if a defense of impredicativity requires platonism in the philosophy of mathematics. I would have thought that a quick glance at intuitionistic mathematics—defensible, highly impredicative, nonplatonistic—would suffice to show that there is no such requirement.[1]

The reader would never know from reading this comment that my essay is almost entirely historical. The question I was asking about realism or platonism (in the essay I prefer the former term) is not "Is realism required for a defense of impredicative methods?," but rather "To what extent did the defenses of impredicative methods in the period that concerned me rest on realism?"[2] And the answer is, to some

---

[1] Review of Sieg, Sommer, and Talcott, pp. 240–241.

[2] Moreover, my story begins not with Poincaré and Russell but with Weyl's *Das Kontinuum*. I don't find a tissue of fallacies in his argument (though of course contestable assumptions).

extent, but not so exclusively as a certain conventional wisdom would have it.

Only briefly at the end do I address the question whether such an assumption is needed for a defense, and there it is explicitly said that it is classical mathematics that is in question.

Still, McCarty's invocation of intuitionism does raise an issue about the historical period I was concerned with. The defenders of impredicativity whom I discuss (Hilbert, Bernays, Ramsey, Carnap, and Gödel) were all concerned to defend classical mathematics. Still, none of them considers intuitionistic mathematics where issues concerning impredicativity are being discussed. I think the reason is clear: intuitionistic mathematics as practiced by Brouwer and those around him was not at the time recognized as impredicative. We might well discern impredicativity in Brouwer's theory of ordinals and its application to the theory of free choice sequences, but such recognition was not common coin in the period I was writing about. In fact, McCarty does not challenge my assertion that "intuitionist analysis, which was just then being developed seriously by Brouwer, was not thought to be subject to the same difficulty" (p. 41). Brouwer's version of intuitionism could be regarded as predicative in an extended sense, which admits what are called generalized inductive definitions.[3] Furthermore, Weyl's version of intuitionism in "Über die neue Grundlagenkrise" has been seen recently to be logically weaker than Brouwer's.[4]

In the last paragraphs of his review, no longer commenting directly on my essay, McCarty raises some points that deserve comment. He is interested in the fate of "predicativism" and rightly notes that it is ignored by mathematicians at large. He does not note that what logicians, in particular Feferman, have been interested in is not exactly predicativism but rather predicativity. How should we understand it, and given some analysis, what can be proved by predicative methods, or can be proved by methods in some way reducible to predicative methods?

In my view, throughout the classical period up to 1931, predicativism remained a less developed view than intuitionism and the viewpoint of

---

[3] See *Mathematical Thought and Its Objects*, §51 and references there; however, I failed to make the connection with Brouwer. Generalized inductive definitions were little studied in the inter-war period. But I don't know of anyone who claimed that they involved a vicious circle.

[4] See van Dalen, "Hermann Weyl's Intuitionistic Mathematics."

the Hilbert school. Concern with predicativity originated in the demand of Poincaré and other French mathematicians that what is talked about should be definable "in a finite number of words." In a sense "predicativism" was stillborn, because Borel, Baire, and Lebesgue used set-theoretic methods that could not satisfy this demand in any rigorous way. The complexity and ambivalence of their views are well described by Gregory Moore in his account of their reactions to Zermelo's formulation of the axiom of choice and proof from it that every set can be well-ordered.[5] Furthermore, at the time the notion of definability was not at all exactly formulated. One respect in which Weyl's *Das Kontinuum* represents an advance is that he did have a much more definite concept of definability to work with.

The rejection of impredicativity obtains motivation from an initially natural conception of what a set is, that it is the extension of a predicate. For then it seems that for each set there should be a predicate that defines it. If a set is defined impredicatively, say by quantifying over all sets of natural numbers, what meaning can we give to the totality of predicates of natural numbers that are supposed to define these sets? The conception in question is reminiscent of Frege's conception of the extension of a concept, and it is only in such terms that he could make sense of discourse about sets. But Frege avoided the problem of predicativity by talking of extensions of concepts rather than of predicates and assuming that concepts obeyed full second-order logic.[6]

Even if the French writers of before 1914 could not make "predicativism" a thorough and comprehensive point of view, their preference for the definable proved *mathematically* fruitful. It led to the development of descriptive set theory, a theory of sets of real numbers that are definable in a definite, though extended, sense. In much of their work and certainly in the field as it developed, an assumption was used that Weyl in 1918 explicitly rejected, that it is legitimate to quantify over arbitrary real numbers and apply classical logic at that point.

---

[5] *Zermelo's Axiom of Choice*, §2.3. A well-known text that illustrates their views and the contrasting, more pro-set-theoretic, view of Jacques Hadamard is Baire et al., "Cinq lettres," translated as appendix 1 to Moore's book.

[6] Since Frege did not comment on the issues about predicativity raised in the first decade of the twentieth century, it is ironic that restricting second-order logic to be predicative was sufficient to save Basic Law V. See Burgess, *Fixing Frege*, ch. 2. In my review of the book (pp. 411–413), I comment briefly on the question whether predicativity should be an issue for Frege.

One might then ask: Could we not accept Weyl's general view, only moving his boundary up one level? That is, the real numbers would be accepted as "extensionally definite," but already sets of real numbers would have to be treated in the way that Weyl proposed for sets of natural numbers. This would be a conception of predicativity relative to the real numbers, on the model of predicativity relative to the natural numbers, which is usually called predicativity *simpliciter*. This would be an instance of the general relativization of the concept of predicativity suggested by Feferman.[7]

Such a view might be argued for on the ground that the real numbers are given by geometric intuition in a way that makes it reasonable to treat them as an extensionally definite totality. In one text Émile Borel admits that "mathematicians have a clear common notion of the continuum," which he says is "acquired through geometric intuition."[8] However, he seems unable to reconcile this with his skepticism about nondenumerable sets. Apart from this attitude, which most of us would not share, it would be difficult to maintain that any geometric intuition gives us the continuum as an extensionally definite totality, that is, a totality of *points*. And that seems necessary for a mathematical program based on this idea.

But if we make such an assumption, reasoning with quantifiers over real numbers becomes acceptable without the limitations imposed by predicativity relative to the natural numbers. This is a powerful assumption, which may not have a philosophical rationale that is at all convincing. However, the standpoint is reasonably well defined mathematically. It is at least roughly what Stephen Simpson calls the "Cinq lettres program," after the famous five letters mentioned above.[9] The notions of Borel set and projective set would be definable from such a standpoint. Simpson maintains that the results of classical descriptive set theory, as it developed before the Second World War, could in principle be proved in such a framework.[10] In fact, since he says that "the basic theory of Borel and analytic sets" can be proved in Friedman's

---

[7] "Predicativity," pp. 620–621.
[8] "Sur les principes de la théorie des ensembles," p. 160. Paul Bernays also says that the conception of the continuum is originally geometric but he never proposes restricting reasoning about sets of real numbers as the standpoint I am suggesting would require. (Cf. §4 of Essay 1 in this volume.)
[9] See Simpson, "Gödel Hierarchy," p. 120.
[10] Ibid.

theory ATR_0, only a small part of what I am suggesting is required for that. Other results belonging to classical descriptive set theory, such as Kondô's theorem, are of the strength of Π_1^1 comprehension.[11] Let's assume that no more is needed for classical (pre-war) descriptive set theory. Then we are still within a fragment of second-order arithmetic that is within the reach of proof theory as developed in the post-war period, so that the assumption I mentioned is stronger than needed also from a philosophical point of view.

That is, however, not true of more recent descriptive set theory, even the part of it that stays within ZFC. Even full second-order arithmetic is not sufficient for the proof of Martin's classical result that Borel sets are determinate. The proof makes essential use of the axiom of replacement, and it had been shown even before by Harvey Friedman that this would be necessary.

In any case, the standpoint in question would have to admit as meaningful many questions that have been explored in modern descriptive set theory that could not be answered within it and indeed required for their solutions assumptions well beyond ZFC.[12]

---

[11] For details on these cases see Simpson, *Subsystems of Second-Order Arithmetic,* §§V.3–4, VI.1–3.

[12] Thanks to Solomon Feferman for comments on two earlier versions.

3

PAUL BERNAYS' LATER PHILOSOPHY

OF MATHEMATICS

1. Mathematics and Philosophy in Bernays' Career

The name of Paul Bernays (1888–1977) is familiar probably first of all for his contributions to mathematical logic. Many of those were in the context of his position as David Hilbert's junior collaborator in his proof-theoretic program inaugurated after the First World War. For those of us starting out in logic in the mid-twentieth century, the monumental *Grundlagen der Mathematik* of Hilbert and Bernays was one of the basic works in mathematical logic that we were obliged to study. That was the more true for those like me who aspired to work in proof theory.

Bernays' collaboration with Hilbert ended with the publication in 1939 of volume II of that work. Because the Nazis had forced his removal from his position in Göttingen in 1933, the collaboration ceased to be face-to-face in 1934, when Bernays moved to Zürich, where he lived for the rest of his life.[1]

Bernays is also known, though less well known, as a philosophical writer. The fact that he had some philosophical training was apparently a reason why Hilbert chose him as his collaborator. Constance Reid tells the story that when Hilbert visited Zürich in the spring of

---

[1] I do not know exactly when Bernays' move took place, but it appears to have been in April 1934. [It is clear from the preface that volume I of Hilbert and Bernays' *Grundlagen*, was completed before Bernays moved. Thereafter there seems to have been no scientific contact between Bernays and Hilbert, although he did correspond in the 1930s with Gentzen and Ackermann. That either of them had a scientific relation to Hilbert during that period is very doubtful, although Gentzen was for a time nominally Hilbert's assistant. Cf. Menzler-Trott, *Logic's Lost Genius*, p. 62 and n.157.]

1917, he went for a walk with two local mathematicians, Bernays and Georg Pólya. The conversation turned to philosophy, and although Pólya was usually rather voluble and Bernays was not, this time Bernays did most of the talking. On the spot Hilbert invited Bernays to Göttingen to work with him, and Bernays accepted.[2] This story is too neat to be quite correct. In fact, in a letter to Reid commenting on a draft of the biography, Bernays writes that the walk indeed occurred in the spring of 1917 but that the offer was made and accepted when Hilbert returned to Zürich in September for his lecture "Axiomatisches Denken."[3]

Although Bernays' self-identification was clearly as a mathematician, and his academic career, such as it was, was in mathematics, he maintained an interest in philosophy throughout his life and continued to write philosophical essays. His writing was not confined to the philosophy of mathematics, but naturally enough that is where its main weight lies. Near the end of his life he put together a collection *Abhandlungen zur Philosophie der Mathematik,* which contains the most important such essays.[4] A more comprehensive collection, *Essays on the Philosophy of Mathematics,* is in preparation and will have the originals and English translations on facing pages, with introductions by various hands.[5]

Bernays' philosophical formation is of interest from a contemporary point of view because it would have to be described as neither

---

[2] Reid, *Hilbert,* pp. 150–151.

[3] Bernays to Reid, November 27, 1968, copy in the Bernays papers in the Archive of the Eidgenossische Technische Hochschule, document Hs 975: 3775. Later correspondence shows that Bernays was concerned about this inaccuracy after the publication of the book. However, the book was later reprinted unrevised. Also in "Kurze Biographie," p. xv, Bernays states that the offer was made when Hilbert came for the lecture, which occurred on September 11, 1917.

The interval would have allowed Hilbert not only to think about the idea of engaging Bernays but to inquire with people who knew his work, such as his *Doktorvater* Edmund Landau, Leonard Nelson (see below), Ernst Zermelo, and in Zürich probably Adolf Hurwitz and Hermann Weyl.

[4] "Sur le platonisme" and "Quelques points de vue" are translated from French, the latter by Bernays himself.

[5] All essays reprinted or translated in *Abhandlungen,* as well as a number of others, and a number of items not cited here, will be included in this collection. Quotations from these essays in the present essay are in the translations for this collection, in some cases based on published translations. Translations of quotations from other writings of Bernays are my own.

"analytic" nor "continental" in the sense in which those terms are now used to classify philosophical styles and tendencies. His principal teacher in philosophy was Leonard Nelson (1882–1927), who would be described as a neo-Kantian but who did not belong to either of the principal neo-Kantian schools, the Marburg school and the southwestern school. However, Bernays writes that earlier in Berlin he heard lectures of Alois Riehl and Ernst Cassirer.[6] Riehl had been one of the founders of neo-Kantianism, and Cassirer was a leading representative in the twentieth century of the Marburg school. Edmund Husserl was a professor in Göttingen during Bernays' studies there and led an active school, but it seems that he and his philosophy had little impact on Bernays. Husserl is not mentioned in Bernays' autobiographical sketch, and the rather few comments about Husserl and phenomenology in Bernays' writings are mostly critical. Nonetheless there are indications that eventually they exercised some positive influence on Bernays. Nelson sought to revive the philosophical approach of J. F. Fries (1773–1843), and Bernays published three early essays in the *Abhandlungen der Fries'schen Schule,* which Nelson had revived as the organ of his school.

Bernays published no more philosophy until several years after his return to Göttingen in 1917, and most of the papers published during his second sojourn in Göttingen reflect rather directly his collaboration with Hilbert. I will make some remarks later about "Die Philosophie der Mathematik und die Hilbertsche Beweistheorie," the most significant and original of these papers.[7]

By his "later" philosophy I have in mind principally what we find in publications after 1945. In the later period he was evidently much influenced by his Zürich colleague Ferdinand Gonseth (1890–1975), with whom he collaborated actively in organizational matters. Remarks of Bernays himself indicate a significant change in his philosophical views between the early 1930s and the post-war period.

> I had come close to the views of [Ferdinand] Gonseth on the basis of the engagement of my thinking *(meinen gendanklichen Auseinandersetzungen)* with the philosophy of Kant, Fries, and Nelson, and so I attached myself to his philosophical school.

---

[6] "Kurze Biographie," p. xiv.
[7] See also §4 of Essay 1 in this volume.

. . . during the period in which these articles were published, my views on the relevant questions have changed almost exclusively in response to new insights gained from research in the foundations of mathematics.[8]

There is a tension between these two remarks if they are applied to the transitional period between 1934 and 1945, since the first suggests that reflections on the philosophy of Kant, Fries, and Nelson, probably differences with them, led to his coming close to Gonseth's views, while the second attributes change in his views to developments in the foundations of mathematics. It seems to me likely that the problems the Hilbert program encountered in the wake of Gödel's theorem, and perhaps other mathematical developments of the time, crystallized some dissatisfactions on Bernays' part with his Kantian legacy. We will see shortly what those dissatisfactions might have been. I will comment on aspects of two essays from the transitional period, the well-known "Sur le platonisme" and the almost unknown "Grundsätzliche Betrachtungen zue Erkenntnistheorie."

Bernays' style of philosophical writing was essayistic, and it is unlikely that he thought of himself as developing a systematic view. I do not find evidence of the drive toward system that one finds in so many of the best philosophers. That is evidently connected with his extremely modest personality and is reflected in his tendency to be a disciple of others, earlier Nelson and Hilbert and later Gonseth. For our purposes, it is most illuminating to single out certain themes that run through his writings. I will concentrate on two, what might be called anti-foundationalism, which summarizes his later epistemological atti-

---

[8] "Kurze Biographie," p. xiv, and preface to *Abhandlungen,* p. vii. In the second Bernays wrote ". . . meine Ansichten . . . sich fast nur insoweit gewandelt haben, als es durch die in der mathematischen Grundlagenforschung gewonnenen neuen Einsichten bedingt wurde." One might also translate this as "My views . . . have changed almost exclusively to the extent that is required by the new insights gained from research in the foundations of mathematics." The remark might be taken to minimize not only philosophical motives for changes in his views, but the changes themselves. I find that difficult to reconcile with the totality of Bernays' philosophical writing. But it also suggests that whatever other reasons for changes there may have been, they were adequately motivated by developments in foundational research. I find that hard to reconcile with Bernays' actual evolution. But he very likely did think that his later views yielded a better understanding of the results of foundational research. (I am indebted here to an anonymous referee.)

tude, including his rejection of the a priori, and structuralism. That leaves out much. A lot of his philosophical writing consists of comment on developments in logical research in foundations, and I will also have little to say about that. His subtle but somewhat elusive conception of intuition and its role is omitted because his most developed presentation belongs to the earlier period, and I am far from sure I understand it. A related issue is Bernays' view that the conception of the continuum is basically geometric; he sees the predicative and intuitionistic theories as striving for a complete arithmetization that does not do justice to the conception and the standard set-theoretical treatment as a workable but not intuitively entirely satisfying compromise. These issues are large enough to be the subject of another essay.[9]

2. Anti-foundationalism

One might introduce Bernays' anti-foundationalism by a probably oversimple representation of the epistemology lying behind the Hilbert program as a kind of indirect foundationalism. It did not seek to make the theories actual mathematics works with rest on evident foundations in the sense that mathematical theorems would be seen to be obtainable by self-evident elementary deductive steps from axioms that are also self-evident, that is would need no further justification and would perhaps be incapable of it. But it offered a justification of mathematical theories by means of a consistency proof, where the method of the consistency proof, Hilbert's finitary method, seems to have the character just described.

In "Die Philosophie der Mathematik" Bernays seems to view the consistency of arithmetic as obtaining such a justification from a finitist proof. However, as far as the justification of arithmetic itself goes, what is accomplished is to give "full, evident certainty . . . that it cannot come to grief through the incompatibility of its consequences" (55); however, Bernays emphasizes that arithmetic and analysis are extremely well confirmed by mathematical experience and application.

However, applying our simple picture of a foundationalist view of a domain of mathematical knowledge to the finitary method is complicated by the fact that Hilbert and Bernays do not view the method as embodied by an axiomatic theory; on the contrary the emphasis is on

---

[9] [But see the comments in §4 of Essay 1.]

the intuitive character of the concepts and claims that the method gives rise to. I would express this by saying that what is attained by finitary concept formation and proof is intuitive knowledge. Still, one could infer from their expositions that the method at least allows what is formalizable in primitive recursive arithmetic. Elsewhere I have attributed to them what I call Hilbert's Thesis:

> A proof of a proposition according to the finitary method yields intuitive knowledge of that proposition. In particular, this is true of proofs in primitive recursive arithmetic (PRA).[10]

The key to an argument for Hilbert's Thesis is the claim that if a function is defined by primitive recursion from functions that can be seen intuitively to be well-defined, then it too can be seen intuitively to be well-defined. Hilbert and Bernays do offer an argument for this, which generalizes one given earlier by Bernays for exponentiation (HB I, 25–27; "Die Philosophie der Mathematik," pp. 38–39). In the paper just referred to I argue that these arguments are the weak point of the case for Hilbert's Thesis.[11] It may be that Bernays already came to doubt them in 1934. In "Sur le platonisme," p. 61, he questions whether it is intuitively evident that the number $67 \exp 257^{729}$ has a decimal expansion, on the ground that it is a number that is "far larger than any occurring in experience."[12] What is significant is not this obvious observation about the number, but his taking it as a reason for questioning whether the knowledge that it has a decimal expansion is intuitive.

It is not certain that intuitive knowledge as Hilbert and Bernays understood it is a case of perfectly evident premises leading to perfectly evident conclusions. It does seem, however, that that is the way Bernays' mentor Leonard Nelson viewed Kantian "pure intuition," and Bernays does not question it in "Über Nelsons Stellungnahme," the exposition of Nelson's philosophy of mathematics that Bernays pub-

---

[10] "Finitism and Intuitive Knowledge, p. 254. [Also *Mathematical Thought and Its Objects* (hereafter MTO), p. 240.] It has been pointed out recently that in practice inferences that went beyond primitive recursive arithmetic were allowed as finitist. See Zach, "The Practice of Finitism." This does not contradict the attribution to Hilbert and Bernays of Hilbert's Thesis.

[11] "Finitism and Intuitive Knowledge," §5, or MTO, §44.

[12] I date the doubts to 1934 because "Sur le platonisme" derives from a lecture given in Geneva on June 18, 1934.

lished just after Nelson's death. That may, however, just reflect the essentially expository character of the essay. Bernays' remark in the 1930 paper could mean only that intuitive finitist proof offers simply the greatest and most reflective certainty that can be attained. A motive for the method was surely that anything proved by it would be acceptable to any of the currents of opinion in foundations at the time, in particular the positions of Brouwer and Weyl.

Whatever may have been Bernays' view in 1928 or 1930, it is clear that by 1934 he was questioning this understanding of finitist mathematics and with it, one might say, the last refuge of foundationalism in the epistemology of mathematics. If he held such a view in the 1920s, it would have to be on the basis of something like Hilbert's Thesis. The remark in "Sur le platonisme" already calls this into question. And in the postscript to the reprint of the 1930 paper in *Abhandlungen,* he writes that "the sharp distinction of the intuitive and the nonintuitive, as it is applied in the treatment of the problem of the infinite, can apparently not be carried through so strictly" (p. 61). That is a philosophical reflection of the mathematical fact that after Gödel's theorem proof-theoretic arguments had to be conducted with assumptions of increasing strength, and it was already clear in the 1930s that the sense in which the method of Gentzen's consistency proof is "intuitive" is not so clear.

What he had in mind in the late remark is indicated by that of 1934, but it is clear that he had carried the matter further. The major step was the introduction in "Quelques points de vue" of the notion of acquired evidence. The German term *Evidenz* and its French equivalent *évidence* are often translated as "self-evidence." This does not fit well the intentions of philosophers such as Brentano and Husserl who have given the concept a prominent role. It certainly does not fit the intention of Bernays. Acquired evidence is the evident character a proposition comes to have for someone who uses it in the context of a developed conceptual scheme,[13] where the sentence involved may have had a use as the scheme developed. So acquired evidence is not something a

---

[13] Bernays uses the term *dialectique* derived from Gonseth. But it appears that what he has in mind is very close to what we would call a conceptual scheme. The relevant sense is the first of two senses of *dialectique* that Bernays distinguishes in "Zum Begriff der Dialektik." He says this can be taken as a generalization of the concept of dialectic as logic, and that one might say that logic is the first example of the scientific fixation of such a dialectic (p. 173).

proposition has by itself; Bernays explicitly says that evidence may be relative to "implicit suppositions" which the conceptual scheme involves.

Bernays says that the evidences in mathematics are almost all acquired and goes so far as to remark:

> What distinguishes this case is that the dialectic is established in our mind in such a penetrating manner that it influences our intuitive imagination, that is to say that it influences the way in which we represent intuitively certain categories of objects. . . . In this way one also understands that intuition can derive notions that surpass the possibilities of a complete effective control and whose conceptual analysis gives rise to infinite structures.[14]

About the concept of natural number (with induction and recursion), he remarks a little later that it is "a full dialectic, which certainly did not exist from the beginning for the mind but which had to be tried out and dared at a certain stage" (ibid., p. 324). That is to say, our conception of natural number had to be developed in a way that involved trial and (very likely) error.

The concept of acquired evidence offers a natural way to interpret the view commonly expressed by set theorists that the axiom of choice is evident, obvious, or follows from the concept of set. Bernays himself gives a more elementary example in a brief description of the obviousness that arithmetic comes to have:

> First we are conscious of the freedom we have to advance from one position arrived at in the process of counting to the next one. But then we take the step of a connection, through which a function that associates a successor with each and every number is posited. Here a *progressus in infinitum* replaces the *progressus in indefinitum*. But it is not immediately obvious that this idea of the infinite number series can be realized; the intellectual experience of its successful realization is then essential for developing a feeling of familiarity, even of obviousness, as acquired evidence.[15]

---

[14] "Quelques points de vue," p. 323.
[15] "Die Mathematik als ein zugleich Vertrautes und Unbekanntes," p. 111.

3. Questioning the a priori

A second aspect of Bernays' anti-foundationalism is his rejection of the idea of a priori knowledge as he had come to understand it from Kant and Nelson. This is indicated already in the paper "Grundsätzliche Betrachtungen zur Erkenntnistheorie," although it is principally directed against the idea of a priori principles in physics (including the idea that the geometry of space is known a priori, which Nelson had continued to hold). The objections he canvasses to the "aprioristic point of view" first of all concern the difficulty of maintaining that view while doing justice to developments in physics. But at the end of the paper, he draws the quite general conclusion that the aprioristic view and "pure empiricism" have in common a presupposition (which Bernays rejects): If reason is essential for empirical knowledge, then it must play its role through principles that are knowable a priori. What is most significant about mathematics is what Bernays does not say: he might have said that mathematics has a priori principles, which, however, do not have implications for physics or other sciences that allow them to come into conflict with new developments. Although his statement of the common presupposition of apriorism and empiricism suggests that he already rejected the view that mathematical knowledge is a priori, his silence could reflect rather uncertainty or suspension of judgment.

In fact it is difficult to pin down what Bernays thought earlier about the a priori character of mathematics in general, beyond finitary mathematics. In a paper evidently addressed to fellow members of Nelson's neo-Friesian school, Bernays describes analysis as rational knowledge, of the same nature as Fries attributed to pure natural science.[16] But the issue seems to have been complicated for him by the Hilbertian conception of the axiomatic method, according to which the status of an axiom is not that of a known truth or even (as in Hilbert's *Grundlagen der Geometrie*) as in itself true or false.[17]

---

[16] "Die Grundgedanken der Fries'schen Philosophie," p. 110. He mentions the use of ideas of totalities and quantification over them.

[17] See the remark on Hilbert's geometry in "Die Bedeutung Hilberts," pp. 95–96, trans. p. 192. Cf. Bernays' discussion remark published with Nelson, "Kritische Philosophie." The polemic against formalism in Nelson's reply may have been one factor convincing Bernays that he had to distance himself from a formalistic interpretation of Hilbert's views and program, as he does in "Die Philosophie der Mathematik," part II §4.

As we shall see, later Bernays is quite definite in rejecting the strictly a priori character of mathematics. Before turning to this, I should explain a difficulty that a contemporary philosopher might have with my regarding the rejection of the a priori character of mathematics as an aspect of Bernays' anti-foundationalism. Philosophers from Gödel to the present have defended the idea of propositions whose justification is a priori but which are neither certain nor unrevisable. That possibility seems not to occur to Bernays, because his model of a priori knowledge comes from a certain interpretation of Kant. Thus he describes the "theory of knowledge a priori" as holding

> ... that we have certain cognitions of natural reality lying in Reason, to be sure first actualized through sensory stimulation, which, when they are brought to full consciousness, can be formulated in a definitive way in the form of general laws. In addition this theory claims that these a priori knowable laws contain the principles for all research in exact natural science and that in particular the method of physical theory formation is fixed in a unique and definitive way.[18]

In some ways this is excessively rigid even as an interpretation of Kant.[19] But Bernays is clearly guided by a general Kantian picture of at least synthetic a priori cognition as determined by the nature of the mind. Since the nature of the mind does not change, it is entirely natural to expect that available synthetic a priori cognition will not change and will thus be unrevisable.[20]

---

[18] "Grundsätzliche Betrachtungen," p. 279.

[19] Bernays goes on to say "so that after the uncovering of these principles there is strictly speaking no further development of physical speculation." He seems to assume that according to Kant a priori principles in physics are limited to mechanics. This leaves no room for what Michael Friedman finds in the *Opus postumum*, a development of the theory to accommodate Lavoisier's discoveries in chemistry. See *Kant and the Exact Sciences,* ch. 5. That development might also consist of "definitive" a priori principles, but Bernays' reading still leaves it a little difficult to understand Kant's continuing concern about empirical laws.

[20] Although the position Bernays takes here is critical of Nelson, one might see Bernays as responding to an internal conflict in Nelson's philosophy, between his orthodox Kantian view of geometry and his view that philosophy should pay close attention to developments in the sciences and should not contradict its results.

Already in "Bemerkungen zur Grundlagenfrage," Bernays manifests directly a dissatisfaction with the idea of the a priori applied to mathematics:

> In general the present-day problem of foundations . . . points to the requirement of *revising our epistemological conceptions (Begriffsbildungen)*. In particular, that widely held opposition, according to which there is in science on one side the self-evident ("logical") and [on the other] accommodation to perception (the "empirical"), appears inadequate for the characterization of the scientific situation. The extension of this dichotomy carried out by Kant with the concept of a priori knowledge does not give a satisfying viewpoint. (p. 87, emphasis Bernays')[21]

Interestingly, Bernays mentions the "problem of foundations," evidently as it has developed in logical research. But it is noteworthy that "Grundsätzliche Betrachtungen" is very unlikely to reflect any influence of Gonseth, whereas the setting of the publication (and probably the writing) of the note of 1939 suggests the contrary.

In the important "Mathematische Existenz und Widerspruchsfreiheit" Bernays is clearer in distancing himself from the idea of an a priori evident foundation of mathematics.

> Instead, one can adopt the epistemological viewpoint of Gonseth's philosophy which does not restrict the character of a duality—due to a combination of rational and empirical factors—to knowledge in the natural sciences, but rather finds it in all areas of knowledge. For the abstract fields of mathematics and logic that means specifically that thought-formations are not determined purely a priori but grow out of a kind of intellectual experimentation. (p. 102)

The intellectual experimentation he refers to is evidently an instance of his notion of intellectual experience *(geistige Erfahrung)*, of which more later.

What he wishes to substitute for the a priori as he understands it is explained a little more fully in the essay "Zur Rolle der Sprache" on

---

[21] Evidently Bernays was thinking of the synthetic a priori. The passage strikes me as rather carelessly written, rather uncharacteristic of Bernays.

Carnap.[22] In effect Bernays proposes replacing the notion of the a priori by the more modest notion of the antecedent (*vorgängig*, translating Gonseth's term *préalable*). He applies this term to "ideas, opinions, and beliefs to which we either consciously or instinctively hold on in our questions, considerations, and methods" (p. 161). But such beliefs might be revised in the development of the science in question. In fact, Bernays writes:

> The scientific method also requires that we make ourselves aware of the antecedent premises, and even make them the object of an investigation, respectively include them in the subject matter of an investigation. (Ibid.)

The antecedent, in contrast to the a priori, is relative either to a state of knowledge or to a discipline.[23] He seems in general terms to be following Gonseth.

Although the final push may have been delivered by Gonseth, Bernays' rejection of the a priori character of mathematics seems clearly to have been motivated by the experience of foundational research in mathematics, where different conceptual constructions competed with none being able to make convincing any claim to definitive character. After introducing the idea of intellectual experimentation, he discerns such experimentation in the different constructions in foundational research and says that even unsuccessful attempts have their value. He sees the competing foundational programs as analogous to competing theories in science ("Mathematische Existenz," p. 102).

---

[22] Although the journal issue in which this essay appeared is dated 1961, its actual publication must have been later, because Bernays wrote to Gödel on December 31, 1961 that he was "presently busy" with the paper (Gödel, *Collected Works* [hereafter CW], vol. IV, pp. 204–205). As the title suggests, the essay is only incidentally concerned with the philosophy of mathematics.

[23] The notion of the antecedent might be instructively compared to Hilary Putnam's notion of the contextually a priori or to the idea of constitutive principles, derived by Michael Friedman from some logical positivist writings. There will be significant differences in both cases. For Putnam, a proposition is contextually a priori in a given epistemic setting only if it is impossible to conceive how it could be false. (He holds that to have been true for Euclidean geometry in the eighteenth century.) This element does not enter into Bernays' conception, nor so far as I know into Gonseth's.

Bernays' notion of intellectual experience *(geistige Erfahrung)* deserves further exploration. The term seems first to occur in Bernays' publications in 1948, in an obscure passage that does not relate the notion to mathematics.[24] But its origin is certainly earlier. In a letter to Gödel of September 7, 1942, Bernays mentions distinguishing "different layers and kinds of evidence,"[25] as he did already in "Sur le platonisme." He goes on to say:

> Here the point of view is to be added that the certainty of a conceptual system (a "dialectic" in the sense of Gonseth) is not given beforehand, but is acquired through use, in the light of a kind of "intellectual experience."[26]

Thus the notion of acquired evidence is also invoked.[27]

The notion is evidently implicitly present in "Mathematische Existenz" and is put into a more general context in some remarks to the third of Gonseth's Zürich conferences, which concern the Gonsethian theme of the intimate relation of rational and empirical factors in knowledge. Illustrating the proposal that experience be thought of more broadly than sense-experience, he says that a field of "experiences of a more general kind" arises in the working out of theories. He goes on to say:

> Even in the domain of mathematical thought one can speak of experience. A proof that proceeds according to the current methods of a mathematical discipline can be considered as a definite construction whose existence is shown, in a way analogous to that in which the result of a numerical computation can be exhibited. Thus there is a sort of intellectual experience with respect to the deductive working out of mathematical ideas.[28]

In another place he again introduces the idea in opposition to the idea of certain a priori knowledge:

---

[24] "Grundsätzliches zur 'philosophie ouverte'," pp. 276–277.

[25] Gödel, CW IV 138–139.

[26] Ibid.

[27] Bernays mentions that he had advocated the notion of intellectual experience in a discussion remark at the first of Gonseth's Zürich conferences in 1938; see Gonseth, *Les entretiens*, pp. 78–79. There is a hint there of the idea of acquired evidence.

[28] "Dritte Gespräche von Zürich," pp. 133–134.

It seems necessary to concede that we also have to learn in the
fields of mathematics and that we here too have an experience
sui generis (we might call it "intellectual experience"). This does
not diminish the rationality of mathematics. Rather, the assumption that rationality is necessarily connected with certainty appears to be a prejudice.[29]

It is probably with what he has earlier called intellectual experimentation in mind that he ends "Die schematische Korrespondenz," the most
Gonsethian of his essays in the philosophy of mathematics, with the
remark:

Gonseth proclaims *ouverture à l'expérience* as a general method;
and as a requirement it is not restricted to research in natural
science, but it is equally important in the field of *intellectual experience.* (p. 188)

What Bernays says about intellectual experience would encourage
the idea that he would admit that mathematical knowledge is a priori
once that thesis is dissociated from the Kantian paradigm. Commenting on Quine's criticism of the analytic-synthetic distinction, Bernays
says that mathematical propositions are justified in a different sense
from physical.[30] But in general he proves elusive on this question. An
essay tantalizingly entitled "Some Empirical Aspects of Mathematics"
offers little enlightenment. He repeats the proposal to replace "the doc-

---

[29] "Bemerkungen zur Philosophie der Mathematik," pp. 174–175.

[30] Ibid., p. 172. It is of some interest that throughout his career Bernays never
seems greatly moved by the analytic-synthetic distinction, although he does comment on it in various places. His Kantian background no doubt predisposed him
against the view held by many philosophers that the a priori character of mathematics could be defended by showing mathematical propositions to be analytic.
Already in "Die Philosophie der Mathemtik, I §2, he criticizes the Fregean logicist
analysis of number on quite different grounds. In "Bemerkungen zur Philosophie
der Mathematik" he criticizes Quine in saying that Carnapian definitions of analyticity do characterize mathematical propositions as distinguished from those of
empirical science, but he seems to mean no more than that they give an extensional
characterization. Bernays discusses Kant's distinction at some length in "Zur Frage
der Anknüpfung an die kantische Erkenntnistheorie," pp. 27–30. Like many other
writers, he says that Kant's distinction is not the same as that discussed in his own
time. Referring to the discussion of the foundations of mathematics, he says that it
"is surely questionable whether a sharp general defined version" of the distinction
has been arrived at (p. 30).

trines of the a priori, the analytical, the self-evident" with Gonseth's of the *préalable* (p. 127) and says that mathematics is antecedent to natural science.[31] But what he says elsewhere about the antecedent certainly does not rule out the possibility that essentially empirical grounds might lead to the modification of mathematical principles. It would be natural to read in that way an expression of agreement with what he (rightly) takes to be a fundamental point in Quine's critique of the analytic-synthetic distinction:

> ... it is inappropriate to divide the validity of a judgment into a linguistic and a factual component. This outlook points in a direction similar to that of the *principe de dualité* of Ferdinand Gonseth.[32]

That Bernays understood Gonseth's principle as having this implication is suggested by the following remark:

> Moreover he [Gonseth] does not divide disciplines into purely empirical and purely rational; rather, he assumes that a kind of "duality" of the empirical and the rational is to be found in all parts [of knowledge].[33]

Gonseth himself emphasized the genesis of mathematical concept formations (especially geometrical) in ordinary experience, and this element does appear in Bernays. But others have thought that a genetic role of this kind is compatible with the science in its developed form having a justification without appeal to the empirical. There is some indication that this was not Bernays' view, but it is far from decisive.

An element of Bernays' epistemology that may be relevant here but is not developed very much is the appeal that he makes at times to phenomenology. He even goes so far as to suggest regarding mathematics

---

[31] This paper is a translation. The translator states that he renders *préalable* as 'preliminary'. However, I continue to follow the Bernays Project translations in rendering it as 'antecedent'. This paper was evidently originally written in French. A copy of the typescript is in the Bernays papers, document Hs 973: 34. (I am indebted to Wilfried Sieg for calling my attention to this document.) That is not surprising since the volume in which the translation appeared is the proceedings of a meeting in Brussels of the Académie Internationale de Philosophie des Sciences.

[32] "Von der Syntax der Sprache," p. 238.

[33] "Überlegungen zu Ferdinand Gonseths Philosophie," p. 119.

as "the theoretical phenomenology of formal structure."[34] He offers phenomenology as an example of objectivity distinct from that referring to "the real objects of nature." His example is colors, about which he says that observations of them "do not lose their objectivity even though we have discovered that colors must be eliminated from physical theories" (ibid.).[35]

In this characterization of mathematics, "theoretical" is important. Bernays may think of mathematical thought as beginning with noting structural (as opposed to qualitative) aspects of the phenomenal, but it quickly idealizes and so goes beyond the strictly phenomenal. Thus he more frequently refers to mathematics as the study of idealized structures.

4. The Stratification of Mathematical Theories

Both Bernays' rejection of the a priori and his relations with Gonseth might lead one to think he holds a holistic view at least structurally analogous to that of W. V. Quine. I say "structurally analogous" because I have already made clear that he did not share Quine's empiricism. There are indeed indications of such a view, but the sketchy idea of phenomenology would suggest a qualification. Another, more developed and more interesting to logicians, is the stratification of mathematics laid out in "Sur le platonisme," which is one of his most influential contributions.

A central idea of that paper, which so far as I know was quite original, is that platonism in mathematics admits of degrees. "Platonism" as he describes it is first of all a methodological tendency, of which his initial illustrations are the use of the law of the excluded middle and the existential form of axioms (in Hilbert's geometry as opposed to that of Euclid, who speaks of figures to be constructed).

---

[34] "Some Empirical Aspects," p. 126. In his essay on Wittgenstein, Bernays criticizes Wittgenstein for rejecting any phenomenology ("Betrachtungen," pp. 124, 141); the second passage overlaps the present one in content.

[35] There follows the most positive comment about Husserl that I have found in Bernays' writings. In general (also here) Bernays criticizes Husserl's claims for intuition, and in an essay about Gonseth he mentions Husserl's transcendental phenomenology as an example of "first philosophy" that he and Gonseth reject ("Charakterzüge," p. 151).

... the tendency of which we are speaking consists in viewing the objects as cut off from all links with the reflecting subject. Since this tendency asserted itself especially in the philosophy of Plato, allow me to call it "platonism." (p. 53)

Platonist mathematical conceptions, he goes on to say, are distinguished by their simplicity and logical strength. They form "representations which extrapolate from certain regions of experience and intuition."

Bernays goes on to make clear that the conceptions he calls platonist can be introduced at some levels of mathematics and not others, and he uses this in a now familiar way to classify stances in foundations. Thus admitting "the totality of natural numbers" allows classical quantificational reasoning about natural numbers but is silent about further steps. Analysis introduces such a conception of set of natural numbers, function of natural numbers, and real number. (It is in this connection that Bernays gives his well-known "quasi-combinatorial" motivation of classical impredicative reasoning about sets of numbers [pp. 53–54]). Set theory, because it involves iteration of this conception, represents a much greater "platonist" commitment, how great depending on what axioms are accepted.

Bernays attributes to Kronecker and Brouwer the rejection of even the initial platonist step of admitting the totality of numbers. Although this is no longer a matter of platonism as he understands it, he notes a stratification here, in that intuitionism goes beyond Hilbert's finitism in reasoning with an abstract concept of proof. The remark about 67 exp 257^{729}, though directed against Brouwer, would equally imply that Hilbert's finitism goes beyond the intuitive.[36] The kind of stratification described by Bernays is a commonplace today, because it has been ratified and refined by many proof-theoretic results and by related enterprises such as the study of hierarchies based on notions of definability. Bernays already takes note in the paper of the first result showing that "platonistic" commitments are not the only dimension on which mathematical conceptions are stratified, the relative consistency proof of classical first-order arithmetic to intuitionistic.

---

[36] Thus Hao Wang places strict finitism, which he calls "anthropologism," lower in this hierarchy than finitism. See "Eighty Years of Foundational Studies," pp. 472–473. He attributes his stratification to Bernays; probably he found it implicit in "Sur le platonisme."

5. Platonism

Philosophers will naturally ask whether and in what sense Bernays was a platonist. Since that word is used in many different ways in philosophy of mathematics and elsewhere in philosophy, we can't be sure of a unique answer. It is clear that the platonism he describes in "Sur le platonisme" is primarily a methodological stance, and furthermore he says that different levels of platonism are appropriate for different domains of mathematics; he suggests for example that the "intuitive concept of number" is the most natural for number theory (p. 64). From his description, it seems clear that it is Hilbert's finitary method that he has in mind, so that he probably means the most elementary number theory rather than more developed number theory such as analytic number theory.

Clearly, however, in saying that "today platonism reigns in mathematics" (p. 56) he does not mean to offer a general objection. What he does reject is what he calls "absolute platonism," characterized as "conceptual realism, postulating the existence of a world of ideal objects containing all the objects and relations of mathematics" (ibid.). This view he considers refuted by the paradoxes.[37] Elsewhere I describe Bernays' view in this paper as methodological platonism.[38]

In later writings Bernays largely avoids the term "platonism"; he may well have thought it gave rise to misunderstanding. There's no doubt that he continued to accept what has been called default realism or default platonism,[39] which amounts to taking the language of classical mathematics at face value and accepting what has been proved by standard methods as true. Some such statements are of the existence of mathematical objects, which would make Bernays a platonist in the sense often used in American philosophy, where 'platonism' is contrasted with nominalism (as espoused by Goodman and later by Field) rather than with constructivism, as was the case in "Sur le platonisme." In fact, a broadly realistic attitude was part of his general approach to

---

[37] This sounds like a view that Gödel would have endorsed from 1944 on. Some of Gödel's difficulties about the notion of concept could be taken to arise from the problem of formulating the view so as to avoid the sort of refutation Bernays has in mind.

[38] Essay 2 in this volume, p. 51.

[39] Tait, "Gödel's Unpublished Papers," pp. 289–290. [See also "Truth and Proof," p. 91, which I should have cited here.]

knowledge in the post-war years. In one description of the common position of the group around Gonseth, he emphasizes that the position is one of trust in our cognitive faculties. He also introduces the French term *connaisance de fait;* the idea is that one should in epistemology take as one's point of departure the fact of knowledge in established branches of science.[40] The stance is similar to the naturalism of later philosophers, though closer to that of Penelope Maddy or to the "substantial factualism" of Hao Wang than to that of Quine. (Wang was certainly influenced by Bernays.)

With respect to the physical world, Bernays' realism is limited by his acceptance of Gonseth's thesis that theories stand in a "schematic correspondence" to reality.[41] This does not directly limit realism about mathematics for Bernays, because according to him mathematics is on the side of the schemata in this correspondence. But he repeatedly describes the structures mathematics is concerned with as idealized.

6. Bernays and Structuralism

This point brings us to the other main theme in Bernays' later philosophy I want to emphasize, his structuralism. In some sense or other, structuralism was part of Hilbert's view of mathematics and seems to have been adopted by Bernays early in their collaboration and held to for the remainder of his career. When he gave a brief characterization of his philosophy, he typically included "the view according to which mathematics is the science of idealized structures."[42] A structuralist outlook is not distinctive of his *later* philosophy, although we shall see that in later writings he took important steps toward working it out. There are nowadays many versions of structuralism, and I will not attempt here to characterize Hilbert's or to relate it to current versions. I will say something later about Bernays' relation to contemporary structuralism.

---

[40] "Grundsätzliches zur 'philosophie ouverte'," p. 275. Generally use of a French term indicates that the concept is taken from Gonseth, but in this case Bernays says that he is making an analogy with a notion of *liberté de fait* introduced by Gonseth in writing on freedom. But naturalism in the sense of the text was certainly Gonseth's view as well.

[41] "Die schematische Korrspondenz," esp. part 1.

[42] Ibid., p. 188. Bernays quotes this in the preface to *Abhandlungen*, p. x.

Bernays' view of existence in mathematics is laid out most fully in "Mathematische Existenz." Its point of departure, the thesis that "existence, in the mathematical sense, means nothing but consistency" is found in Poincaré's writings and in Hilbert's early writings and was in a rough way a motivating idea of Hilbert's program. Bernays sees the thesis as opposed to a traditional form of platonism according to which mathematical entities have "ideal being" independent of being thought of and of being "determinations" of something real.[43] He criticizes this thesis as doing no methodological work. Although he does not immediately address the vexed question what is meant by 'independent', it appears that he gives it quite a strong interpretation. This is indicated by the fact that so much else of what he says is compatible with platonism characterized in roughly this way. To begin with he rejects the nominalist program of eliminating reference to what he calls theoretical *(ideell)* entities, which in the context we might call abstract entities.[44]

Bernays goes on to sketch four interpretations of existence statements about theoretical entities, none of which he takes to imply independent existence. (c), existence relative to a structure, will be Bernays' own central concept. I will leave aside (a) and (b) because Bernays says that they involve "a kind of contentual reduction" (p. 95). (d) is described as follows:

> Existence of theoretical entities may mean that one is led to such entities in the course of certain reflections. For example, the statement that there are judgments in which relations appear as subjects expresses the fact that we are also led to such "second-order" judgments (as they are called) when forming judgments. (p. 95)

---

[43] As we shall see Bernays wishes to avoid this view; that may be a reason why he eschews the label 'platonism' for his own view. But he doesn't use the term at all in this paper. The reason might be that he does not want to repudiate the views of "Sur le platonisme," although he may have thought that some of the formulations of that paper suggested something beyond the methodological stance he was concerned with.

[44] Like Husserl, Bernays uses the terms *ideal* and *ideell*. Following the translations of *Essays*, I render the first as 'ideal' and the second as 'theoretical'; however, its meaning is often closer to 'abstract'. Bernays' usage is not particularly close to Husserl's, and there is no particular reason to think it derived from the latter.

Since "being led" is in the sense of the objectively appropriate, Bernays denies that there is reduction in this case. Why, then, does he deny that it involves independent "ideal" existence? He remarks: "The existence statement is kept within the particular conceptual context, and no philosophical (ontological) question of modality which goes beyond this context is entered into" (ibid.). That suggests that in all these cases he has in mind something formally analogous to Carnap's distinction of internal and external questions, and an affirmation of "independent existence" would be a metaphysician's affirmative answer to an external question of existence.[45]

In this general discussion sense (c) of existence for theoretical entities is one of the key concepts of the paper. It is existence relative to a structure (in a broad sense), what he calls relative existence *(bezogene Existenz)*. But his route to the development of this idea goes by way of a more direct discussion of the equation of mathematical existence and consistency. Applied in general to existence statements individually, of individual objects or of objects satisfying some condition, the thesis is criticized on generally familiar grounds (pp. 96–99). Bernays suggests that its appeal may have rested on a too simple understanding of consistency.

Bernays proceeds to consider the case of existence axioms of an axiomatic theory. Here there is an obvious objection to equating existence with consistency, that consistency is a property of the system of axioms as a whole (p. 99). But since the axiom system may be regarded as a description of a structure, the existence claims of the axioms can be understood as statements of relative existence. Thus Bernays comes to the point, much less familiar at the time than it is now, that typical

---

[45] When writing the present essay, Bernays is unlikely to have known Carnap's "Empiricism, Semantics, and Ontology," the *locus classicus* of the distinction of internal and external questions; however, the idea is certainly foreshadowed in earlier writings of Carnap. In his essay "Zur Rolle der Sprache" on Carnap, Bernays does not comment on the distinction. However, in "Von der Syntax der Sprache," also largely about Carnap, he comments that it is in general incorrect to pose the question "Is there such and such?" in an "absolute sense" and that here the criticism by logical empiricism is justified. As in the present essay, he maintains that questions of existence are sensible only in a specific conceptual context. That suggests that he viewed Carnap's distinction sympathetically.

A relation of Bernays' and Carnap's thought at this point is already proposed by Heinzmann, "Paul Bernays et la philosophie ouverte," p. 27.

existence statements in mathematics are in the context of a structure. (His example is Euclidean geometry.)

The philosophical question is in his view only shifted by this observation, since the question then arises of the existence of the structure. Here, there is some justification for equating existence with consistency, as in the case of non-Euclidean geometry (p. 100). But typically consistency is made out in such cases by the construction of a model, which leads to a potential regress of relative existences. This is the point where Hilbert shortly after 1900 saw the necessity of a syntactic consistency proof as a way to end such a regress.

At this point Bernays does not turn so quickly to proof theory. He writes:

> We finally reach the point at which we make reference to a theoretical framework *(ideeller Rahmen)*. It is a thought-system that involves a kind of methodological attitude; in the final analysis, the mathematical existence posits relate to this thought-system. (p. 100)

He says that mathematical experience has tested the consistency of this framework to such an extent that there is "de facto no doubt about it," and that is what is needed for the existence posits made within the framework, which are presumably of relative existence.

Bernays is not as explicit as one would wish as to what this framework might be. He probably thinks there are different options even within classical mathematics; a little later he speaks of "indeterminacies" in demarcating a framework (p. 101). The somewhat obscure remarks about the number system (ibid.) are clearly meant to be illustrative. His point is perhaps that it is not so clear as appears at first sight what positing the number system consists in. Although in the first instance it is a matter of the existence of objects, conceptual understanding and with it logic are essentially involved.

Bernays' thesis that mathematical practice operates within a framework that is not uniquely determined is another point of contact between his thought and that of Carnap. For Carnap, a framework is essentially linguistic and specifiable precisely with the methods of mathematical logic, a formal system at least in a generalized sense, since the characterization may involve infinitary rules. Bernays' description of the framework as a "thought-system" implies that he thinks of it in a more informal and perhaps mentalistic way. However, he clearly thinks that a framework can be delineated more or less precisely and that axiomatization and formalization are means of attaining greater precision.

The importance of the question how he precisely thinks of the concept of framework is somewhat diminished by a point on which he places some emphasis, that a precisely delineated framework for mathematics intends "a certain domain of mathematical reality" that is at least to a certain degree independent of the "particular configuration" of the framework (p. 102). What he has specifically in mind is the fact that different axiomatizations and constructions (e.g., of the real numbers) can be equivalent even if they differ in their ontology. Bernays seems to recognize that the concept of relative existence loses some of its sharpness by its being viewed in this way (p. 104). He adds the observation that mathematical reality is not fully exhausted by a delimited framework, what Gödel called the inexhaustibility of mathematics.

Bernays' understanding of typical mathematical existence statements in terms of relative existence makes his view a structuralist view in a general sense. However, he does not pursue an idea central to later structuralist views of mathematical objects, which are based on the idea that such objects have no more in the way of a "nature" than is given by the basic relations of a structure to which they belong.[46]

Bernays mentions two difficulties for his perspective on mathematical existence, that it might reinstate the "ideal existence" that he has rejected, and that it might be a form of relativism. He considers what he has already said about the independence of mathematical reality of the particulars of a framework as a sufficient answer to the second objection. About the first, he makes the observation that in natural science the central assertions of existence are of the "factually real," and other talk of existence, for example of laws, "appears as mere improper existence" (p. 104). There is no such contrast in mathematics. "It is not a question of being *(Dasein)* but of relational, structural connections and the emergence (being induced) of theoretical entities from other such entities" (ibid.). To me that leaves the rejected sense of independent existence somewhat elusive. I don't believe that Bernays would have taken the Carnapian line that, to the extent that they make sense

---

[46] Thus I would no longer affirm what I wrote in "The Structuralist View," p. 303, that the paper gives a "clear general statement" of such a view.

at all, answers to external questions are simply pragmatic recommendations of one type of framework over another.

I have gone into some detail about this paper in part because I consider it the best of Bernays' post-war philosophical papers and because it is little known, although it is surely one of the most important contributions to mathematical ontology of its time, not as sharply formulated as the contemporary writings of Carnap and Quine but more sensitive to actual mathematics. Even to this day it has excited little comment in the literature.[47]

What is distinctive about Bernays' structuralism? As I have said, he does not say that mathematical objects have no more of a nature than is given by the basic relations of a structure to which they belong. It is not clear how much he considered that view.[48] His paper develops what could be regarded as a major difficulty: that basic structures of arithmetic, analysis, and geometry can be constructed in ways that are fundamentally equivalent but differ in the choice of basic relations. A point on which contemporary structuralisms differ from each other concerns how to understand the concept of structure and what role to give the concept. Bernays talks freely of structures in his philosophical writing but does not say how he understands the existence of structures in relation to the discussion in "Mathematische Existenz." That might make it tempting to attribute to him the view that structures are primitive mathematical entities, as seems to be the case for Stewart Shapiro. But he does not commit himself on this question, and I do not know what he would have thought of the views on this question that have been advanced in the contemporary discussion of structuralism.

---

[47] I speculate on the reasons for this neglect in my introduction to the paper in *Essays,* on which the present discussion draws.

[48] Wilfried Sieg, in "Beyond Hilbert's Reach?," p. 378, points out that in "Die Philosophie der Mathematik," p. 21, Bernays sketches the logical move characteristic of if-thenism or deductivism, the simplest of the forms of structuralism in this sense aspiring to eliminate reference to mathematical objects, and, by way of motivating concern with consistency, notes the problem of vacuity (the nonexistence of a system of objects satisfying given axioms) that this proposal faces. (Cf. my "The Structuralist View," pp. 311–313 [or MTO, §11].) Sieg may overinterpret in writing that Bernays "presents the standard account" of if-thenism. First, he does not propose it as a method for eliminating reference to mathematical objects. Second, although he does describe the knowledge that a proof from certain axioms yields as expressible as a statement of pure logic, he also describes it as a statement about predicates, without saying anything about what sort of entities predicates are.

7. Concluding Remarks

What can we say about Bernays' achievement as a philosopher? The present discussion cannot be the basis for a complete answer to this question, both because of the period it concentrates on and because of its selectiveness as to topics. Bernays played a major role in developing and articulating the philosophical views accompanying Hilbert's program, but that has been largely outside our purview. I have also done little to explore the question of his debt to others, notably Nelson and Gonseth.

Bernays' virtues as a writer on philosophy of mathematics are evident: contact with actual mathematics, especially mathematical logic, combined with familiarity with the issues that concern philosophers and sensitivity to the difficulties that philosophical positions are prone to. Probably the last is the most difficult for a mathematician to achieve.[49]

In the aspects of his work I have canvassed in this essay, one can discern some philosophical contributions that have proved enduring: his stratification of mathematical conceptions in "Sur le platonisme," the observation in that paper that points to a distinction between finitism and intuitionism on one side and "strict finitism" on the other (although he made no attempt then or later to articulate strict finitism as a foundational stance), possibly the "quasi-combinatorial" description of the conception of arbitrary subset of an infinite set. Others have not been widely attended to but stand up well in the light of discussion in the philosophy of mathematics, for example the idea of acquired evidence and the accompanying idea of intellectual experience, and the discussion of mathematical ontology in "Mathematische Existenz" with its thoughtful exploration of issues that would nowadays surround different forms of structuralism.

On the other hand the essayistic mode of writing and the absence of a systematic drive mean that many ideas are not very much worked out. His judiciousness in discussing different views and his loyalty to earlier associations sometimes makes his stance border on eclecticism. Fortunately for us this tendency is more noticeable in general philosophical

---

[49] Bernays' membership in the schools of Nelson and later Gonseth probably sharpened his abilities in this respect by engaging him in philosophical debates.

discussion than when Bernays is focused on mathematics.[50] One can hardly question the claim that Bernays was an acute and well-informed commentator on issues in the foundations of mathematics, with a high degree of philosophical literacy. I have claimed some genuine philosophical contributions, but their extent might be disappointing given the amount Bernays wrote. I will not attempt to assess whether that disappointment would be overcome by exploring the issues I have not taken up here.[51]

---

[50] What seems to me an example is the attempt in "Zur Frage der Anknüpfung" to reconstruct Kantian transcendental idealism with the help of Gonseth's idea of "schematic correspondence" of theories and reality.

[51] This essay results in considerable measure from my (somewhat marginal) participation in the project of preparing a collected edition of Bernays' papers in the philosophy of mathematics. [See now *Essays*.] In §6 I draw on one of the introductions I wrote for that edition. I have learned much about Bernays from the other participants, especially (in fact over a longer period) Wilfried Sieg. Section 7 was prompted by comments made by Albert Visser at the Athens meeting [Logic Colloquium 2005, where the first version of the essay was presented]. I am also indebted to an anonymous referee and to correspondence with Miriam Franchella, although I have not gone into many questions about Bernays' relation to Nelson and Gonseth suggested by her "Paul Bernays' Philosophical Way."

4

Kurt Gödel was born on April 28, 1906, in Brünn, Moravia, then part of Austria and now Brno, Czech Republic. His father held a managerial position in a textile firm, and Gödel grew up in prosperous circumstances. He was baptized in his mother's Lutheran faith. He attended schools in his home city and in 1924 entered the University of Vienna, initially to study physics but eventually studying mathematics. He also attended lectures in philosophy and sessions of the Vienna Circle of logical positivists. He completed the Dr. Phil. in 1929 and qualified as a *Privatdozent* (unsalaried lecturer) in 1933. He retained that position until it was abolished after the Nazi annexation of Austria in 1938, but lectured only intermittently, in part because of health problems and in part because of visits to the Institute for Advanced Study, in 1933–1934, fall 1935 (when he had to return early), and fall 1938 (after which he taught for a semester at Notre Dame). In 1938 he married Adele Nimbursky.

In January 1940, in the face of concerns about the war and uncertainty about his academic position, he left Vienna with his wife and took up a position at the Institute, where he remained for the rest of his life. He became a permanent member in 1946 and professor in 1953. He retired in 1976. During that time he traveled little and never returned to Europe. He had many health problems, both physical and mental, and his health declined seriously from early in 1976. He died in Princeton on January 14, 1978.

Gödel's career was as a logician institutionally in mathematics. Although he was interested in philosophy throughout his life and it became a major preoccupation from about 1942 on, he published little philosophy, and his impact on philosophy probably rests more on his mathematical work, especially the incompleteness theorems of 1931, than on his actual philosophical writing, even when his publications are augmented by posthumous materials. His enduring fame rests primarily on four contributions:

1. In his dissertation he proved the completeness of first-order quantificational logic, that is, that if a formula of this logic is not formally refutable then there is a model in which it is true, in fact one where the domain of the quantifiers is the natural numbers. Gödel generalized his result to prove that the same holds for denumerable sets of formulae and along the way proved the compactness theorem: If every finite subset of a set of formulae has a model, so does the whole set. These theorems could be described as the fundamental theorems of model theory. Gödel's presentation is notable for its conceptual clarity. The results were published in "Die Vollständigkeit."

2. In his epoch-making paper "Über formal unentscheidbare Sätze," Gödel proved his two incompleteness theorems.[1] First, he described for a system of arithmetic based on the simple theory of types a formula F that "says of itself that it is unprovable" and proved that if the system is consistent, then F is unprovable. If in addition the system is ω-consistent (cannot prove there is a number satisfying some condition while refuting it for each individual number), then ¬F is also unprovable, and thus the system is incomplete. The method was that of arithmetization of syntax, coding symbols of the system by natural numbers so that formulae and proofs could be coded as sequences and so also as numbers, so that properties such as being a proof of a given formula could be represented by predicates expressible in the system. Gödel showed that this could be carried out for first-order arithmetic, so that the incompleteness result applies to it as well. Second, if the system satisfies some further conditions, then if it is consistent the formula expressing the consistency of the system is unprovable. In this sense the consistency of the system cannot be proved in the system itself.

3. In the mid-1930s Gödel turned his principal attention to set theory. Between 1935 and 1937 he proved of the standard systems of set theory, such as that of Zermelo-Fraenkel (ZF), that if they are consistent, they remain so after the addition of the axiom of choice and the generalized continuum hypothesis, a generalization of Cantor's early

---

[1] [This is not quite accurate, since although the second incompleteness theorem is stated in the 1931 paper, only an indication of the proof is given. A full proof first appeared in Hilbert and Bernays, *Grundlagen*, vol. II.]

conjecture that the power of the continuum is the least uncountable cardinal.[2] Thus the latter cannot be refuted by standard set-theoretic principles. The proof proceeded by constructing an inner model of "constructible sets," essentially an extension of Russell's ramified hierarchy to allow levels of sets indexed by all the ordinals. He proved that it is consistent with ZF that all sets are constructible and that this hypothesis implies the axiom of choice and the generalized continuum hypothesis. Apart from the great interest of the results themselves, Gödel's methods proved fundamental to later research in set theory.

4. In the late 1940s, prompted by reflection on Kant's conception of the transcendental ideality of time, Gödel discovered solutions of the field equations of general relativity in which the matter of the universe is in a state of uniform rigid rotation and in which there are closed timelike curves. The latter raised the theoretical possibility of time travel, but what interested Gödel philosophically were solutions in which one could not define an objective global time. Gödel published his results in two papers, as well as a philosophical comment,[3] all reprinted in volume II of the *Collected Works* (abbreviated as CW II). His initial models appear not to be physically realistic, but the later ones may be. The work was influential in the development of general relativity and cosmology. The connection with Kant is explored in some manuscripts on relativity theory and Kantian philosophy, probably earlier drafts of "A Remark." Two of them are published in CW III.

Especially in the 1930s, Gödel obtained other significant results in mathematical logic, many published in short papers. Several concern intuitionistic logic and its relation to classical logic and modal logic. Most foundationally significant is the translation of classical logic into intuitionistic, which implies that if intuitionistic first-order arithmetic HA is consistent, then so is classical (PA). (See "Zur intuitionistischen Arithmetik.") His interpretation of HA by primitive recursive functionals of finite type, discovered in 1941 but only published in 1958, introduced a method widely used in post-war work in proof theory.

Gödel's major results in logic noted above are pillars on which later mathematical logic rests; in fact (1) and (2) were decisive steps in the

---

[2] Gödel announced these results in 1938. His full proof appears in *The Consistency of the Continuum Hypothesis*.

[3] "A Remark about the Relationship between Relativity Theory and Idealistic Philosophy."

process by which mathematical logic became a mature subject, which came to be pursued more and more independently of the grand foundational programs that had dominated the 1920s. In the long run, this process had an impact on philosophy in that logic became a tool that could be used by philosophers of widely varying interests and views.

One major step remained to be taken in the 1930s for mathematical logic to reach maturity. That was the analysis of computability and the proof of undecidability results, in particular the undecidability of first-order logic. Gödel participated in this development; for example in his Princeton lectures of 1934 he presented the concept of general recursive function, one of the equivalent notions that came to be identified with that of computable total function on the natural numbers. However, the decisive steps and results were obtained by others, particularly Alonzo Church and A. M. Turing. Gödel recognized the importance of Turing's analysis. In particular, it enabled him to state his own first incompleteness theorem in what he regarded as full generality, since it enabled a general definition of formal system (see CW I 369).

The philosophical impact of Gödel's logical work can be further discussed here largely in connection with Gödel's own philosophical thought. One immediate impact must be mentioned, however: In the 1920s David Hilbert founded proof theory, a branch of mathematical logic, and proposed to resolve epistemological questions about mathematics by formalizing fundamental mathematical theories and giving a proof of their consistency by a syntactic analysis of formal proofs. Proofs of consistency were to be carried out by the finitary method, which corresponded to the most elementary constructive methods in arithmetic. Hilbert's own statements and the actual methods of proof-theoretic work in his school encouraged the conclusion that finitary methods could be captured within a restricted part of PA. Then the second incompleteness theorem would imply that they were not sufficient even to prove the consistency of PA. This was a very serious blow to Hilbert's program, and although work in the type of proof theory it inaugurated has continued to this day, its philosophical ambitions have had to become more modest.

Gödel's own philosophical work reflected his highly rationalistic temperament as well as his training in mathematics and physics. Although the Vienna Circle, especially Carnap, stimulated his interest in logic and foundations, he reacted against their views. During most of his career positive influence came more from historical figures, especially Leibniz and Kant, than from the philosophical movements of his own day (apart from programs in the foundations of mathematics). He was aloof from the philosophical debates of the post-war period and indeed sought to avoid controversy of any kind.

Nearly all of Gödel's philosophical publications and much of his other philosophical writing naturally concern logic and mathematics. Gödel is widely known for defending a platonist view and a philosophical conception of mathematical intuition. On the first point, he could be described as a realist; in an unsent reply to a questionnaire in 1975 he described himself as a "conceptual realist" even since 1925 (see CW IV 447). In letters of 1967–1968 to Hao Wang, he argued that such a view contributed significantly to his principal discoveries in logic.[4] With respect to the incompleteness theorem, what he thought important was the heuristic role of the concept of objective truth, independent of provability. Remarks from the 1930s suggest that it was only over time that he came to the full-blown platonism he avows in "Russell's Mathematical Logic" (hereafter RML) and later writings.[5]

Gödel nonetheless showed sympathy for idealism and sometimes described his philosophy as idealistic. Although Kant was an important influence on him, he criticized Kant's transcendental idealism as too subjectivist. But he was able to combine ringing expressions of realist conviction with idealist sympathies (see for example the letter to Gotthard Günther of June 30, 1954, in CW IV).[6] Concern to reconcile these tendencies was probably a factor leading him to embark in 1959 on a study of Husserl's works and to be favorable thereafter to Husserl's views.[7]

Gödel's platonism has several elements. One, which may underlie his reaction to the Vienna Circle, is the insistence that mathematical statements have a "real content" as opposed to being tautologies or

---

[4] See Wang, *From Mathematics to Philosophy*, pp. 8–11, as originally sent in CW V, 396–399, 403–405.

[5] [See Essay 7 in this volume and its Postscript, also Davis, Review of Dawson and "What Did Gödel Believe?"]

[6] [Further to this subject see my "Gödel and Philosophical Idealism," which, however, does not consider Gödel's appropriation of Husserl.]

[7] See "The Modern Development" and Føllesdal's introductory note to it in CW III, also Tieszen, "Kurt Gödel's Path," and van Atten and Kennedy, "On the Philosophical Development of Kurt Gödel." [See now also Tieszen, *After Gödel*, and Hauser, "Gödel's Program Revisited."]

reflections of convention or the use of language. Such a view need not be realism in a strong sense, and Gödel admits it is compatible with intuitionism.

A second is the conviction of the inexhaustibility of mathematics, in the sense that any formal system or even definite conceptual framework can in principle be expanded to yield new statements and theorems. This was one of Gödel's reasons for insisting on the generality that the Church-Turing thesis gave to the first incompleteness theorem. However, he already observed in 1931 that his undecidable sentence becomes provable by ascending to higher types. In a lecture of 1933 he describes in very clear terms what is now called the iterative conception of set. He argues that any procedure for generating sets in this way leads, once clearly understood, to a procedure that generates more sets. In the usage of Michael Dummett, the concept of set is "indefinitely extensible."

A third element of his platonism, which emerges only in RML and "What Is Cantor's Continuum Problem?" (hereafter WCCP), is his robust defense of set theory, taking its language at face value and thus accepting its ontology of sets. He wrote that the iterative conception captures Cantor's set theory "in its whole original extent and meaning" and that it "has never led to any antinomy whatsoever." In 1947 he already conjectured that the continuum hypothesis (CH) is independent (as Paul Cohen proved in 1963) but holds that set theory, however incomplete, describes a "well-determined reality"; independence of CH would not make idle the question of its truth. Gödel was already then interested in strong axioms of infinity (also called large cardinal axioms) and hoped that axioms of that type might settle CH. Such axioms have subsequently played an important role in set theory, but the independence results have extended to those proposed so far, so that it is only in conjunction with axioms of another kind that they might settle CH. Even the question whether CH has a determinate truth-value remains open.

The fourth and most problematic aspect of Gödel's platonism is that his realism extends to concepts as well as mathematical objects. Concepts are objects associated with predicates.[8] One might expect a

---

[8] [Although this is the view of RML, in late years he either denied that concepts are objects or was uncertain about the question. See the Postscript to Essay 5 in this volume.]

rationalist like Gödel to be a realist about properties and relations. But he had more specific reasons. Concepts are important to his epistemology as objects of intuition (see below). Set theory avoids paradox by allowing that some predicates do not have sets as extensions, so that concepts are not reducible to sets or objects too much like sets. Finally, he took examples of conceptual analysis in mathematics, yielding sharp and essentially unique concepts (such as that of computability), to indicate that such concepts are not man-made.

Though Gödel thought the iterative conception of set free of antinomies, he did regard the antinomies as a serious problem for the foundations of logic. Late in life he maintained that the "intensional paradoxes" (which presumably concerned the notion of concept and related notions) remained unsolved. He seems to have hoped for a type-free theory of concepts, and the problem of devising such a theory that is both strong and consistent remains difficult. But it is not clear how much effort he himself put into devising a formal theory of concepts, or how high a priority it was for him. He was, however, undoubtedly dissatisfied with his own thought about concepts.

Gödel's epistemological views on mathematics have two distinctive features. First, already in RML he proposed that axioms in mathematics might be accepted not on the ground of intrinsic evidence but on the basis of their consequences. As he later put it, "There might exist axioms so abundant in their verifiable consequences, shedding so much light on a whole field, and yielding such powerful methods of solving problems that, no matter whether or not they are intrinsically necessary, they would have to be accepted at least in the same sense as any well-established physical theory" (WCCP, 1964 version; CW II 261). In 1944 this was a bold proposal (though with a precedent in Russell), but in time it has come to be widely accepted that mathematical evidence is to some degree a posteriori in this way. That it is to some degree empirical is a further thesis, which Gödel did not embrace.

Second, Gödel held that what he calls mathematical intuition is an essential source of evidence in mathematics; he famously wrote that "we do have something like a perception of the objects of set theory, as is seen from the fact that the axioms force themselves upon us as being true" (CW II 268). In his earlier philosophical publications there are only tenuous indications of this view, but it is fully avowed in the 1964 version of WCCP and developed earlier in the unpublished "Is Mathematics Syntax of Language?" Intuition, as Gödel understands it, is rational evidence. A proposition is a deliverance of intuition if it is rationally evident by itself (i.e., not inferred). The conception is not Kantian; intuition in the relevant sense is not spatio-temporal and does not have the limits of application to mathematics that Kant's and later Kantian conceptions of intuition give rise to. It probably derives from conceptions of rational evidence in early modern philosophers such as Descartes and Leibniz. Gödel argues that mathematical intuition has to be admitted by someone who does not hold a reductionist view such as he attributed to Carnap, although his rather expansive view of its scope could not be shared by someone who does not accept higher set theory. Gödel does not consider an empiricist view such as Quine's.

The above quotation speaks of something like a perception of the *objects* of set theory. Distinctive of Gödel's view is that he thought rational evidence of this kind analogous to perception. One argument is from the inexhaustibility of mathematics, so that in the development of mathematics new intuitions will continue to arise. Another is that we can "perceive" concepts more or less clearly, e.g., he thought the concept of computability was perceived much more clearly after Turing's analysis. It follows that there is no reason to think mathematical intuition infallible. When axioms of set theory "force themselves on us as being true," the objects in question are primarily *concepts,* in particular the concept of set itself, and of course the concept of membership. It is doubtful that intuition of particular sets plays any significant role in the evidence of the axioms.

Gödel saw a parallelism both ontologically and epistemologically between mathematics and physical theory, and that allowed him to find a place for a posteriori evidence in mathematics. In this respect his view resembles Quine's. But he insisted that mathematics has no implications for physical reality and argued that basic mathematical principles are analytic. However, he meant by this that they follow from the concepts in them, so that the claim is quite compatible with realism. It is probable that intuition as described above amounts to seeing how a proposition is made true by the concepts constituting it. Of certain strong axioms of infinity Gödel thought this had not yet been made out, so that whatever plausibility they might have, we do not know them by intuition.

One idealistic aspect of Gödel's philosophy is his rejection of materialism and mechanism. In the Gibbs lecture of 1951 he considered the implication of his incompleteness theorem for a mechanist view of the mind. Unlike J. R. Lucas,[9] who argued in 1961 that the theorem itself implied the falsity of mechanism, Gödel argued for a disjunction: "Either . . . the human mind infinitely surpasses the powers of any finite machine, or else there exist absolutely unsolvable diophantine problems," i.e., problems expressible as whether a Gödel sentence is true or false (CW III 310). But his view of the power of reason made him favor the first disjunct. Some of his thoughts on this subject appeared in his lifetime in Wang, op. cit., pp. 324–326, in a section revised by Gödel.

As remarked above, Gödel's work in general relativity originated in reflections on Kant's view of time. In "A Remark," he expresses the view that general relativity implies that change is "an illusion or appearance due to our special mode of perception" (CW II 202). Although he admits that in some models an objective global time can be defined, he took his own models to imply that whether this is true depends on the mean motion of matter, and he seems to have thought that too contingent. But the issue concerned the place of time in the overall scheme of things. Gödel's overall view of physical knowledge was cautious but otherwise not more anti-realistic than his view of mathematical knowledge. Without going into the matter, he rejects drawing such conclusions from quantum mechanics. In discussing Kant's idealism in the light of relativity, he suggests that the former should be modified, "i.e. it should be assumed that it is possible for scientific knowledge, at least partially and step by step, to go beyond the appearances and approach the world of things" (CW III 244).[10]

Gödel held a wide spectrum of philosophical views and aspired to general system in philosophy. The outstanding characteristic of his world view is rationalistic optimism, evident in fourteen theses with which he once summarized his philosophy in a manuscript (Wang, *A Logical Journey,* p. 316). He described his philosophy as "rationalistic, idealistic, optimistic, and theological" and as "a monadology with a central monad" (ibid., p. 290). The kinship of this general outlook with Leibniz's is obvious. Gödel thought it should be possible to found metaphysics scientifically, with the help of the axiomatic method (and later phenomenology).

---

[9] "Minds, Machines, and Gödel."
[10] [On the context of this remark, see "Gödel and Philosophical Idealism," pp. 178–179.]

Gödel's more general philosophical reflection is to be found almost entirely in shorthand notebooks and in the remarks in his conversations with Wang. He himself thought he had not worked out his philosophy enough for a more systematic exposition. One concrete result, however, was a modal-logical version of the ontological proof written out in 1970, related to Leibniz's version of the argument.[11]

What is Gödel's significance as a philosopher? Although his views are and will remain controversial, he is a major philosopher of mathematics, where his platonist and rationalist view were developed in the context of reflection on his own mathematical work, and the data he appeals to, particularly the incompleteness theorems and the nature of higher set theory, are a challenge for any view. Outside this area, the views in the philosophy of time developed along with his cosmological models, his views on minds and machines, and his version of the ontological argument, have excited interest and will probably continue to do so. His general world view seems to be of interest mainly as an expression of a genius of a very particular temperament. Whether others sympathetic with that temperament will find his sketches promising remains to be seen.[12]

---

[11] See Gödel, "Ontological Proof." [Naturally this has generated a lot of discussion, some of it reflected in Robert Merrihew Adams's introductory note in CW III.]

[12] [I am much indebted to John Dawson, Warren Goldfarb, Peter Koellner, Wilfried Sieg, and Mark van Atten for their comments and suggestions. Some have saved me from errors or infelicities. Unfortunately this acknowledgment could not appear in the original published version.]

5

This paper was written for *The Philosophy of Bertrand Russell,* a volume of Paul Arthur Schilpp's series The Library of Living Philosophers. In his letter of invitation of November 18, 1942, Schilpp proposed the title of the paper and also wrote, "In talking the matter over last night with Lord Russell in person, I learned that he too would not only very greatly appreciate your participation in this project, but that he considers you the scholar par excellence in this field." Gödel sent in the manuscript on May 17, 1943. There followed a lengthy correspondence about stylistic editing proposed by Schilpp and Gödel's own deliberations concerning revision.[1] Before Gödel had submitted the final version, Russell had completed his reply to the other papers and decided that under the circumstances he would not reply to Gödel's. When Gödel finally sent in the revised version on September 28, he wrote to Russell attempting to change his mind about not replying.[2] He undertook to disabuse Russell of his impression that what Gödel said would not be controversial, and emphasized his criticisms of Russell. However, Russell confined himself to the following brief note:

> Dr. Gödel's most interesting paper on my mathematical logic came into my hands after my replies had been completed, and at a time when I had no leisure to work on it. As it is now about eighteen years since I last worked on mathematical logic, it would have taken me a long time to form a critical estimate of Dr. Gödel's

---

[1] [Much of this correspondence, including the letter quoted above, is published in *Collected Works* (hereafter CW), vol. V, pp. 217–232. Note that all citations of Gödel's works are in the pagination of CW; "Russell's Mathematical Logic" (hereafter RML) is cited merely by page number.]

[2] [For this letter see CW V 207–208.]

opinions. His great ability, as shown in his previous work, makes me think it highly probable that many of his criticisms of me are justified. The writing of *Principia Mathematica* was completed thirty-three years ago, and obviously, in view of subsequent advances in the subject, it needs amending in various ways. If I had the leisure, I should be glad to attempt a revision of the introductory portions, but external circumstances make this impossible. I must therefore ask the reader to give Dr. Gödel's work the attention that it deserves, and to form his own critical judgment on it.[3]

Gödel subsequently submitted "A Remark about the Relationship between Relativity Theory and Idealistic Philosophy" to the volume on Einstein in Schilpp's series and accepted an invitation to contribute to the Carnap volume. Several drafts of this paper on Carnap survive in Gödel's *Nachlaß,* but it was never actually submitted.[4] Still later, Gödel declined an invitation to contribute to the Popper volume.

This paper was reprinted twice,[5] with only editorial changes in the text. In an opening footnote added in Benacerraf and Putnam, Gödel clarifies the difference between the use of the term "constructivistic" in that paper and the more usual uses; this remark is revised and expanded in the reprint in Pears.

The paper is notable as Gödel's first and most extended philosophical statement. It holds mainly to the form of a commentary on Russell; as such it has been quite influential. However, Gödel is not reticent in expressing his own views. In the present note I shall give more emphasis to what it reveals about the thought of its author.

The organization of the paper is difficult for the present commentator to analyze. He cannot but endorse Hermann Weyl's remark that the paper is "the work of a pointillist: a delicate pattern of partly disconnected, partly interrelated, critical remarks and suggestions."[6] Nonetheless, Gödel's paper might be divided as follows:

---

[3] Schilpp, *The Philosophy of Bertrand Russell,* p. 741.
[4] [Versions III and V have appeared in CW III.]
[5] In Benacerraf and Putnam, *Philosophy of Mathematics* (both 1st and 2nd eds.), and in Pears, *Bertrand Russell.*
[6] Review of Schilpp, p. 210.

I follow this division in the remainder of this note, where I use a number in parentheses without a date to indicate a page number in Gödel's paper.[7]

The Gödel archive contains reprints of the paper (designated below as A–D) and a loose page (designated as E) containing annotations to it. All changes on A–E are listed in the textual notes for the paper at the end of CW II. Many of the annotations are textual emendations, for the most part either of a stylistic nature or for greater explicitness. A few indicate changes of view on specific points. There is no way of knowing whether any of these changes represents a final position for Gödel. They are hardly reflected at all in the reprints cited in note 3. E contains some remarks on Bernays's review of the paper.

1. Introductory Remarks

Early on, Gödel remarks on Russell's "pronouncedly realistic attitude" and the analogy with natural science expressed by Russell's remark, "Logic is concerned with the real world just as truly as zoology, though with its more abstract and general features."[8] He also mentions another, epistemological analogy in Russell's view that the axioms of logic and mathematics do not have to be evident in themselves but can obtain justification from the fact that their consequences agree with what has been found evident in the course of the history of mathematics (121). Gödel remarks: "This view has been largely justified by subsequent

---

[7] [As noted above, in the pagination of CW.]
[8] *Introduction to Mathematical Philosophy*, p. 169.

developments, and it is to be expected that it will be still more so in the future."

The essay as a whole might be seen as a defense of these attitudes of Russell against the reductionism prominent in his philosophy and implicit in much of his logical work. It was perhaps the most robust defense of realism about mathematics and its objects since the paradoxes had come to the consciousness of the mathematical world after 1900. Bernays's earlier defense of realism (for example, in "Sur le platonisme") was more cautious. Gödel begins to develop this theme when he turns to Russell's approach to the paradoxes (see especially §§3–5 below).

2. The Theory of Descriptions

This discussion is noteworthy. Gödel indicates (122 and note 5) a formal argument for Frege's thesis that the signification (his translation of Frege's *Bedeutung*) of two sentences is the same if they agree in truth-value. The argument collapses intensional distinctions on the basis of simple assumptions about signification. A similar argument for the same conclusion, from somewhat different assumptions, occurs in Church's "Carnap's *Introduction to Semantics*," pp. 299–300. Such collapsing arguments have been prominent since in philosophical discussions of meaning and reference, modality and propositional attitudes. Gödel concedes that Russell's theory of descriptions avoids Frege's conclusion and allows a sentence to signify a fact or a proposition.[9] He expresses the suspicion that it only evades the problem (123).

---

[9] Church (Review of Quine, "Whitehead") observes that Russell's theory of descriptions eliminates apparent violations of the substitutivity of identity in intensional contexts. The universal substitutivity of identity is one of the assumptions on which Gödel's and Church's collapsing arguments turn. Church's observation is the basis for the reply to Quine's criticism of modal logic in Smullyan, "Modality and Description." Quine's use of the argument in criticizing modal logic is its most widely known and influential use. However, it does not occur in either of his two early papers on this theme ("Notes on Existence and Necessity" and "The Problem of Interpreting Modal Logic"). The earliest occurrence of a collapsing argument of the Gödel-Church type that I have been able to find in Quine's writings is *From a Logical Point of View*, p. 159; see also "Three Grades," pp. 163–164. Both these arguments are essentially the same as Church's.

[Gödel's own argument and its significance are analyzed at length in Neale, "The Philosophical Significance of Gödel's Slingshot."]

3. The Paradoxes and the Vicious Circle Principle

Gödel says that Russell "freed them [the paradoxes] from all mathematical technicalities, thus bringing to light the amazing fact that our logical intuitions (i.e., intuitions concerning such notions as: truth, concept, being, class, etc.) are self-contradictory" (124). Many readers have been puzzled by the contrast between this statement and the defense of the concept of set in "What Is Cantor's Continuum Problem?,"[10] where he says that the set-theoretical paradoxes "are a very serious problem, but not for Cantor's set theory" (1947 version, p. 180), revised in 1964 to ". . . problem, not for mathematics, however, but rather for logic and epistemology" (p. 258).[11] Closer examination makes the passages not difficult to reconcile. In the later paper Gödel says that the concept of set in contemporary mathematics, including Cantor's set theory, can be taken to be what we call the iterative conception of set, according to which sets are obtained by iterated application of the formation of sets of previously given objects, beginning with some well-defined objects such as the integers. It is this conception that has "never led to any antinomy whatsoever" and whose "perfectly 'naive' and uncritical working" has "so far proved completely self-consistent" (WCCP, 1964 version, p. 259).

In the present paper, Gödel is following Russell in being concerned with the foundations of logic in a larger sense; note that in the above quotation our logical intuitions are said to be "concerning such notions as: truth, concept, being, class, etc." Though it is mentioned (134), the iterative conception of set is kept in the background, perhaps more so than was optimal for the purpose of defending realism. But, as we shall see, Gödel's purpose was not limited to defending realism about sets or the objects of classical mathematics.

---

[10] Cited hereafter as WCCP, with the date 1947 or 1964 to indicate which version is being cited.

[11] Perhaps Gödel thought it necessary to clarify the difference of his concerns from those of RML. [We might note that Zermelo discovered Russell's paradox at about the same time. He may have had a clearer sense of its significance for set theory than Russell, but he did not show interest in or appreciation of its general logical significance, which is what Gödel's remark points to. On what Zermelo accomplished see pp. 166–168 of Ulrich Felgner's introductory note to "Untersuchungen über die Grundlagen der Mengenlehre" in Zermelo's *Collected Works*.]

After short remarks about two proposals that Russell discussed briefly in "On Some Difficulties," the "theory of limitation of size" and the "zig-zag theory," Gödel turns to the vicious circle principle. The masterly analysis of the ambiguities of this principle and the criticism of the principle itself constitute probably the most influential piece of direct commentary on Russell in the essay and also supply a major argument for Gödel's own position. Gödel's main criticism, which had already been intimated by Ramsey,[12] is that the strongest form of the principle, that no totality can contain members *definable only in terms of* this totality, is true only if the entities whose totality is in question are "constructed by ourselves." Gödel adds: "If, however, it is a question of objects that exist independently of our constructions, there is nothing in the least absurd in the existence of totalities containing members, which can be described . . . only by reference to the totality" (127–128). Earlier Gödel observed that this form of the principle is not satisfied by classical mathematics or even by the system of *Principia Mathematica* with the axiom of reducibility. He considered this "rather as a proof that the vicious circle principle is false than that classical mathematics is false" (127).

These remarks lead Gödel into the declaration of his realistic point of view. But before discussing this we should note his characterization of the position that would justify the strong vicious circle principle, which he calls constructivistic or nominalistic. He seems to regard this viewpoint as involving the eliminability of reference to such objects as classes and propositions (127–128). His model is clearly Russell's no-class theory.

4. Gödel's Realism

Classes and concepts, according to Gödel, may be understood as real objects "existing independently of our definitions and constructions" (128). It should be stressed, as it has not been in previous commentaries on this paper, that Gödel's realism extends not only to *sets* as described in axiomatic set theory, but also to what he calls *concepts:* "the properties and relations of things existing independently of our definitions and constructions." He is clearly referring to both in his often-quoted remark that "the assumption of such objects is quite as legitimate as

---

[12] "The Foundations of Mathematics," p. 41.

the assumption of physical bodies" and that "they are in the same sense necessary to obtain a satisfactory system of mathematics as physical bodies are necessary for a satisfactory theory of our sense perceptions" (128). These remarks have excited critical comment.[13] To deal in an adequate way with the questions they raise would be quite beyond the scope of a note of this kind. However, we should state some of these questions.

(i) What does Gödel mean by "real objects" and "existing independently of our definitions and constructions"? A question of this kind arises about any form of philosophical realism. If realism about sets and concepts is to go beyond what would be asserted by a non-subjectivist form of constructivism, this reality and independence will have to amount to more than objective existence. In the present paper, Gödel does not undertake very directly to clarify his meaning, though something can be learned from his discussion of the ramified theory of types (see below).[14] Of course, the general question "What is realism?" has been much debated in quite recent times, largely through the stimulation of the writings of Michael Dummett.[15]

A point that needs to be stressed, however, is that Gödel saw his realism in the context of concrete problems and as motivating mathematical research programs. This is perhaps most evident in WCCP, with its defense of the view that the continuum hypothesis is definitely true or false even though probably (and by 1963 certainly) independent of the established axioms of set theory. Further reflection shows

---

[13] For example, Chihara, *Ontology and the Vicious Circle Principle*, pp. 61, 75–81; Chihara, "A Gödelian Thesis," part I; Dummett, "Platonism," p. 204.

[14] In "A Remark about the Relationship between Relativity Theory and Idealistic Philosophy," Gödel argues that the general theory of relativity calls in question the objectivity òf time and change. He sees this as a confirmation of idealistic views, in particular Kant's. He does not attempt to draw any parallel with a possible anti-realist view of mathematics.

Of course, Kant did not "deny the objectivity of change" (ibid., p. 202) if what is meant by the latter is the existence of an objective temporal order that is the same for all observers with our forms of intuition. Evidently Gödel thinks of the dependence of the temporal ordering of events on the position and state of motion of the observer according to relativity theory as parallel to the dependence of the very temporality of the experienced world on the constitution of our cognitive faculties according to Kant.

[Further to these issues, see my "Gödel and Philosophical Idealism."]

[15] For example, the essays in *Truth and Other Enigmas*, especially "Truth," "Realism," and "The Reality of the Past."

that it is very much present in the paper under discussion, where Gödel criticizes ideas of Russell that obstructed the transfinite extension of the hierarchies of simple and ramified type theories. The remarks about his own theory of constructible sets (136; see below) are an illustration of the "cash value" of realism for Gödel.[16]

(ii) How does Gödel understand the parallel between the objects of mathematics and "physical bodies"? It would be tempting to suppose that Gödel views sets and concepts as postulated in a theory to *explain* certain data.[17] This is suggested by the parallel itself between the necessity of sets and concepts for a "satisfactory system of mathematics" and the necessity of physical bodies for a "satisfactory theory of our sense perceptions"; it is also in line with Gödel's approval, noted above, of Russell's suggestion that mathematical axioms can be justified by their consequences. But more direct evidence for this interpretation is lacking.[18]

Gödel does, however, use the notion of "data" with reference to mathematics. Indeed, one point of parallelism between mathematics and physics is simply that "in both cases it is impossible to interpret the propositions one wants to assert about these entities as propositions about the 'data'" (128). But at this point he refrains from saying what plays the role of data in the case of mathematics. We shall return to this matter in connection with his discussion of the ramified theory.

(iii) By "concepts" Gödel evidently means objects signified in some way by predicates. The notion "property of set," which he counts among the primitives of set theory (WCCP 1947, p. 182 n.17, or 1964, p. 260 n.18), is clearly a special case of this notion. Why he should have considered "property of set" a primitive of set theory is clear enough from

---

[16] Cf. note 48a of "Über formal unentscheidbare Sätze," which Burton Dreben called to my attention.

[17] This interpretation is assumed by Dummett ("Platonism," p. 204) and in some of Chihara's criticisms (for example, "A Gödelian Thesis," pp. 214–215).

[18] Gödel does not use the language of explanation in the two passages where he is most explicit about the justification of mathematical axioms through their consequences (WCCP 1964, pp. 261, 269), although in the former he does describe a (hypothetical) situation in which axioms "would have to be accepted in at least the same sense as any well-established physical theory." In spite of its lack of direct support, the interpretation in terms of explanation is difficult to refute.

In these passages, Gödel seems to me to be considerably more cautious about justification of axioms by their consequences than he appears to be in the passage noted above in the text.

the role of classes in set theory and from the generalization with respect to predicates contained in the axioms of separation and replacement. Gödel therefore did not lack mathematical motivation for adding something like concepts to his ontology.[19]

But what sort of theory of concepts did Gödel envisage? What consequences for the theory does realism about concepts have, once the existence of sets as "real objects" is granted? The use of the notion of class that is standard in set theory is predicative relative to the universe of sets. In the above-cited note, he seems to envisage impredicative theories of properties based on the simple theory of types, which he also mentions as a theory of concepts in the present paper (130). In the note he remarks that such theories are not deductively stronger than extensions of the axioms referring to sets.

It is clear that Gödel takes his realism about concepts to justify an impredicative theory, and he suggests strongly that he would prefer a stronger theory than the simple theory of types. He claims (130) that impredicative specifications of properties do not themselves lead to absurdity and that a property might "involve" a totality of properties to which it belongs, thus contradicting the second of the three forms of the vicious circle principle he has earlier distinguished.[20] He also

---

[19] This view may be reflected in Gödel's choice of a theory with class variables as the framework for *The Consistency of the Continuum Hypothesis*. Note the remark on p. 34 (p. 2 of the original), "Classes are what appear in Zermelo's formulation . . . as 'definite Eigenschaften'."

[20] Whether the simple theory of types conflicts with this form of the principle depends on how it is interpreted. It seems clear, for example from the emphasis Gödel places on the claim (128) that classical mathematics, and in particular *Principia* with reducibility, does not satisfy the *first* form, that Gödel thinks that the extensional simple theory of types with its higher-order variables interpreted to range over sets does satisfy the second form (and the third as well). Whether an intensional form of the simple theory, in which the variables range over properties and relations, satisfies the second form will depend on the underlying notion of intension. Gödel's remark (130) that "the totality of all properties (or of all those of a given type) does lead to situations of this kind," in which the second form is violated, makes it clear that he envisages a notion of property that would lead to an interpretation of the simple theory of types where the second form of the principle is violated. Gödel's annotations to reprint A, however, call this into question. In RML, the violation of the second form of the principle is said to arise because a universal quantification over properties of a given type contains these properties as constituents of their content (130). This is questioned in A on the ground that universal quantification "does not mean in the same way as conjunction does."

remarks: "Nor is it self-contradictory that a proper part should be identical (not merely equal) to the whole, as is seen in the case of structures in the abstract sense" (130). Of course, on the set-theoretic conception of structure, such identity (as opposed to isomorphism) does not obtain and would indeed be self-contradictory, at least if one takes "proper part" in its obvious meaning of a substructure whose domain is a proper subset of the whole. Gödel is evidently thinking in terms of an informal notion of structure according to which isomorphic "structures" (in the set-theoretic sense) are the *same* structure. But it is a problem to construct a theory in which this sameness is interpreted as identity.[21]

Gödel seems to regard the simple theory of types as the best presently available solution to the paradoxes for a theory of concepts, but "such a solution may be found . . . in the future perhaps in the development of the ideas sketched on pages 132 and 150" (130). The former refers to his remarks on Russell's "zig-zag theory," the latter to the frequently quoted but enigmatic suggestion that a concept might be assumed "significant everywhere except for certain 'singular points' or 'limiting points', so that the paradoxes would appear as something analogous to dividing by zero." Evidently he has in mind a type-free theory; he mentions as attempts the work of the early 1930s on theories based on the λ-calculus (in particular Church, "A Set of Postulates"), which he views as having had a negative outcome, in view of Kleene and Rosser, "The Inconsistency." Gödel does not return to this theme in later publications, except for the brief remark that "the spirit of the modern abstract disciplines of mathematics, in particular the

---

It should be noted that Gödel's first remark in E on Bernays's review of the paper is "Misunderstanding of my interpretation of type theory for concepts." Presumably he is attributing such a misunderstanding to Bernays. I am unable to determine in what the misunderstanding consists.

[21] In the language of category theory, we could say that an object A is a proper part of an object B if there is a monomorphism of A into B that is not epi; this does not exclude identity. The alternative, to say that A is a proper part of B if there is a monomorphism of A into B but there is not an epimorphism, does of course exclude identity but does not fit what Gödel says. Of course, the relevant difference between the construal of structures as tuples of a domain and relations on it, and the language of categories, is that the former forces, while the latter does not, a distinction between isomorphism and identity. This is a much more superficial difference than the matter of the self-applicability of categories (see below).

theory of categories, transcends this [iterative] concept of set, as becomes apparent, e.g., by the self-applicability of categories" (WCCP, 1964, p. 258 n.12; not in the 1947 version.) However, he evidently thought that Mac Lane's distinction between large and small categories captured "the mathematical content of the theory" as it then stood. But the program of constructing a strong type-free theory has attracted others, with inconclusive results so far.[22]

Gödel's remarks about realistic theories of concepts in the present paper have an inconclusive character; no available theory satisfies him. In later publications, as we have noted, he is virtually silent on the subject. The question arises whether Gödel himself worked on the project of constructing a theory that would answer to his conception. Whether he did is not known, but the absence of more definite information would suggest the conjecture that he never formulated such a theory to his own satisfaction. It is to be hoped that transcriptions of Gödel's shorthand notebooks will shed light on these questions.[23]

5. The Ramified Theory of Types

Returning to Russell, Gödel begins his discussion of the ramified theory by remarking on Russell's "pronounced tendency to build up logic as far as possible without the assumption of the objective existence of such entities as classes and concepts" (131). He reads the contextual definitions of locutions involving classes in *Principia* as a reduction of classes to concepts,"[24] but reasonably enough finds matters not so clear when it comes to concepts and propositions. Influenced especially by the introduction to the second edition, Gödel finds in *Principia* a program according to which all concepts and propositions except logically simple ones are to "appear as something constructed (i.e., as something

---

[22] For a survey see Feferman, "Toward Useful Type-Free Theories, I" and its sequel. [The sequel has apparently not appeared.]

[23] [On this point see the Postscript to this essay.]

[24] Gödel's reading is misleading in that he clearly understands concepts to be *objects,* while the "ambiguity" that Russell attributes to prepositional functions is close to Frege's "unsaturatedness." But, from his later comment on the notion of propositional function (136–137), it is clear that this does not result from misunderstanding but is rather a conscious assimilation of Russell's conceptual scheme to his own. Cf. note 38 below.

not belonging to the 'inventory' of the world)" (132). This program offers an intrinsic motivation for the ramification of the type hierarchy, but does not yield a theory strong enough for classical mathematics, for well-known reasons: the impredicative character of standard arguments in analysis, and the question whether, to construct number theory, one can replace the Frege-Russell definition of the predicate "natural number" by one in which the second-order quantifier is restricted to a definite order (134–135).[25]

It is not as clear as it might be how Gödel sees the realization of the program he attributes to Russell even to obtain ramified type theory without reducibility. The introduction to the second edition of *Principia* proceeds on the basis of the Wittgensteinian idea that "functions of propositions are always truth functions, and that a function can only occur in a proposition through its values" (p. xiv). But it is hard to see how propositions involving quantifiers are to be interpreted as truth functions of atomic propositions unless infinitary propositional combinations are allowed, as Ramsey in effect proposed; Gödel's comment on that is that one might as well adopt the iterative conception of sets as pluralities (134). Gödel says that Russell "took a less metaphysical course by confining himself to such truth-functions as can actually be constructed" (134). But what are the allowed means of construction? Gödel apparently has in mind an interpretation of the ramified theory in which the higher-order variables range over *predicates,* that is, the linguistic expressions that "express" the propositional functions that the quantifiers range over on the naive reading. This is indicated by Gödel's later remark that for propositional functions to be "defined (as in the second edition of *Principia*) to be certain finite . . . combinations (of quantifiers, propositional connectives, etc.)" (135) would presuppose the notion of finiteness and therefore arithmetic, and by the earlier characterization of the constructivist view he is trying to explicate as a form of nominalism (127–129).

Such an interpretation is certainly possible and well known, provided that at the level of individuals one has elementary syntax. But this reading of Gödel still leaves some puzzles. Translating a statement about propositional functions of order n as one about predicates,

---

[25] Gödel leaves the latter question open. A negative answer is claimed without proof in Wang, "Ordinal Numbers" (see *A Survey*, p. 642). A full treatment, with a proof, is given in Myhill, "The Undefinability of the Set of Natural Numbers."

namely, predicates containing quantifiers only for propositional functions of order $< n$, requires the notion of satisfaction, or at least truth, for the language with quantifiers of order $< n$. It is surprising that Gödel nowhere remarks on this fact, particularly since in this context we would have to suppose satisfaction or truth introduced by an inductive definition, another obstacle to applying the idea in reducing arithmetic to logic. Moreover, such an interpretation entirely eliminates quantification over the sort of entities Gödel calls concepts, at least in the absence of such locutions as propositional attitudes, and therefore does not leave "the primitive predicates and relations such as 'red' or 'colder' " as "real objects" (132). But to suppose that Gödel would have remarked on the latter point if he had had the present interpretation in mind may be to attribute to him a Quinean distinction of "ontology" and "ideology" that is foreign to him.

The difficulties faced by this last interpretation suggest to me that Gödel did not distinguish clearly in his own mind between a nominalist theory of concepts in which such entities are *eliminated*, and a theory in which every concept in the range of a quantifier is "signified" by an expression for it that is antecedently understood, but in which reference to concepts is not actually eliminated, because, presumably, one does not give a contextual definition of quantifiers over them. The latter sort of theory might now be realized by a substitutional interpretation of quantifiers over concepts; given such an interpretation, the most that can be asked in establishing the "existence" of a concept is the construction of a meaningful expression "signifying" it.

Something more should be said about Gödel's use of the term "data." In the discussion of the ramified theory it refers to that on the basis of which classes and concepts are constructed, or perhaps to what is allowed as primitive in a theory in which reference to classes and concepts is eliminated. (See for example p. 132 n.33.) Gödel does not say here what epistemological force this might have. The analogy with sense perception may be limited to the context of interpreting Russell, who was interested in a program that would represent the objects of physics as "logical constructions" from sense-data. In Gödel's own epistemological view of mathematics, what corresponds most closely to sense perception is something quite different, namely elementary arithmetical evidence (see p. 121).

Like many other commentators, Gödel found that in the *first* edition of *Principia* the constructivistic attitude was fatally compromised by the axiom of reducibility, but his description (133) of what survived is worth noting.[26] However, Gödel's remarks about the axiom of reducibility show lack of sensitivity to the essentially intensional character of Russell's logic; the fact that every propositional function is *coextensive* with one of lowest order does not imply "the existence in the data of the kind of objects to be constructed" (132), if the objects in question are concepts or propositional functions rather than classes. It is similarly misleading to say that, "owing to the axiom of reducibility, there always exist real objects in the form of primitive predicates, or combinations of such, corresponding to each defined symbol" (133). Russell himself was closer to the mark in saying that the axiom accomplishes "what common sense effects by the admission of classes."[27] The ramified theory with reducibility would fit well with a conception according to which classes are admitted as "real objects," but the conception of propositional functions (concepts) is [relative to classes] constructivistic. This insensitivity is quite common in commentators on Russell, but is somewhat surprising in Gödel, since his own "concepts" are evidently intensions of a kind, and his "Über eine bisher noch nicht benützte Erweiterung" shows a very subtle and fruitful handling of intensional notions.[28]

---

[26] Gödel mentions here the treatment of propositional connectives as applied to propositions containing quantifiers, presumably in *9 of *Principia,* which he says "proved its fecundity in a consistency proof for arithmetic." The connection between *9 and Herbrand's work suggests that he has in mind Herbrand, "Sur la non-contradiction," which, however, covered only first-order arithmetic with quantifier-free induction. A proof within Herbrand's framework that covers full first-order arithmetic became known only some years later; see Dreben and Denton, "Herbrand-Style Consistency Proofs," and Scanlon, "The Consistency of Number Theory." Could Gödel have seen at this time how to extend Herbrand's proof? The errors in the proof of the fundamental theorem in Herbrand's dissertation would have been an obstacle. At the time Gödel was at least aware of a difficulty; see van Heijenoort, *From Frege to Gödel,* p. 525. Another possibility, suggested by van Heijenoort, is that Gödel was thinking of the proof he published in "Über eine bisher noch nicht benützte Erweiterung," which he had discovered not long before writing the present paper. This now seems to me somewhat more likely.

[It has now been established van Heijenoort's conjecture is correct. . . . It is also known that Gödel saw a gap in Herbrand's proof and worked out a way to correct it.]

[27] "Mathematical Logic as Based on the Theory of Types," p. 167 of the reprint in van Heijenoort, *From Frege to Gödel.*

[28] Of course, the theory of intensional equality in that paper is very different from more usual constructions in intensional logic, and in the present paper he suggests that concepts might obey extensionality (129).

Gödel concludes his discussion of the ramified theory with well-known remarks in which he views his own theory of constructible sets as an extension of the hierarchy of orders, now within the framework of ordinary (impredicative) mathematics, to arbitrary transfinite orders (135–136). After what I have said about Gödel's treatment of the axiom of reducibility, I should call attention to his characterization of his theorem that every constructible set of integers has order $< \omega_1$ as a "transfinite theorem of reducibility"[29]—thus, when set against Russell, a striking application of his realistic point of view.

Gödel remarks that even from the predicative standpoint an extension of the hierarchy of orders is possible and, moreover, demanded by the theory. This remark may be the first suggestion of a program that was pursued by several logicians from 1950 on, of extending and analyzing the resources of predicative mathematics by means of ramified theories with transfinite orders. The first work of this kind sought to give a better reconstruction of classical mathematics than earlier predicative work by constructing transfinite ramified theories.[30] It left open the question what ordinal levels can be admitted in such a construction.

Gödel already offers a hint in remarking that one can extend the hierarchy of orders "to such transfinite ordinals as can be constructed within the framework of finite orders" (136). It seems evident that such a procedure might be iterated; this gives rise to the notion of autonomous iteration that is prominent in later analyses of predicativity. The first such proposal is made in Wang, "The Formalization": given an interpreted language Σ_α (in his setting, ramified set theory with ordinal levels $< \alpha$), one extends it by admitting as new levels ordinals β such that a well-ordering of type β is *definable* in Σ_α (*A Survey*, p. 579). In this case, since Wang's theory Σ_0 could define all recursive orderings, the iteration closes after one step. This was shown by Clifford Spector, who constructed a sequence of systems similar to Wang's, indexed by the recursive ordinal notations, and showed that any well-ordering definable in one of the systems has as order type a recursive ordinal; he

---

[29] In reprint A, Gödel amended the text in a way that omits this phrase.

[30] Lorenzen, "Die Widerspruchsfreiheit der klassischen Analysis," "Maß und Integral," and *Einführung in die operative Logik*; Wang, "The Formalization of Mathematics."

also showed that the sets of natural numbers definable in some system of the sequence are exactly the hyperarithmetic sets.[31]

Kleene's work on the hyperarithmetic hierarchy related it to a transfinite ramified hierarchy, and a number of technical results suggested the thesis, advanced with some reservations by G. Kreisel ("La prédicativité," p. 373), that a set of natural numbers is predicatively definable if and only if it is hyperarithmetic. Kreisel seems to have been more confident of the "only if" than of the "if" part of this thesis.[32] The main reservation about the latter concerned the use in the definition of a hyperarithmetic set of the notion of a recursive ordinal; should one not demand, for an ordinal to count as predicatively obtained, that an ordering of that type be *predicatively recognized* to be a well-ordering?[33] This question leads into the analysis of predicative *provability*, in which another type of autonomous iteration (first suggested in Kreisel, "Ordinal Logics") plays a role. In the context of ramified theories, one constructs a progression of theories (in the precise sense of Feferman, "Transfinite Recursive Progressions") such that levels up to an ordinal given by a primitive recursive ordering are admitted if at an earlier stage that ordering has been *proved* to be well-founded. This led to a precise characterization of the predicatively provable statements of ramified analysis.[34] The same idea is applied to the admission of stages in the progression of theories, and it is therefore applicable to

---

[31] Spector, "Recursive Ordinals and Predicative Set Theory."

[32] Ibid., pp. 387–388. See also his "The Axiom of Choice and the Class of Hyperarithmetic Functions," p. 318, and Feferman, "Systems of Predicative Analysis," p. 10.

[33] Kreisel, "La prédicativité," p. 387. Accepting this demand seems to involve giving up the attempt to characterize predicative definability independently of predicative provability. In fact the independent discussion of predicative definability seems to have petered out when Feferman's and Schütte's results became known.

That a definite meaning can be given to the notion "predicatively definable set of natural numbers" without using impredicative concepts seems very doubtful, and Kreisel already stressed (ibid.) that he was approaching the analysis of predicativity with the help of impredicative notions, that of a well-ordering in particular. (See also Lorenzen's reasons for rejecting the question how far his iterative construction of language strata can be carried, *Einführung in die operative Logik,* p. 189.) Kreisel's attitude was in line with Gödel's realism and may have been influenced by it. But it should be noted that the notions involved can still be understood constructively.

[34] Feferman, "Systems of Predicative Analysis"; Schütte, "Predicative Well-Orderings" and "Eine Grenze für die Beweisbarkeit."

118

unramified theories, which allowed Feferman to extend his characterization to the usual language of analysis in "Systems" and to set theory.[35] The results of Feferman and Schütte all point to the conclusion that an ordering that can be predicatively proved to be a well-ordering is of type less than a certain recursive ordinal Γ_0, the first strongly critical number.[36]

6. The Simple Theory of Types

What is of greatest interest in Gödel's discussion of the simple theory is his questioning of it as a theory of concepts and the hint for a possible type-free theory commented on above (§4). It should be noted that he motivates his suggestion by means of Russell's idea that propositional functions have limited ranges of significance (137–138).

7. Analyticity

Gödel now turns to "the question whether (and in which sense) the axioms of *Principia* can be considered to be analytic." In a first sense— roughly, reducibility by explicit or contextual definitions to instances of the law of identity—he says that even arithmetic is demonstrably not analytic because of its undecidability. This sense is of interest because it seems to be directly inspired by Leibniz. If infinite reduction, with intermediary sentences of infinite length, is allowed (as would be suggested by Leibniz's theory of contingent propositions), then all the axioms of *Principia* can be proved analytic, but the proof would require "the whole of mathematics . . . e.g., the axiom of choice can be proved to be analytic only if it is assumed to be true" (139). This remark anticipates later arguments criticizing the thesis that mathematics is analytic, such as Quine, "Carnap and Logical Truth." Gödel continues: "In a second sense a proposition is called analytic if it holds 'owing to the meaning of the concepts occurring in it', where this meaning may perhaps be undefinable (i.e., irreducible to anything more fundamental)." In this sense, Gödel affirms the analyticity of the axioms of the first edition of *Principia* other than the axiom of infinity

---

[35] "Predicative Provability in Set Theory" and "Theories of Finite Type."
[36] Feferman, "Systems of Predicative Analysis," part II; see also Schütte, *Proof Theory*, ch. 8.

for two interpretations, "namely if the term 'predicative function' is replaced either by 'class' (in the extensional sense) or (leaving out the axiom of choice) by 'concept'" (139). The first is the sort of interpretation suggested by remarks of Russell such as that in "Mathematical Logic as Based on the Theory of Types" quoted above.[37] The second prompts the remark that "the meaning of the term 'concept' seems to imply that every propositional function defines a concept." Gödel's intuitive notion of concept seems in that respect to have resembled Frege's notion of extension.[38]

For analyticity in this sense, Gödel sees the difficulty that "we don't perceive the concepts of 'concept' and 'class' with sufficient distinctness, as is shown by the paradoxes" (139–140). But, rather than following Russell's reductionism, the actual development of logic (even by Russell in much of his work) has consisted in "trying to make the meaning of the terms 'class' and 'concept' clearer, and to set up a consistent theory of classes and concepts as objectively existing entities" (140). In spite of the success of the simple theory of types and of axiomatic set theory, "many symptoms show only too clearly, however, that the primitive concepts need further elucidation." But surely Gödel must have seen matters differently with respect to the two notions of class and concept: in the former, he seems to insist in WCCP 1947, there is a well-motivated theory that is quite satisfactory as far as it goes; what is lacking in our "perception" of the notion of set is intu-

---

[37] Gödel's remarks about the axiom of reducibility commented on above might suggest that he had in mind an extensional interpretation that would collapse the ramification of the hierarchy. But that in the absence of extensionality such collapse does not occur is made clear in Church, "Comparison." It is straightforward to construct possible-worlds models of Church's formulation where reducibility holds but orders do not collapse.

[Against this suggestion about Gödel's meaning is the remark that if propositional function variables are taken to range over concepts, then the axiom choice is not analytic.]

[38] Except for Frege's commitment to an extensional language, Gödel's "concepts" closely resemble the *objects* that Frege says are signified by such phrases as "the concept *horse*." In the Fregean context, they are hardly distinguishable from the extensions. Indeed, Gödel's suggestion (150; see above) that one might assume a concept significant everywhere except for certain "singular points" recalls Frege's unsuccessful proposal for a way out of the paradoxes. (See Frege, *Grundgesetze*, 2:262, and Quine, "On Frege's Way Out.")

ition regarding the truth of axioms that would decide such questions as the continuum problem. In the latter, he had no theory to offer that answered to his intuitive notion, and it does not appear that such a theory has been constructed since.

The seriousness with which Gödel takes the notion of analyticity in this section has not attracted the attention of commentators. It seems to show a greater engagement with the ideas of the Vienna Circle than has usually been attributed to Gödel, although there is no doubt that the disagreement is deep. With the Vienna Circle and many other analytic philosophers before the impact of Quine's criticism, Gödel believes firmly that mathematical propositions are true by virtue of "the meaning of the concepts occurring" in them, though others might have said "words" instead of "concepts," but he denies that mathematics is true by convention (as perhaps it would be if it were analytic in his first sense) or that its truth is constituted by linguistic rules that we lay down or embody in our usage.[39] Thus, he says, this position does not contradict his view that "mathematics is based on axioms with a real content," since the *existence* of the concepts involved would have to be an axiom of this kind (p. 139 n.47), presumably if one were undertaking to derive the truth of the axioms from their being in some way implied by the concepts.[40]

---

[39] Gödel's annotation in reprint A to note 47 seems to reject the possibility that by virtue of meaning every mathematical proposition can be "reduced to a special case of $a = a$." This thesis is mentioned in the remarks in E on Bernays's review. Bernays appears to argue against it, using Frege's distinction of sense and signification (Gödel's translation of *Bedeutung*): since the transformations that reduce a mathematical proposition to an identity will in general not preserve sense, the "meaning" by virtue of which the reduction proceeds can only be signification (p. 78 of his review). On the latter reading the thesis reduces to triviality.

There is no way of knowing whether this argument influenced Gödel. It seems to me to be unconvincing. That P can be reduced to Q by virtue of the sense of P need not imply that the sense of Q is the same as that of P. [Gödel made this distinction in a letter to Bernays of August 11, 1961, CW IV 104–105.]

[40] Gödel evidently elaborated further on this issue in the unfinished paper, "Is Mathematics Syntax of Language?," undertaken in the mid-1950s for Schilpp, *The Philosophy of Rudolf Carnap*. (See Wang, "Some Facts," p. 658.) From a very brief examination, it seems to reinforce the points of agreement and disagreement with the Vienna Circle noted in the text.

[This theme is discussed further in Essay 6 of this volume, which in particular takes account of the Syntax paper.]

8. Concluding Remarks

A sort of coda to this intricate paper is formed by Gödel's closing remarks noting that mathematical logic had not yet come close to fulfilling the hopes of "Peano and others" that it would contribute to the solution of problems in mathematics, and attributing this to "incomplete understanding of the foundations": "For how can one hope to solve mathematical problems by mere analysis of the concepts occurring, if our analysis so far does not even suffice to set up the axioms?" (152). His suggestion that the hopes expressed by Leibniz for his *characteristica universalis* might, after all, be realistic is one of his most striking and enigmatic utterances. It is known that Gödel was for a time much occupied with the study of Leibniz and that he regarded Leibniz as the greatest influence on him of the philosophers of the past. But the strongly Leibnizian flavor of the last pages of the present paper recedes in his later writings (except perhaps the remarks to the Princeton Bicentennial Conference), and about the substance of his reflections on Leibniz little is known.[41]

---

[41] But see also Wang, "Some Facts," p. 657 n.8. [Evidently much more has become known on this subject from Gödel's *Nachlaß* and from remarks in Gödel's conversations with Hao Wang reported in Wang, *A Logical Journey*. In particular, see Robert Merrihew Adams's introductory note to "Ontological Proof" in CW III, as well as the texts in appendix B of that volume. Rather brief comments on Gödel's appropriation of Leibniz occur in §VI of my "Gödel and Philosophical Idealism."]

I am much indebted to John W. Dawson Jr. for assistance, and to Burton Dreben, Wilfried Sieg, Hao Wang, and the editors (especially Solomon Feferman) for comments and suggestions. Without the work of the late Jean van Heijenoort on Gödel's annotations to A–E, my own brief comments on them would not have been possible.

At the time this essay was written (the mid-1980s) Gödel's notion of concept and his realism about them had hardly been discussed. Some of the documents in which it is developed, such as "Is Mathematics Syntax of Language?," had hardly been studied.[1] At the end of §4 above, I conjecture that Gödel never formulated to his own satisfaction a formal theory of concepts. Even now I do not know of evidence from his notebooks that bears on the question what kind of theory, if any, he had in mind. However, two letters to Gotthard Günther indicate that in the late 1950s he thought of concepts in a type-free way.[2] Remarks reported by Hao Wang in *A Logical Journey* (LJ) tend to show that later he thought the same; for example Gödel insists that concepts may apply to themselves, e.g., in 8.6.15 and 8.6.20. He also remarks more than once on the absence of a theory of concepts; remark 8.6.16 says, "At the present stage, the program of finding axioms for concepts seems wide open."

In "Russell's Mathematical Logic," Gödel definitely regards concepts as objects. That appears not to have been his view in late years. In remark 7.3.12 of LJ, it is denied that concepts are objects. In *Reflections*, p. 307, Wang also states that this was Gödel's later view. But these are the only definite statements to that effect that I have found, and only the first purports to be a quotation. Gödel may have been uncertain about the question. Wang's exposition "Sets and Concepts" is silent on the point, although it begins with statements that clearly

---

[1] I saw "Is Mathematics Syntax of Language?" only when my work on the note was essentially complete.

[2] I refer to the letters of April 4, 1957, *Collected Works*, vol. IV, pp. 526–529, and of January 7, 1959, ibid., 534–535.

imply that concepts are unities (§1.1). A little later it is denied that classes are unities (§1.3, cf. remark 8.6.5 of LJ).[3] One might ask how concepts can be unities and yet not be objects.

In §5 above I wrote,

> Gödel's remarks about the axiom of reducibility show lack of sensitivity to the essentially intensional character of Russell's logic; the fact that every propositional function is *coextensive* with one of lowest order does not imply "the existence in the data of the kind of objects to be constructed" (132), if the objects in question are concepts or propositional functions rather than classes. It is similarly misleading to say that, "owing to the axiom of reducibility, there always exist real objects in the form of primitive predicates, or combinations of such, corresponding to each defined symbol" (133). (p. 116)

This remark was severely criticized by Philippe de Rouilhan.[4] Rouilhan says that the charge that Gödel missed *(a manqué)* the intensional character of Russell's logic is "absurd."[5] I am a little puzzled by the interpretation Rouilhan places on my remarks. Since I think that on the main issue Rouilhan is right, I shall not go into his specific comment but rather try to say what I ought to have said.

What are "the kind of objects to be constructed"? Evidently what I had in mind was a reading of this phrase as referring to the complex propositions or propositional functions that arise in Russell's ramified logic. The problem is that the entities of lowest order are also propositions and propositional functions, and since they are what any construction proceeds from, they are "in the data." If we ignore the intensional character of Russell's logic, then (assuming the axiom of reducibility) the complex ones are effectively identical to the basic ones. But at least a

---

[3] Wang reports Gödel as taking talk of classes to be a *façon de parler;* he seems to have envisaged a contextual definition in terms of concepts. See remarks 8.6.6–9 of LJ. So, in contrast to concepts as well as sets, they are not genuine entities.

[4] *Russell et le cercle des paradoxes,* p. 274. I believe it was Warren Goldfarb who called my attention to this passage.

William Demopoulos also criticizes this remark ("The 1910 *Principia*'s Theory," p. 174). He says that I have missed the "essentially epistemological purpose of reducibility." The claim is very likely true, but since the issue concerns Russell rather than Gödel, I do not pursue the matter here.

[5] That he *missed* it is a little stronger than the claim I actually made.

more charitable reading of Gödel's remark would be that what are to be constructed are entities serving the purposes of classical mathematics, in which case his claim is that Russell is assuming that there are "in the data" entities that accomplish this, namely the propositional functions of lowest order that are coextensive with the ones of higher order that figure in one or another mathematical argument.[6] That reading of Gödel's remark makes it quite compatible with the remark I quote from Russell, that the axiom of reducibility accomplishes what common sense effects by the admission of classes.

Leaving aside these particular passages in Gödel's essay, Rouilhan has a stronger defense of the essay against the charge of lack of sensitivity to the intensional character of Russell's logic. In connection with the discussion of what he calls the constructivistic or nominalistic standpoint and the first form of the vicious circle principle, Gödel introduces the concept of a "notion": "a symbol together with a rule for translating sentences containing a symbol into such sentences as do not contain it, so that a separate object denoted by the symbol appears as a mere fiction" (128). Different definitions "can be assumed to define two different notions" (128–129). A little later he writes:

> As Chwistek has shown,[7] it is even possible under certain assumptions admissible within constructive logic to derive an actual contradiction from the unrestricted admission of impredicative definitions. To be more specific, he has shown that the system of simple types becomes contradictory if one adds the "axiom of intensionality" which says (roughly speaking) that to different definitions belong different notions. This axiom, however, as has just been pointed out, can be assumed to hold for notions in the constructivistic sense. (129)

Gödel is attributing to Chwistek something very close to the Russell-Myhill paradox, presented by Russell in a letter to Frege of September 29, 1902, and in appendix B of the *Principles of Mathematics*. This

---

[6] In the second of my quotations from Gödel above, if one reads 'predicate' as I normally would, as a linguistic item, the remark would have no connection with the axiom of reducibility. It is likely that Gödel meant 'propositional function' or something equivalent.

[7] Leon Chwistek, "Die nominalistische Grundlegung der Mathematik." Gödel probably refers to pp. 369–370 of the paper, but I have not succeeded in understanding Chwistek's argument.

shows that the simple theory of types becomes inconsistent if an assumption is added that makes the entities intensions with refined identity criteria.[8]

Although Gödel is not directly discussing Russell's text at this point, he is clearly discussing an issue that arises for Russell's logic as an intensional logic. Rouilhan comments on the Russell-Myhill paradox in two places, with only brief mention of Gödel.[9] It is no doubt for that reason that he does not take either of those occasions to criticize the first sentence of the passage from my essay quoted above. But it does serve to reinforce his point.

---

[8] I discuss this paradox in the context of the Frege-Russell correspondence in *From Kant to Husserl,* pp. 151–153.

[9] *Russell et le cercle des paradoxes,* p. 228; "Russell's Logics," pp. 346–347.

6

QUINE AND GÖDEL ON ANALYTICITY

This essay is more about Gödel than about Quine. My excuse is that it concerns an aspect of Quine's philosophy that is well known, while the related aspect of Gödel's philosophy is comparatively little known; indeed, most of the texts where it is presented have remained unpublished until now.[1] I do hope, moreover, that seeing some of Quine's arguments in comparison with Gödel's will put some points in Quine's philosophy into sharper relief.

I

Let me remind you of some points about Quine's discussion of analyticity. It is useful to distinguish three strands.

1. All but the last section of "Two Dogmas of Empiricism" is devoted to criticism of possible explications of analyticity; Quine argued that these used notions that are just as problematic as the notion "analytic" itself. The debate about analyticity in the 1950s tended to focus on these arguments and on the demand, whose satisfaction would have been a reply to many of them, for a behavioral criterion of synonymy. The analyticity of *logic* (and, given the tendency shared at the time by Quine to subsume mathematics under logic, of mathematics as well) was not at issue in these arguments, since for the most part they concerned definitions of analyticity where it is an immediate consequence of the definition that logical truths are analytic and, furthermore, no questions were asked about the notion of logical truth itself.

---

[1] [At the time this essay went to press, the publication of volume III of Gödel's *Collected Works* (hereafter CW) was expected quite soon, and it in fact occurred in the year when the essay appeared (1995).]

2. Another strand is a criticism of the view, identified with the Vienna Circle, that logic and mathematics are distinguished both from natural science and more dubious endeavors such as metaphysics by their analytic character. The first steps in this criticism are to be found in the early essay "Truth by Convention." It is not the dominant strand in "Two Dogmas." Quine's most developed presentation of his criticism of this view is in "Carnap and Logical Truth," but he has added to it in subsequent writings, particularly *Philosophy of Logic*. Here, the nature of the enterprise precludes taking the status of logical truth for granted.

3. The last section of "Two Dogmas" begins a more fundamental critique of notions of meaning, continued in *Word and Object* and later writings.

Kurt Gödel was also a critic of the Vienna Circle's view of logic and mathematics. Though it was clear from writings published in his lifetime that his own philosophy was very different, his more extended criticisms are to be found in writings that he himself did not publish, "Some Basic Theorems of the Foundations of Mathematics and Their Philosophical Implications," delivered as the Gibbs lecture to the American Mathematical Society in 1951, and the unfinished essay "Is Mathematics Syntax of Language?"[2]

Gödel has occasionally been linked with Quine as a "platonist," but commentators have seldom seen a close affinity of their views. This is entirely understandable given the difference of their general philosophical outlooks, particularly the strongly naturalistic tendency in Quine and the strongly anti-naturalistic tendency in Gödel.[3] No doubt for this reason, not much attention has been paid to a noticeable similarity in the *arguments* that Gödel and Quine direct against the positivist view of logic and mathematics. In this essay, I intend to look at this similarity a little more closely. Both conclude from rather similar considerations that the Carnapian thesis of the analyticity of logic and mathematics will not do the philosophical work claimed for it. I shall

---

[2] The *Nachlaß* contains six drafts of this essay, which Gödel agreed to write for *The Philosophy of Rudolf Carnap* but never submitted. Versions III and V appear in CW III. Cited below as Syntax, by version number and page in CW for those versions, by manuscript page otherwise. The Gibbs lecture will be cited hereafter as Gibbs.

[3] The latter is expressed most sharply in the shorthand draft from about 1961 of a lecture that was never given, "The Modern Development of the Foundations of Mathematics in the Light of Philosophy."

try to explain how these considerations nevertheless lead Gödel and Quine to some quite different conclusions. Gödel's views raise the question whether one can have a realist view of mathematics, certainly opposed to that of Carnap and importantly differing from that of Quine, and still acknowledge the data on which Quine's criticism of mathematical conventionalism rests. Relying to some extent on Gödel, Hao Wang seems to claim not only this but that from this realist point of view an interesting sense in which mathematics is analytic survives.[4] Some of the fascination that Gödel's discussion of these matters exerts indeed relies on the hope that we might extract from it a notion of analyticity that will help in the philosophy of mathematics. I believe, however, that Gödel's ideas do not offer a response to some arguments that a Quinian would offer against a revival of the idea of mathematics as analytic.

Before going into these matters it is worth pointing out some similarities in Quine's and Gödel's personal relations to the Vienna Circle. Gödel attended many of the meetings, and his teacher Hans Hahn was one of the leaders. Gödel credited the Circle with arousing his interest in the foundations of mathematics.[5] Carnap as well as Hahn was important to his development, although he never agreed with either's philosophy. Quine also attended the Circle for a brief period in 1932–1933, and he has recorded on several occasions the great impact made on him just afterward by Carnap. Although he proved to be Carnap's most influential critic, he never forgot his debt and always treated Carnap with high respect.[6]

Both Quine and Gödel seem to have come to the Circle with convictions opposed to its dominant views. Gödel claimed later to have been a "conceptual realist" as early as 1925.[7] Burton Dreben argues that

---

[4] "Two Commandments of Analytic Empiricism."

[5] In the so-called Grandjean questionnaire; see Wang, *Reflections on Kurt Gödel,* p. 17 [or CW IV 447; cf. also 443. On the influence of Carnap on the young Gödel and Gödel's influence on Carnap, see the Gödel-Carnap correspondence and Warren Goldfarb's introductory note to it (CW IV 337–359), as well as Awodey and Carus, "Gödel and Carnap" and the references there to other writings of theirs, and Goldfarb, "On Gödel's Way In."].

[6] Quine gives an account of his relations with Carnap in "Homage to Rudolf Carnap."

[7] Also in the Grandjean questionnaire, Wang, *Reflections on Kurt Gödel,* pp. 17–18 [or CW IV 447; cf. also 444. On the problems that this claim poses, see Essay 7 in this volume and its Postscript.]

Quine too brought to Vienna convictions opposed at least to those of Carnap, in that he took truth and existence literally with reference to logic and mathematics, rejecting Carnap's "dualism" according to which logic and mathematics are empty of content.

> Quine never believed this. He is and always has been a faithful monist. Philosophy, logic, and mathematics are continuous with (natural) science. I repeat, for Quine truth is truth and existence is existence.[8]

Dreben points out that in contrast, for Carnap it was a permanent lesson of Wittgenstein's *Tractatus* that logic and mathematics (as well as, in a different way, metaphysics) have no factual content. Except perhaps for one specific force of the word 'factual', this view is a major target of Gödel's criticism.

II

In works published in his lifetime Gödel addresses the question whether mathematical statements are analytic in only one place. Near the end of "Russell's Mathematical Logic,"[9] he considers the question "whether (and in which sense) the axioms of *Principia* can be considered to be analytic" (p. 138).

The first sense Gödel considers is "the purely formal sense that the terms occurring can be defined . . . in such a way that the axioms and theorems become special cases of the law of identity and disprovable propositions become negations of this law" (p. 139). If the definitions (explicit or contextual) "allow one actually to carry out the elimination in a finite number of steps in each case," then analyticity in this sense implies the existence of a decision procedure and therefore cannot obtain.

Gödel's formulation recalls Leibniz's definition of necessary truth. It is not especially close to those either used by the Vienna Circle or directly criticized by Quine. That analytic truths are "true by definition" was, however, a frequent slogan in the period around 1930. But the Vienna Circle envisaged no such impoverished logical context but rather something closer to Frege's definition of an analytic truth as one that

---

[8] Dreben, "Quine," p. 87.
[9] Cited hereafter as RML. The reader is reminded that writings of Gödel are cited in the pagination of CW.

can be proved by means of general logical laws and definitions, where the "general logical laws" would be codified in a system like that of *Principia Mathematica.*

Nonetheless Gödel begins with this narrow notion both his remarks in RML and a somewhat parallel discussion in Gibbs, which is, however, more explicitly directed at a conventionalist view of mathematical truth. In both places he distinguishes different notions of analyticity which yield different answers to the questions whether mathematical truths are analytic.[10] In proceeding in this way, he was virtually alone before the impact of Quine's critique of analyticity.[11]

The narrow view must have interested Gödel because of its relevance to Leibniz and Kant. Also, it makes the analyticity of arithmetic in his view definitely refutable, since if the reduction procedure the idea envisages existed for first-order arithmetic, it would yield a decision procedure.[12] Interestingly, he wrote on a reprint of RML that arithmetic is demonstrably not analytic in Kant's sense, thus making clear that he took this narrow notion to capture Kantian analyticity.[13]

---

[10] In RML, the question is specifically whether the axioms of *Principia* are analytic.

[11] Compare Quine's remark, "Kant's intent . . . can be restated thus: a statement is analytic when it is true by virtue of meanings and independently of fact" ("Two Dogmas," p. 21). Thus at this point Quine did not distinguish between Kant's notion of analyticity and that of truth by virtue of meaning. (On Gödel's discussion of the latter notion, see §III below.)

This seems far from obviously correct as an interpretation of Kant. But it raises the question: Where does the notion of "truth by virtue of meanings" come from? Quine writes (in 1951) as if it were quite standard. But it is not so easy to trace its origin. It appears to be in informal characterizations of analyticity in the Vienna Circle (as Burton Dreben suggested to me). But the matter deserves investigation. In "Two Dogmas" Quine does not deal directly with this notion, but rather with more precise attempts to explicate it, of which "truth by semantical rules" is the most direct expression.

[12] Göran Sundholm suggested to me that the narrower notion of analyticity might be at work in a more positive way in his consistency proof for intuitionistic arithmetic by primitive recursive functionals of finite type and his reflections on it. See "Über eine bisher noch nicht benützte Erweiterung."

An important feature of the system T in this work is the decidability of intensional equality at higher types, where equality "means that the two functions have the same procedure of computation" (CW II, 276 n.7, a note to the expanded English version). Thus truth of an identity of such functions implies reducibility to a self-identity by the defining equations of the functions involved.

[13] The remark "Th[eorie] der natürlichen Zahlen nachweislich nicht analytisch im Kantischen Sinn" is referred to p. 150 of the text. See CW II 319.

The rest of what Gödel has to say about these matters is more perspicuously presented in Gibbs; I will now follow that source. There he discusses conventionalism after having criticized the idea that mathematics and mathematical objects are our free creation. After some initial remarks, he concludes that this idea is vague and proposes to disprove

> the most precise, and at the same time most radical, formulation that has been given so far. It is that which interprets mathematical propositions as expressing solely certain aspects of syntactical (or linguistic) conventions, . . . According to this view, mathematical propositions, duly analyzed, must turn out to be as void of content as, e.g., the statement "All stallions are horses." (Gibbs 315–316)[14]

Gödel first interprets this by the same narrow view of analyticity considered in RML and offers the same argument from undecidability as a disproof. But, he says, the more refined versions do not fare better:

> The weakest statement that at least would have to be demonstrable, in order that this view concerning the tautological character of mathematics be tenable, is the following: Every demonstrable mathematical proposition can be deduced from the rules about the truth and falsehood of sentences alone (i.e. without using or knowing anything else except these rules). (Gibbs 316)

He offers as examples the standard rules about truth-functional connectives and points out that for some statements (tautologies) their truth follows simply from the rules:

> Now it is actually so, that for the symbolisms of mathematical logic with suitably chosen semantical rules, the truth of the mathematical axioms is derivable from these rules. (Gibbs 317)

Gödel refers at this point to the attempt by Ramsey[15] (already discussed in RML) to show the truths of the simple theory of types tau-

---

[14] It should be remarked that a portion of text beginning in the above quotation and extending to p. 319 was marked by Gödel as to be left out. Since it was not crossed out in the manuscript (as were other passages), it is reasonable to suppose Gödel's intention was merely to leave it out of his oral presentation. But other conjectures about Gödel's intent are possible.
[15] "The Foundations of Mathematics."

tologies, and to §34 of Carnap's *Logical Syntax of Language*,[16] where Carnap gives a characterization in this form of the analytic and synthetic sentences of his richer Language II and proves (theorem 34i.21) that every sentence of the language that is demonstrable is analytic. Gödel continues:

> However (and this is the great stumbling block) in this derivation the mathematical and logical concepts and axioms themselves must be used in a special application, namely as referring to symbols, combinations of symbols, sets of such combinations, etc. Hence this theory, if it wants to prove the tautological character of the mathematical axioms, must first assume these axioms to be true. So while the original idea of this viewpoint was to make the truth of the mathematical axioms understandable by showing that they are tautologies, it ends up with just the opposite, i.e. the truth of the axioms must *first* be assumed and then it can be shown that, in a suitably chosen language, they are tautologies. (Ibid.)

This argument is of a general form that also occurs in Quine's criticism of conventionalist views. It goes roughly as follows. Suppose a philosophical view attempts to show truths of a certain class epistemologically unproblematic by showing that they all have a certain property P (e.g., analyticity or tautologousness). It is then argued that in order to show that truths of this class have the property P, statements or principles of the very class in question have to be assumed.

Quine is probably the best known user of this kind of argument. But it can be traced back at least to Henri Poincaré's well-known criticisms of logicism (and of Hilbert's early views). In the former case Poincaré argued, roughly, that the reduction of mathematical induction to higher-order logic could be shown to be correct only by using mathematical induction. Gödel in fact notes the affinity of his own argument and Poincaré's.[17]

---

[16] London: Kegan Paul, 1937. This is a section that had not been in the German edition but had been published as an article under the revealing title "Ein Gültigkeitskriterium für die Sätze der klassischen Mathematik."

[17] Gibbs 319 n.28. Gödel defends Frege (but not Hilbert) against Poincaré's charge. The matter has been the subject of a lot of more recent discussion; see for example my own *Mathematics in Philosophy*, pp. 167–171, 173–175.

Quine's use of arguments of this type belongs to the second of the above-noted strands of his critique of analyticity; thus it has much the same target as Gödel's. In "Truth by Convention," you will recall, Quine considers the view that elementary logic is true by convention. But the conventions have to be general. Thus to conclude, of particular sentences, that they are true (by the conventions) one has to use the very logical inferences that the idea of truth by convention was to justify.

In "Carnap and Logical Truth," Quine discusses Carnap's claim that logical and mathematical truth is "syntactically specifiable." First, the logical truths of Language II are syntactically specifiable in a vocabulary consisting of names of signs, concatenation, and

> the whole logico-mathematical vocabulary itself. . . . But *this* thesis would hold equally if "logico-mathematical" were broadened to include physics, economics, and anything else under the sun; Tarski's routine of truth-definition would still carry through just as well. No special trait of logic and mathematics has been singled out after all. (p. 125)

Quine also makes the observation appealed to by Gödel that, to show the truth (syntactical or otherwise) of the theorems of Language II or another formalization, a stronger formalization is needed.

Gödel also makes a closely related observation: If one can divide the *empirical* facts into two parts A and B so that B "implies nothing in A," one could then, parallel to the argument that mathematical truth is syntactically specificable and so void of content, argue in the same way that the propositions expressing B are void of content (Gibbs 320). As we shall see, he does not, and given his views would not, press the point in a direction followed by Quine, of questioning whether the distinction between logical and mathematical vocabulary on one side, and vocabulary such as that belonging to empirical science on the other, is a significant one. Rather he argues that if an opponent of the above argument objects that "certain observable facts B" are being disregarded, it can be replied that the same is being done when it is asserted that mathematics is void of content.

Gödel undertakes to reinforce his point against conventionalism by inquiring whether the considerations he has advanced yield an "outright disproof" of it. This would be so if the above form of argument could be shown to be applicable independently of the mathematical language and its interpretation. Not exactly this, but something he thinks close enough can be proved:

> Namely, it follows by the metatheorems mentioned [the incompleteness theorems] that a proof of the tautological character of the mathematical axioms is at the same time a proof for their consistency and cannot be achieved with any weaker means of proof than are contained in the axioms themselves. (Gibbs 317)

This strikes me as a genuinely weaker conclusion, at least until one has looked at consistency proofs in detail, because it implies at most that *some* strong set of assumptions is needed to show a given system of mathematics tautological or true by convention. It would not show any circularity in particular cases. Gödel goes on to say, however, that it follows with "practical certainty" that to prove the consistency of classical arithmetic and stronger systems *abstract concepts* are needed. What he had in mind are higher-order notions, such as would occur in a consistency proof by truth-definition in the context of a system with higher-order logic, or perhaps Gödel's own proof of the consistency of arithmetic using computable functionals of finite type, or intentional notions such as arise in the interpretation of intuitionistic methods as referring to proofs or provability. (Cf. Gibbs 318 n.27.) He says of such concepts that they are "certainly not syntactical."[18]

The upshot of this discussion is that there is something like a common logical core in Quine's and Gödel's attack on the views they

---

[18] Gibbs 318. Outside the context of a criticism of a view like Carnap's, Gödel pursued this and related points in "Über eine bisher noch nicht benützte Erweiterung."

This criticism might be compared with that in Michael Friedman, "Logical Truth and Analyticity in Carnap's *Logical Syntax of Language*." Friedman argues that if a syntactic metalanguage is to formulate "a neutral metaperspective from which we can represent the consequences" of different standpoints as formulated as different "linguistic frameworks" in Carnap's sense, it has to be very weak, in effect primitive recursive arithmetic. But then Gödel's theorem shows that analyticity for the stronger languages is not captured.

Friedman's conception of what is syntactical is close to that of Gödel in the remark cited in the text. It is doubtful that Carnap accepted it at the time of *Logical Syntax*; at all events, he later gave up the view that analyticity is a syntactical concept. [Friedman changed his views between the publication of this essay and its reprinting in *Reconsidering Logical Positivism*; on p. xvi he states that his "corrected views" are expressed in essay 9 of that volume.]

attribute to Carnap. The underlying claim is that mathematics can't be treated as void of content because there is no noncircular argument to the claim that mathematical statements have the kind of simulacrum of truth that this view supposes them to have, since in the argument that they have it they are treated as being simply *true*.

Before I turn to the point at which Quine's and Gödel's treatments diverge, I want to point out that in saying that one can't show that such a principle as the axiom of choice is analytic without assuming it to be true (RML 139), Gödel is making a point of which Carnap is perfectly well aware. In fact, he points to it specifically in §34 of *Logical Syntax* with respect to both the axiom of choice and the principle of mathematical induction. He remarks that if *S* is the formal sentence expressing the axiom of choice, a proof using choice that *S* is analytic is not circular. Such a proof is not meant to show that the axiom is "materially true":

> They only show that our definition of 'analytic' effects at this point what it is intended to effect, namely, the characterization of a sentence as analytic if, in material interpretation, it is regarded as logically valid. (p. 124)

The question whether the axiom of choice "should be admitted into the whole of the language of science" is not decided by such an argument; it is rather a "matter of choice."

III

The background of Quine's argument in §VII of "Carnap and Logical Truth" is skepticism about the importance of the distinction between the "logico-mathematical" and the "empirical" as a classification either of vocabulary or of truths. As Gödel goes on, both in RML and the Gibbs lecture, so far from pressing his arguments in this particular direction, he parts company with Quine on this point and in a way sides with Carnap: "It is correct that a mathematical proposition says nothing about the physical or psychical reality existing in space and time, because it is true already according to the meaning of the terms in it, irrespective of the world of real things" (Gibbs 320). In RML this is offered, unsurprisingly enough, as a different sense of analyticity, in which it is true that the axioms of *Principia* (other than Infinity) are analytic. There his formulation is puzzling and possibly confused: "In a second sense a proposition is called analytic if it holds 'owing to the meaning of the concepts occurring in it', where this meaning may perhaps be undefinable" (p. 151).

Even for a "conceptual realist," there seems to be a doubling of intensions into "concepts" and their "meaning," which may not do any work. It may reflect some uncertainty on Gödel's part about the notion of proposition. In the formulation quoted above from Gibbs, we would naturally read "terms" as referring to expressions.[19] Gödel might have been tempted to think of the constituents of a proposition in some other way and to have hesitated as to whether concepts are constituents of propositions.[20]

To try to sort this out would take us too far afield. Clearly Gödel thought that mathematical statements are in some way true in virtue of meaning and that this justifies regarding mathematics as fundamentally different from empirical science, although he was at pains to reject the Wittgenstein-Carnap characterization of this difference as lying in the fact that the former lacks "real content."[21]

Gödel's argument has a certain Quinian flavor: "Nothing can better express the meaning of the term "class" than the axiom of classes [i.e., the comprehension schemata of type theory] and the axiom of choice" (RML 139; cf. Gibbs 321).

At this point we encounter Gödel's notion of concept, a difficult, I would even say obscure, notion, but one central to his philosophy. Recall that Gödel said he had been a *conceptual* realist since 1925. It seems that from the beginning Gödel's realism was not just what we would describe as realism about mathematical objects and mathematical truths. That it extends to what he calls concepts is already clear in RML.[22] In Gibbs he writes:

---

[19] Indeed he says that the axioms of set theory are valid "owing to the meaning of the term 'set'" (Gibbs 321).

[20] But in Syntax V he says forthrightly, "Mathematical propositions are true in virtue of the *concepts* occurring in them" (p. 357).

[21] In one place Gödel distinguishes the kind of content possessed by "conceptual truths" (including mathematical) from "factual" content (Syntax V, p. 360). So although insisting on the content, even the "real" content, of mathematics, he denies that mathematics has *factual* content. This agrees with the usage of Edmund Husserl, for whom the factual contrasts not with the conventional or tautological but with the essential.

[22] I hope I have made this clear in Essay 5 of this volume, especially pp. 110–113.

What is wrong, however, is that the meaning of the terms (that is, the concepts they denote) is asserted to be something man-made and consisting merely in semantical conventions. The truth, I believe, is that these concepts form an objective reality of their own, which we cannot create or change, but only perceive and describe. (Gibbs 320)

The same view is expressed in "Is Mathematics Syntax of Language?":

Mathematical propositions, it is true, do not express physical properties of the structures concerned [in physics], but rather properties of the *concepts* in which we describe those structures. But this only shows that the properties of those concepts are something quite as objective and independent of our choice as physical properties of matter. This is not surprising, since concepts are composed of primitive ones, which, as well as their properties, we can create as little as the primitive constituents of matter and their properties. (Syntax V, p. 360)

In RML and probably also in the unpublished essays, concepts are objects associated with predicates.[23] Gödel is undecided about their identity conditions. They play a role like that of meanings, but he does not rule out the possibility that they might be extensional (RML 129).

That mathematics is analytic in this sense might be glossed as meaning that mathematical truths are true by virtue of the relations of the concepts denoted or expressed by their predicates. One might ask why the same is not true of other general statements such as 'All men are mortal'. It seems in accord with Gödel's view to regard this as true if and only if the concept *man* stands in a certain relation (presumably inclusion) to the concept *mortal*. We all recognize a difference between such a case and that of the axioms of set theory. Gödel himself has expressed it by saying that the latter express "the very meaning of the

---

[23] Wang reports that later Gödel was inclined to distinguish concepts from objects along Fregean lines (*Reflections*, p. 307). [This remark is not accurate. Although Wang clearly says in the place cited that Gödel's later view was that concepts are not objects, the place cited does not say that he distinguished them "on Fregean lines." I think Wang may have said something to that effect in conversation with me, but there is no written record known to me of his attributing a Fregean view to Gödel. Further to the question whether Gödel regarded concepts as objects, see the Postscript to Essay 5 in this volume.]

term 'set' " (Gibbs 321). Quine challenges us to show that this is more than a rough and ready distinction. In some cases of interest to Gödel, such as elementary logic, Quine's view is that we translate others so that these principles come out true, but this rests on the fact that we find them obvious and therefore doesn't tell us what the obviousness consists in.

I will suggest below a Gödelian answer to this sort of objection. It involves another conception about which what Gödel says leaves a lot of unanswered questions: mathematical intuition. Two relatively clear passages suggest that the primary meaning of mathematical intuition is a form of intuition of *truths,* what I have elsewhere called intuition *that,* contrasted with intuition of objects. In one famous passage he seems to pass, without any justification, from the first to the second:

> But, despite their remoteness from sense experience, we do have something like a perception of the objects of set theory, as is seen from the fact that the axioms force themselves on us as being true. I don't see any reason why we should have less confidence in this kind of perception, i.e. in mathematical intuition, than in sense perception, which induces us to build up physical theories and to expect that future sense perceptions will agree with them.[24]

By "the objects of set theory" it is clear that Gödel means not only sets but also certain concepts. Some light on what he has in mind is shed by the following:

> The similarity between mathematical intuition and a physical sense is very striking. It is arbitrary to consider "this is red" an immediate datum, but not so to consider the proposition expressing modus ponens or complete induction (or perhaps some simpler propositions from which the latter follows). For the difference, as far as it is relevant here, consists solely in the fact that in the first case a relationship between a concept and a particular object is perceived, while in the second case it is a relationship between concepts. (Syntax V, p. 359)

---

[24] "What Is Cantor's Continuum Problem?," 1964 version, p. 268. Wang informs me that Gödel expressed to him the view that the *de dicto* and the *de re* are inextricably intertwined; presumably this means one could not have intuition of truths without intuition of objects or at least concepts.

Gödel does not formulate the "proposition expressing modus ponens."[25] But it would have to be a generalization about propositions, involving probably also the notion of following from or valid inference. Gödel's procedure seems to be to begin with evidences whose form is propositional, such as set-theoretic axioms, or statements to the effect that certain inferences are valid (or however else principles like modus ponens and induction are exactly formulated). "Mathematical intuition" is then in the first instance just the insight that these statements report. Much of Gödel's use of the term serves simply to defend against conventionalism or empiricism the view that there is such a thing as mathematical *Evidenz*. The latter view is far less controversial than Gödel's view appears to be. But it already distinguishes his position both from Carnap's and from Quine's. Carnap's view that mathematics and logic are "without factual content" implies that there is nothing for such *Evidenz* to be about. Quine's holistic empiricism rejects the idea of autonomous mathematical evidence.

The choice of modus ponens and induction as elementary examples indicates that in this respect Gödel does not distinguish mathematics sharply from logic. It is also evidently the case that he does not distinguish sharply between cases where there is ontological commitment (as most of us would certainly think even of elementary set-theoretic axioms), cases such as mathematical induction where there is such commitment in the background but the principle in question does not bring in any specific ontological commitment of its own, and cases like modus ponens, where (I at least would wish to say) ontology is not involved at all in particular instances, and arises only through the general problem of formulating principles of logic.

Gödel's position becomes both more problematic and more controversial when he goes on to read these evidences as involving perceptions of "concepts." This view, and the lack of stress on the two distinctions I have just mentioned, are mutually reinforcing. From the remark about 'this is red' I infer that even the most elementary perception of a truth involves perception of concepts, since there "a relationship be-

---

[25] The examples of modus ponens and induction also occur in Syntax III, note 34 (p. 347), but Gödel is not more precise about the former. The "simpler propositions" from which mathematical induction follows appear from this place (and also Gibbs 321) to be axioms of set theory or second-order logic. Gödel's hesitancy is due to the fact that such a reduction of mathematical induction depends on a nonconstructive standpoint (presumably because of the impredicativity involved).

tween a concept and a particular object is perceived." I am supposing that if one perceives a relation between *a* and *b*, one also perceives *a* and *b*.[26] If we accept this principle, and the view that predicates designate concepts, then if we consider (to fix ideas) any statement that can be formulated as a first-order sentence, it follows that perceiving it to be true will involve perception of concepts. This would be an explanation of Gödel's puzzling transition from the axioms of set theory "forc[ing] themselves on us as being true" to our having "a kind of perception of the objects of set theory." In the apparently harder case of modus ponens, Gödel is talking of the "proposition expressing modus ponens"; that too will involve predication, even though modus ponens itself is a rule of propositional logic.

We need to be clear, however, that even if these suggestions are correct they only describe situations where propositions are "perceived" to be true. I don't think that is meant by Gödel to be a term applicable to any situation where one attains propositional knowledge. That suggests an answer to the above Quinian objection. Perception of concepts is insufficient to verify "all men are mortal" or to verify axioms from which it follows. If it is "perceived" to be true at all (which I am inclined to think Gödel would not say), perception not just of the concepts but of their instances, actual men, is involved. One could say that it is not analytic because it is not a priori. It is well known that Gödel allows that set-theoretic axioms might be justified not directly but by their consequences. In one place he contrasts this with seeing them to be true by virtue of the meaning of their terms:

> For these axioms there exists no other rational (and not merely practical) foundation except either that they (or propositions implying them) can be perceived to be true (owing to the meaning of the terms or by an intuition of the objects falling under them) or that they are assumed (like physical hypotheses) on the ground of inductive arguments, e.g. their success in the applications. (Syntax III 346–347; emphasis Gödel's)

---

[26] This view of predication clearly opens Gödel to the third-man type of argument Frege uses against the view that predicates designate objects. It is not surprising that Gödel came to hold a Fregean view himself (see note 23 above [but note the added correction]). It is also not surprising that he should have found the paradoxes (in particular Russell's paradox) to be an unsolved problem for the foundations of logic.

Note that such an "inductive" justification is contrasted with "perceiving" axioms to be true. Here as elsewhere Gödel makes clear that arguments of this kind include arguments based on the *mathematical* consequences of the axiom in question. It then appears that axioms might be justified in a way that is from an epistemological point of view a priori (though presumably not conclusive) but does not show them to be true by virtue of the relevant concepts.[27]

One might object that the reply I have offered to the objection that Gödel does not distinguish axioms of set theory from "all men are mortal" only explains how the latter is not *known* by "perception" of concepts; it does not explain how it fails to be *true* by virtue of the relations of concepts. And if the latter obtains, should that not make it analytic in the sense we are considering? Whether Gödel could reply to this rejoinder, it does indicate a weakness of the very rudimentary theory of meaning with which he operates. There is really very little explanation in his writings known to us of the relation between language and the concepts designated by predicates, or how that relation arises in the use of language or is manifested in it. Gödel may seem to maintain his version of the thesis that mathematics is analytic by virtue of a museum theory of meaning. Unlike the version assailed by Quine, however, any museum postulated by Gödel is not a mental museum, as his statements about the mind-independence of concepts make clear. The real reproach one might make against Gödel, in my view, is that he barely has a theory of meaning at all. He relies on the idea that predicates denote concepts but does not tell us much of anything about this relation. What we can gather about it is derived from remarks about mathematical intuition, but that mathematical and logical evidence involves "perception" of concepts seems to be a consequence of the assumption that predicates designate concepts, rather than a justification of it (on phenomenological or other grounds).

There is one indication that Gödel himself felt frustration about the role the notion of concept played for him; he may well have been aware of the thinness of his own theory of concepts. Early in 1959, he

---

[27] An example would be very strong axioms of infinity such as that there is a measurable cardinal. In a 1966 note to "What Is Cantor's Continuum Problem?" Gödel wrote of such axioms, "That these axioms are implied by the general concept of set in the same sense as Mahlo's has not been made clear yet" (CW II, 260 n.20).

replied to an inquiry from Paul Arthur Schilpp about the state of his promised contribution to *The Philosophy of Rudolf Carnap*. He says, very apologetically, that the paper will not be forthcoming. One thing he says in his explanation is very revealing:

> It is easy to allege very weighty and striking arguments in favor of my views, but a complete elucidation of the situation turned out to be more difficult than I had anticipated, doubtless in consequence of the fact that the subject matter is closely related to, and in part identical with, one of the basic problems of philosophy, namely the question of the objective reality of concepts and their relations.[28]

Did Gödel do more towards creating a theory of concepts which would be more satisfactory to him, and which would overcome some of the disappointment we feel when we see that no real theory of meaning backs up the claim that mathematics is analytic in the sense he claims it is? This question has two sides, a logical and what we might call an epistemological. In RML, Gödel makes a number of remarks about a possible formal logic of concepts; the simple theory of types is offered as a reasonably satisfactory theory, but he says one might reasonably look for a stronger theory, presumably a type-free theory.[29] We don't know how much more he might have done to develop such a theory.[30] It is doubtful that it could have played much of a role in his criticism of conventionalism; hence we cannot conclude

---

[28] Letter to P. A. Schilpp, February 3, 1959, CW V 244.

[29] See, in my introductory note in CW II, Essay 5 in this volume, pp. 111–113, and passages referred to there.

[30] The Fregean turn mentioned by Wang (see note 23 above) might suggest that he became more satisfied with the simple theory of types than at the time of RML. Wang informs me that that is not so: Gödel thought there should be a general concept *concept*, which of course the theory of types would not admit. Cf. also Wang, *Reflections*, p. 194.

[On the alleged "Fregean turn," see the addendum correcting note 23 above. The correspondence with Gotthard Günther, which took place during the time when Gödel was working on Syntax, indicates a definite interest in a type-free theory of concepts and the hope that Günther might develop an idea of his own in that direction. See in particular Gödel's letters of April 4, 1957, and January 7, 1959 (CW IV 524–529, 532–535) and §3 of my introductory note to their correspondence.]

much from the absence of indication in Gibbs and Syntax of how Gödel might have envisaged it.

By the epistemological side, I mean the question of what Gödel might have done to develop views that address the questions about knowledge of language and meaning that have been central to the philosophy of language in the last generation. (Here at last we meet the third strand of Quine's critique of analyticity.) It would be easy to conclude from our discussion so far that Gödel did not think very much about these problems and does not offer any challenge to a view of meaning like Quine's. The latter conclusion may, however, be hasty: Although I have not dwelt on it more than briefly, Gödel's case for autonomous mathematical evidence is well developed, and he at least indicates that something of this kind is at work in some quite elementary cases. The empiricist framework of Quine's theory does not give mathematical evidence any role of its own in the theory of language, and it is possible that focusing on this might change our views about some things, although it is not clear that results would ensue that would be essentially different from what Quine derives from the obvious character of at least part of elementary logic.

There are two reasons of a more documentary character for being dissatisfied with the picture of Gödel's views on mathematical intuition and perception of concepts that we have presented so far. First, Gödel has rather specific examples in mind when he talks of perception of concepts, in particular the concept of set, and the existence of what he in one place calls "an open series of extensions" of the axioms of set theory,[31] and the concept of mechanical procedure, where it was surprisingly possible to arrive at an analysis which captured what Gödel considered to be *the* concept, which turned out to have a certain absolute character.[32] Whether it is right to regard either the reflection that convinces us of set-theoretic axioms or our understanding, after Turing, of the notion of mechanical procedure, as "intuition" in more than a weak sense, or as "perception of concepts," these are data that challenge any theory of meaning and evidence in mathematics.

The second reason is that the documents on which we have relied nearly all date from no later than early 1959, and it was in that year

---

[31] "What Is Cantor's Continuum Problem?," 1964 version, p. 268.

[32] See the 1946 "Remarks before the Princeton Bicentennial Conference on Problems in Mathematics."

that he began the serious study of Husserl.[33] One thing he may have obtained from Husserl is a less naive picture of the relation of evidence and meaning than seems to be at work as late as Syntax.[34] Some idea of how he viewed Husserlian phenomenology can be obtained from "The Modern Development" (see note 7). There he says that phenomenology offers a systematic method for clarification of meaning "that does not consist in giving defininitions," where the clarification

> consists in focusing more sharply on the concepts concerned by directing our attention in a certain way, namely, onto our own acts in the use of these concepts, onto our powers in carrying out our acts, etc. (CW III 382, trans. 383)

This is in itself not very helpful, although one could turn to Husserl's extensive writings to clarify it.[35] In defense of the phenomenological approach, however, Gödel offers the following remark:

> If one considers the development of a child, one notices that it proceeds in two directions: it consists on the one hand in experimenting with the objects of the external world and with its [own] sensory and motor organs, on the other hand in coming to a better and better understanding of language, and that means—as soon as the child is beyond the most primitive designating [of objects]—of the basic concepts on which it rests. With respect to the development in this second direction, one can justifiably say

---

[33] Wang, *Reflections*, pp. 12, 28. Note that the above cited letter to Schilpp is dated February 3, 1959; it is extremely likely that at that point Gödel gave up work on Syntax.

Wang, presumably interpreting something Gödel himself said, says that he began to study Husserl "probably to look for a deeper foundation of human knowledge in everyday life" (*Reflections*, p. 12).

[34] It has been speculated that the supplement to the 1964 version of "What Is Cantor's Continuum Problem?" reflects this influence. It does contain elements not present in earlier philosophical writings. [On this subject see §5 of Essay 7 in this volume, as well as the Postscript to Essay 7.]

[35] Indeed, given that so little documentation of Gödel's reflections on Husserl has been available, one might well turn to Husserl for help in understanding Gödel. A beginning is made in Richard Tieszen, "Kurt Gödel and Phenomenology." [The available documentation has greatly expanded since this essay was written, and a number of writers have written on Gödel's appropriation of Husserl. Worthy of mention are Mark van Atten and Juliette Kennedy, Dagfinn Føllesdal, Kai Hauser, and Tieszen.]

that the child passes through states of consciousness of various heights, e.g., one can say that a higher state of consciousness is attained when the child first learns the use of words, and similarly at the moment when for the first time it understands a logical inference.

This is the only place known to me where Gödel even begins to enter the circle of ideas where Quine's deeper discussion of meaning moves. It offers only a hint of a way to proceed, that we might look at the development of the "understanding of language" as having some dimensions that are independent of responses to stimulation of the child's sense-organs, and that in particular we might see the development of logical inference as one step of this kind. But it would be a long way from this to a real Gödelian theory of meaning and to a clarification of the epistemological basis for his use of the notion of concept.[36]

---

[36] I wish to thank Solomon Feferman, John W. Dawson Jr., and Warren Goldfarb, fellow editors of Gödel's unpublished works, for access to papers at various stages of editing. The paper has benefited from comments by members of the audience at the San Marino conference [on Quine, in 1990] and also at presentations of later versions at Carnegie-Mellon University and Columbia University. I am particularly grateful to Goldfarb, George Boolos, Burton Dreben, James Higginbotham, Wilfried Sieg, Hao Wang, and the editors for comments and discussion.

Unpublished materials by Gödel are from the papers held at the Firestone Library of Princeton University and are quoted by kind permission of the Institute for Advanced Study.

[I discuss Gödel's conception of analyticity and its development further in "Analyticity for Realists."]

In his published comments on the papers in the volume in which this essay appeared, Quine writes, "I am glad to find Gödel agreeing, however, that mathematics has content. Indeed it has ontological content. Unlike logic in my narrow sense, it has its special objects."[1] Quine's remark tempts the conclusion that it is only by virtue of its having objects that mathematics has content. I don't know if this was Quine's considered view; his holism would suggest the contrary. His eye at the time was on the contrast of mathematics with logic, and he had of course long insisted that logic does not have objects of his own, and shortly before this passage he sketches an argument to the effect that first-order logic is in a sense analytic.

A number of essays have appeared that discuss the force of the arguments by Gödel that are discussed in this essay. The authors that interest me are scholars of Carnap, and their main concern is with the question of the force of Gödel's arguments against Carnap's views, either in *Logical Syntax* or later.[2] It was not my own purpose to make such an assessment. Among other things, that would have required me to offer a serious interpretation of Carnap. However, these writings raise some questions about what is said in my essay about *Gödel's* arguments.

---

[1] "Reactions," p. 352.

[2] I have in mind Warren Goldfarb's introductory note to Syntax in *Collected Works,* vol. III: Thomas Ricketts, "Carnap's Principle of Tolerance"; Goldfarb and Ricketts, "Carnap and the Philosophy of Mathematics"; Friedman, "Tolerance and Analyticity"; and some essays by Steve Awodey and A. W. Carus, in particular "How Carnap Could Have Replied to Gödel" and "Gödel and Carnap." There are differences among these writers in the interpretation of Carnap.

In my essay Gödel's appeal to his own incompleteness theorem is mentioned only briefly. But in these writings, arguments appealing to the incompleteness theorem seem to be regarded as his central ones. This difference could, I believe, be partly accounted for by the difference of aims between my essay and these writings: between uncovering similarities (and also differences) between Gödel's arguments and Quine's and assessing the force of Gödel's arguments as aimed at Carnap's views, particularly in *Logical Syntax*. Gödel's argument is naturally aimed at the view that "mathematics is syntax of language." In the well-organized presentation in Syntax V, Gödel singles out three theses, of which the first is a Gödelian reading of that thesis:

> Mathematical intuition, for all scientifically relevant purposes, in particular for drawing the conclusions as to observable facts occurring in applied mathematics, can be replaced by conventions about the use of symbols and their applications. (III 356)

It is in his argument against that claim that Gödel deploys the incompleteness theorem. The idea is, roughly, that a mathematical theory (which for the sake of argument can be assumed to be a system of conventions) cannot be acceptable if it implies false empirical propositions. But this will be avoided only if the theory is consistent. But to prove this for the theories that are relevant to science, we require (by the second incompleteness theorem) some assumption that is not embodied in the theory.

Awodey and Carus point out that Carnap could object that it is too much to ask that we should have a *proof* of consistency.[3] It is sufficient that the theory should *be* consistent. Gödel might reply that that is still a substantive mathematical assumption. It's possible, however, that a defender of Carnap could have a counter-reply. One that occurs to me is more Wittgensteinian: The mathematical statement of consistency belongs to a mathematical model of mathematical proof. It is meant to imply that in the real world, no mathematician will derive $0 = 1$ or, in application, some manifestly false empirical statement. The model fits well enough so that if a theory is consistent, we can be pretty sure of that, or rather, if a mathematician does derive $0 = 1$, there will be some concrete indication that he has not followed the rules that define the system whose consistency is in question. But in the case of

---

[3] "How Carnap Could Have Replied to Gödel," pp. 207–208.

inconsistency, the idealization inherent in the model is closer to home. It might be that a system is inconsistent by virtue of the existence of a derivation of $0 = 1$ or some other contradiction that is so long that we have no reason to believe that any human mathematician will ever arrive at a proof of an actual contradiction, even with the help of the most powerful computers.

Goldfarb and Ricketts point out that in the course of this argument Gödel says, "What must be known is that the rules, by themselves, do not imply the truth or falsehood of any proposition expressing an empirical fact" (Syntax V, CW III 357).[4] They interpret Gödel to require that the "empirical facts" be fixed prior to mathematics. They don't claim that that was Gödel's own view of the relation of mathematics to empirical science, and they point out that it is not the view of Carnap in *Logical Syntax*.[5] I'm not sure that this remark of Gödel does express such a view. But Goldfarb seems to attribute to Carnap a view that is virtually opposite: A linguistic framework is defined only when its rules of logic (including mathematics) are fixed, and it is only in a determinate framework that we have a definite notion of "empirical fact."[6] It then seems that Gödel could not make a reply that seems natural to me: that the most that his argument could require is that there be some limited domain of fact that would be independent of mathematics. It would be analogous to Quine's observation sentences. It appears that that would not have been acceptable to the Carnap of *Logical Syntax*, but it does have some intuitive plausibility.

Goldfarb observes, correctly in my view, that Gödel frames the issue as a straightforward epistemological issue (ibid., p. 330). It appears that Carnap's conception of philosophy is so different that, although good reasons can be given why he should be unmoved by Gödel's arguments, he does not offer a direct response in terms close to those in which Gödel poses the problems.

This state of affairs makes it difficult to discuss another point on which Gödel takes issue not only with Carnap but with the Vienna

---

[4] Goldfarb and Ricketts, "Carnap and the Philosophy of Mathematics," p. 64; also Goldfarb, introductory note to Syntax, p. 328.
[5] Goldfarb says that a view of that kind was expressed earlier by logical positivists (Carnap included); see his introductory note, ibid.
[6] Ibid.

Circle, apparently as he encountered it in his student days. That is the point that Gödel expresses by saying that mathematical statements have a "real content." In the second of the three theses he states at the beginning of Syntax V, he uses the term "void of content" to describe what he opposes. Although Quine does not use that sort of language, this is a point where Gödel's arguments parallel Quine's, as the paper points out and as Goldfarb seems to agree (ibid., p. 332). Both question whether Carnap has sufficient reason to distinguish logic and mathematics as "without content" and laws of nature as having empirical content. Of course they differ in the implications of this claim. In a way Quine argues that mathematics does have empirical content, while Gödel regards positivism as begging the question by rejecting any idea of non-empirical content.

Goldfarb points out that the background of positivist views on this issue was the idea derived from Wittgenstein's *Tractatus* that to have sense, a proposition must exclude possibilities. Necessary propositions therefore lack sense. This idea has a descendant in possible-worlds semantics, of which Carnap in his later years was a pioneer. The usual conclusion is that necessary propositions, since they are true in all possible worlds, have the same meaning. The problem this poses, for example for the analysis of propositional attitudes, was already recognized in discussions of the 1940s and 1950s framed to some extent by logical empiricism. In more recent work there are many competing semantic paradigms, and different concepts of necessity. It is doubtful that Wittgenstein's original idea has survived.

The difference in Quine's and Gödel's directions that was more emphasized in the essay is that Gödel holds to an analytic-synthetic distinction but ties it to his own conceptual realism. His position has the weakness that he did not develop a theory of meaning or an independent theory of concepts that would even satisfy him. Nonetheless his position has a strong point that Quine's and Carnap's both lack. That is that in crucial respects his views developed out of his experience as a mathematician.

To me, a view something like Gödel's on this point has always seemed to be the default position. It is independent of Gödel's conceptual realism and other strongly "platonist" views. Very powerful considerations would have to be advanced in order to overthrow it. It is close to the realism that William Tait describes as the default position.[7]

I will make one comment on an argument that seems to me to be common to Gödel and Quine. Gödel puts the matter as follows; the context is the observation that mathematical propositions do not imply any empirical facts:

> By this method, if a division of the empirical facts into two parts, A and B, is given such that B implies nothing in A, a language can be constructed in which the proposition expressing B would be void of content. And if your opponent were to say: "You are arbitrarily disregarding certain observable facts B," one may answer: "You are doing the same thing, for example, with the law of complete induction, which I perceive to be true on the basis of my understanding (that is, perception) of the concept of integer. (Gibbs, CW III 320)

Gödel's idea is apparently that one might divide the empirical statements into two parts, where if the statements in one part implied none of the statements in the other, one could by parity of reasoning with that of Carnap declare that those of the first part are "void of content." Quine uses similar considerations to argue that the division of a language into analytic and synthetic statements can be arbitrary.

My comment is: Could one not raise the same issue within what is normally taken as the sphere of logic and mathematics? Consider a formal language L where the logic consists of second-order arithmetic with first-order comprehension; it might be the theory ACA_0 of reverse mathematics, or possibly some extension of it that does not provide for more sets of natural numbers. Carnap's Principle of Tolerance tells us that only pragmatic considerations can decide between admitting as our logic such a theory as opposed to, say, full second-order arithmetic. But could not L contain rules, not part of the logic, that imply the existence of sets of integers that are definable in second-order arithmetic but not first-order definable, so that the statement of the existence of such a set would not be analytic? Then L would in effect declare that a weaker mathematics is "without factual content" while admitting as

---

[7] "Truth and Proof," p. 91. Cf. §3 of Essay 12 in this volume.

derivable statements that would intuitively be recognizable as mathematical but which do have factual content.

Whether or not this idea can be worked out in a way that is at all persuasive, it is of a piece with the considerations that have been advanced to the effect that the extreme pluralism about mathematics implied by the Principle of Tolerance is highly counterintuitive and does not fit the intuition of mathematicians.[8]

---

[8] Such a view has been urged by Peter Koellner in lectures. In an unpublished paper, he has carried out a more thorough critical examination of Carnap's philosophy of mathematics. It should be admitted that Gödel does not focus on the Principle of Tolerance, and he may not have grasped its central role in *Logical Syntax*.

7

PLATONISM AND MATHEMATICAL INTUITION

IN KURT GÖDEL'S THOUGHT

The best known and most widely discussed aspect of Kurt Gödel's philosophy of mathematics is undoubtedly his robust realism or platonism about mathematical objects and mathematical knowledge. This has scandalized many philosophers but probably has done so less in recent years than earlier. Bertrand Russell's report in his autobiography of one or more encounters with Gödel is well known: "Gödel turned out to be an unadulterated Platonist, and apparently believed that an eternal "not" was laid up in heaven, where virtuous logicians might hope to meet it hereafter."[1] On this Gödel commented:

> Concerning my "unadulterated" Platonism, it is no more unadulterated than Russell's own in 1921 when in the *Introduction to Mathematical Philosophy* ... he said, "Logic is concerned with the real world just as truly as zoology, though with its more abstract and general features." At that time evidently Russell had met the "not" even in this world, but later on under the influence of Wittgenstein he chose to overlook it.[2]

---

[1] Russell, *Autobiography 1914–1944*, p. 356.

[2] From a draft reply to a 1971 letter from Kenneth Blackwell, quoted in Wang, *Reflections on Kurt Gödel*, p. 112. [See now *Collected Works* (hereafter CW), vol. IV, pp. 316–317.] The quotation is from Russell, *Introduction*, p. 169. Gödel was fond of this particular quotation from Russell. In commenting on it in "Russell's Mathematical Logic" (hereafter RML), however, he stated erroneously (II 121, n.3) that it had been left out in later editions of *Introduction*. See Blackwell, "A Non-Existent Revision." Evidently Russell himself did not pay close attention to Gödel's footnote. The specific issue about "not" is not pursued elsewhere in Gödel's writings, and I shall not pursue it here. Gödel also remarks that Russell's statement gave the impression that he had had many discussions with Russell, while he himself recalled only one.

One of the tasks I shall undertake here is to say something about what Gödel's platonism is and why he held it.

A feature of Gödel's view is the manner in which he connects it with a strong conception of mathematical intuition, strong in the sense that it appears to be a basic epistemological factor in knowledge of highly abstract mathematics, in particular higher set theory. Other defenders of intuition in the foundations of mathematics, such as Brouwer and the traditional intuitionists, have a much more modest conception of what mathematical intuition will accomplish. In this they follow a common paradigm of a philosophical conception of mathematical intuition derived from Kant, for whom mathematical intuition concerns space and time as forms of our sensibility. Gödel's remarks about intuition have also scandalized philosophers, even many who would count themselves platonists. I shall again try to give some explanation of what Gödel's conception of intuition is. It is not quite so intrinsically connected with his platonism as one might think and as some commentators have thought. I hope to convince you that even though it is far from satisfactory as it stands, there are at least genuine problems to which it responds, which no epistemology for a mathematics that includes higher set theory can altogether avoid. I will suggest, however, that Gödel aims at what other philosophers (in the tradition of Kant) would call a theory of reason rather than a theory of intuition. Gödel is, however, evidently influenced by a pre-Kantian tradition that does not see these two enterprises as sharply distinct and that admits "intuitive knowledge" in cases that for us are purely conceptual.[3]

In connection with these explanations I shall try to say something about the development of Gödel's views. Late in his career, Gödel indicated that some form of realism was a conviction he held already in his student days, even before he began to work in mathematical logic. Remarks from the 1930s, however, indicate that at that time his realism

---

Note that Gödel's writings are cited in the pagination of CW. Where it is clear that it is Gödel who is being cited, we give only the volume and page number.

[3] It is possible that Gödel was influenced by the remarks about intuitive knowledge in Leibniz's "Meditations on Knowledge, Truth, and Ideas." Knowledge is intuitive if it is clear, that is gives the means for recognizing the object it concerns, distinct, that is if one is in a position to enumerate the marks or features that distinguish an instance of one's concept, adequate, that is if one's concept is completely analyzed down to primitives, and finally one has an immediate grasp of all these elements.

fell short of what he expressed later. But it appears in full-blown form in his first philosophical publication, "Russell's Mathematical Logic" (RML). The strong conception of mathematical intuition, however, seems in Gödel's published writings to come out of the blue in the 1964 supplement to "What Is Cantor's Continuum Problem?" Even in unpublished writings so far available it is at most hinted at in writings before the mid-1950s. In what follows I will trace this development in more detail.

1. Realism in General and in Gödel

Speaking quite generally, philosophers often talk as if we all know what it is to be a realist, or a realist about a particular domain of discourse: realism holds that the objects the discourse talks about exist, and are as they are, independently of our thought about them and knowledge of them, and similarly truths in the domain hold independently of our knowledge. One meaning of the term "platonism" which is applied to Gödel (even by himself) is simply realism about abstract objects and particularly the objects of mathematics.[4]

The inadequacy of this formulaic characterization of realism is widely attested, and the question what realism is is itself a subject of philosophical examination and debate. One does find Gödel using the standard formulae. For example in his Gibbs lecture of 1951, he characterizes as "Platonism or 'Realism'" the view that "mathematical objects and facts (or at least something in them) exist independently of our mental acts and decisions" (III 311) and that "the objects and theorems of mathematics are as objective and independent of our free choice and our creative acts as is the physical world" (III 312 n.17). In RML—as I have said the first avowal of his view in its mature form—he does not use this language to characterize Russell's (earlier) "pronouncedly realistic attitude" of which he approves, but he does in his well-known criticism of the vicious circle principle, where he says that the first form of the principle "applies only if the entities involved are constructed by ourselves. . . . If, however, it is a question of objects that exist independently of our constructions, there is nothing in the least

---

[4] For a general discussion of mathematical platonism, see Maddy, "The Roots of Contemporary Platonism."

absurd in the existence of totalities containing members which can be described . . . only by reference to this totality" (II 127–128).[5]

Gödel is concerned in the Russell essay to argue for the inadequacy of Russell's attempts to show that classes and concepts can be replaced by "constructions of our own" (II 140), and the Gibbs lecture contains arguments against the view that mathematical objects are "our own creation," a view maybe more characteristic of nineteenth-century thought about mathematics than of that of Gödel's own time.

Rather than exploring how Gödel himself understands these characterizations, I will note some points that are more distinctive of Gödel's own realism. Introducing the theme in RML, he quotes the statement from Russell quoted above and then turns to an "analogy between mathematics and natural science" he discerns in Russell:

> He compares the axioms of logic and mathematics with the laws of nature and logical evidence with sense perception, so that the axioms need not necessarily be evident in themselves, but rather their justification lies (exactly as in physics) in the fact that they make it possible for the "sense perceptions" to be deduced; which of course would not exclude that they also have a kind of intrinsic plausibility similar to that in physics. I think that . . . this view has been largely justified by subsequent developments, and it is to be expected that it will be still more so in the future. (II 121)

In other places, as is well known, Gödel claims an analogy between the assumption of mathematical objects and that of physical bodies:

> It seems to me that the assumption of such objects [classes and concepts] is quite as legitimate as the assumption of physical bodies and there is quite as much reason to believe in their existence. They are in the same sense necessary to obtain a satisfactory system of mathematics as physical bodies are necessary for a satisfactory theory of our sense perceptions. (II 128)

---

[5] Cf. also: "For someone who considers mathematical objects to exist independently of our constructions and of our having an intuition of them individually . . ." ("What Is Cantor's Continuum Problem?," 1964 version, II 258). This paper will be cited as WCCP, with the date 1947 or 1964 according to which version is being cited.

In the 1964 version of the Continuum Problem paper the question of the "objective existence of the objects of mathematical intuition" is said (parenthetically) to be "an exact replica of the question of the objective existence of the outer world" (II 268).

Thus a Gödelian answer to the question what the "independence" consists in is, for example, that mathematical objects are independent of our "constructions" in much the same sense in which the physical world is independent of our sense-experience. Gödel does not address in a general way what the latter sense is, although some evidence of his views can be gleaned from his writings on relativity. The main thesis of "A Remark on Relativity Theory" is that relativity theory supports the Kantian view that time and change are not to be attributed to things as they are in themselves. But this thesis is specific to time and change; it is perhaps for that reason that he is prepared in one place to gloss the view by saying that they are *illusions,* a formulation that Kant expressly repudiates.[6] Gödel is not led by the considerations he advances to reject a realist view of the physical world in general; for example he does not suggest that space-time is in any way ideal or illusory. In fact, he frequently reproaches Kant for being too subjectivist.[7] But he is quite cautious in what little he says about how far we can be realists about knowledge of the physical world. But in his discussion of Kant, he clearly thinks that modern physics allows a more realistic attitude than Kant held; for example he remarks that "it should be assumed that it is possible for scientific knowledge, at least partially and step by step, to go beyond the appearances and approach the things in themselves."[8]

---

[6] II 202; Kant, *Critique of Pure Reason,* B69.

[7] E.g., WCCP 1964, II 268. However, he interprets Kant's conception of time as a form of intuition as meaning that "temporal properties are certain relations of the things to the perceiving subject" ("Some Observations," version B2, III 231), and he finds that there is at least a strong tendency of Kant to think that, interpreted in that way, temporal properties are perfectly objective.

[8] "Some Observations," version C1, III 257; cf. version B2, III 240. Of course it is quantum mechanics that has been in our own time the main stumbling block for realism about our knowledge in physics. Gödel says little on the subject; what little he does say (e.g., ibid., version B2, n.24 and n.25) indicates a definitely realistic inclination without claiming to offer or discern in the literature an interpretation that would justify this.

2. The Development of Gödel's Realism

I now want to approach the question of the meaning of Gödel's realism by inquiring into its development. One distinctive feature of Gödel's realism is that it extends to what he calls concepts (properties and relations), objects signified in some way by predicates. These would not necessarily be reducible to sets, if for no other reason because among the properties and relations of sets that set theory is concerned with are some that do not have sets as extensions.[9] It may be that this feature arose from convictions with which Gödel started. In an (unsent) response to a questionnaire put to him by Burke D. Grandjean in 1975, Gödel affirmed that "mathematical realism" had been his position since 1925.[10] In a draft letter responding to the same questions, Gödel wrote, "I was a conceptual and mathematical realist since about 1925."[11] The term "mathematical realism" occurs in Grandjean's question; the term "conceptual" is introduced by Gödel.

Gödel's response to Grandjean would suggest that he was prepared to affirm in 1975 that the realism associated with him was a position he had held since his student days. Moreover, in letters to Hao Wang quoted extensively in the latter's *From Mathematics to Philosophy*, Gödel emphasized that realistic convictions, or opposition to what he considered anti-realistic prejudices, played an important role in his early logical achievements, in particular both the completeness and the incompleteness theorems.[12]

Before I turn to these statements, let me mention the remarks of Gödel from the 1930s, to which Martin Davis and Solomon Feferman have called attention, that do not square with the platonist views expressed in RML and later. We have the text of a very interesting general lecture on the foundations of mathematics that Gödel gave to the

---

[9] Thus "property of set" is counted as a primitive notion of set theory (WCCP 1947, II 181 n.17, or 1964, II 260 n.18). This notion corresponds to Zermelo's notion of "definite property" (cf. Gödel, *The Consistency of the Continuum Hypothesis*, II 34).

[10] Wang, *Reflections on Kurt Gödel*, pp. 17–18.

[11] Ibid., p. 20.

[12] Köhler, "Gödel und der Wiener Kreis," contains interesting suggestions about the influences on Gödel as a student that might have encouraged realistic views. They are not specific enough as regards mathematics to bear on an answer to the questions of interpretation considered in the text. In discussing Gödel's relations with the Vienna Circle, Köhler writes as if he already held at the beginning of the 1930s the position of RML and later writings. The evidence does not support that.

Mathematical Association of America (MAA) in December 1933. Much of it is devoted to the axiomatization of set theory and to the point that the principles by which sets, or axioms about them, are generated naturally lead to further extensions of any system they give rise to. When he turns to the justification of the axioms, he finds difficulties: the non-constructive notion of existence, the application of quantifiers to classes and the resulting admission of impredicative definitions, and the axiom of choice. Summing up he remarks,

> The result of the preceding discussion is that our axioms, if interpreted as meaningful statements, necessarily presuppose a kind of Platonism, which cannot satisfy any critical mind and which does not even produce the conviction that they are consistent.[13]

It is clear that Gödel regards impredicativity as the most serious of the problems he cites and notes (following Ramsey) that impredicative specification of properties of integers is acceptable if we assume that "the totality of all properties [of integers] exists somehow independently of our knowledge and our definitions, and that our definitions merely serve to pick out certain of these previously existing properties" (ibid.). That is clearly a major consideration prompting him to say that acceptance of the axioms "presupposes a kind of Platonism."[14]

The other remarks are glosses on his work on constructible sets and the consistency of the continuum hypothesis. In the first announcement of his consistency results Gödel says,

> The proposition A [i.e., $V = L$] added as a new axiom seems to give a natural completion of the axioms of set theory, in so far as it determines the vague notion of an arbitrary infinite set in a definite way.[15]

Acceptance of $V = L$ as an axiom of set theory would not be incompatible with the *philosophical* realism Gödel expressed later, although it would be with the mathematical views he expressed in connection with

---

[13] "The Present Situation in the Foundations of Mathematics," III 50.

[14] The cautious and qualified defense of a kind of platonism in Bernays, "Sur le platonisme," was delivered as a lecture about six months later. We think of one of the influential tendencies in foundations of the time, logicism after Frege and Russell, as a platonist view. That was not the way its proponents saw it in the 1930s.

[15] "The Consistency of the Axiom of Choice," II 27.

the continuum problem. But regarding the concept of an arbitrary infinite set as a "vague notion" certainly does not square with Gödel's view in 1947 that the continuum problem *has* a definite answer.[16]

Another document from about this time indicates that, after proving the consistency of the continuum hypothesis and probably expecting to go on to prove its independence, Gödel did not yet have the view of the significance of this development that he later expressed. In a lecture text on undecidable diophantine sentences, probably prepared between 1938 and 1940, Gödel remarks that the undecidability of the sentences he considers is not absolute, since a proof of their undecidability (in a given formal system) is a proof of their truth. But then he ends the draft with the remarkable statement:

> However, I would not leave it unmentioned that apparently there do exist questions of a very similar structure which very likely are really undecidable in the sense which I explained first. The difference in the structure of these problems is only that variables for real numbers appear in this polynomial. Questions connected with Cantor's Continuum Hypothesis lead to problems of this type. So far I have not been able to prove their undecidability, but there are considerations which make it highly plausible that they really are undecidable. ("Undecidable Diophantine Propositions," III 175)[17]

---

[16] Martin Davis notes that in *The Consistency of the Continuum Hypothesis* Gödel refers to $V = L$ as an axiom, indicating that he still held the view expressed in the above quotation from the 1938 announcement. (See his introductory note to "Undecidable Diophantine Propositions," in CW III, at p. 163.) It would confirm, however, only the first of the two distinguishable aspects of the view of 1938.

[17] *Note added in proof (December 1994).* William Tait raises the question what Gödel means by "the sense which I explained first" in this passage. This is not completely clear, but it is very probable that Gödel refers to the discussion at the beginning of the lecture of Hilbert's conviction of the solvability of every well-posed mathematical problem and intends that a proposition undecidable in this sense would contradict Hilbert's conviction.

Some additional confirmation of my reading of this passage and that from "The Consistency of the Axiom of Choice" quoted on p. 159 is offered by the remark in his 1939 Göttingen lecture that "it is very plausible that with A [i.e., $V = L$] we are dealing with an absolutely undecidable proposition on which set theory bifurcates into two different systems, similar to Euclidean and non-Euclidean geometry" (III 155).

It is hard to see what Gödel could have expected to "prove" concerning a statement of the form he describes other than that it is consistent with and independent of the axioms of set theory, say ZF, and that this independence would generalize to extensions of ZFC by axioms for inaccessible cardinals in a way that Gödel asserts that his consistency result does. There seems to be a clear conflict with the position of WCCP 1947; it's hard to believe that at the earlier time he thought that exploration of the concept of set would yield new axioms that would decide them. Moreover the statement is a rather bold statement. I don't think it can be explained away as a manifestation of Gödel's well-known caution in avowing his views.

Let me now turn to the most informative documents about Gödel's early realism, the letters to Wang. There he explains the failure of other logicians to obtain the results obtained by him as due to philosophical prejudices, in particular against the use of non-finitary methods in metamathematics, deriving from views associated with the Hilbert school, according to which non-finitary reasoning in mathematics is justified "only to the extent to which it can be 'interpreted' or 'justified' in terms of a finitary metamathematics."[18] This is applied to the completeness theorem, of which the main mathematical idea was expressed by Skolem in 1922. Gödel also asserts that his "objectivistic conception of mathematics and metamathematics in general" was fundamental also to his other logical work; in particular "the highly transfinite conception of 'objective mathematical truth', as *opposed* to that of 'demonstrability'" is the heuristic principle of his construction of an undecidable number-theoretic proposition (ibid., p. 9).

It should be pointed out that only one of the examples Gödel gives essentially involves impredicativity and thus conflicts sharply with the view of "The Present Situation": his own work on constructible sets. Where the conflict lies is of course in accepting the conception of the constructible sets as an intuitively meaningful conception, but it's on this that Gödel lays stress rather than on the fact that at the end of the process one can arrive at a finitary relative consistency proof. Gödel is said to have had the idea of using the ramified hierarchy to construct a model quite early; whether by the time of the MAA lecture he had seen

---

[18] Wang, *From Mathematics to Philosophy* (hereafter FMP), p. 8. [See also CW V 397. CW prints the letters as they were originally received by Wang; some editing intervened before their publication in FMP.]

that it "has to be used in an *entirely nonconstructive way*" (Wang, op. cit., p. 10) is not clear. It seems to have been only in 1935 that he had a definite result even on the axiom of choice.[19]

It seems we cannot definitely know whether Gödel in December 1933 already thought the "kind of Platonism" he discerned more acceptable than he was prepared to say. But it seems extremely likely that, with whatever conviction he embraced impredicative concepts in first developing the model of constructible sets in the form we know it, his confidence in this point of view would have been increased by his obtaining definite and important results from it. The remarks from 1938 show that there was already a further step to be taken; one possible reason for his taking it may have been reflection on the consequences of $V = L$ for descriptive set theory, which could have convinced him that $V = L$ is false. But it should be pointed out that the idea that some mathematical propositions are absolutely undecidable is one that Gödel still entertained in his Gibbs lecture in 1951, and in itself it is not opposed to realism.[20]

---

[19] Wang writes (*Reflections*, p. 97):

From about 1930 he had continued to think about the continuum problem. . . . The idea of using the ramified hierarchy occurred to him quite early. He then played with building up enough ordinals. Finally the leap of taking the classical ordinals as given made things easier. It must have been 1935, according to his recollection in 1976, when he realized that the constructible sets satisfy all the axioms of set theory (including the axiom of choice). He conjectured that the continuum hypothesis is also satisfied.

Seen in light of the remarks in "The Present Situation," the "leap of taking the classical ordinals as given" was a decisive step in the development of Gödel's realism about set theory. Wang's remarks (evidently based on Gödel's much later recollection) suggest, but do not explicitly say, that this leap was taken close enough to 1935 to be probably later than December 1933. On the other hand Feferman conjectures that the rather casual treatment in the 1933 lecture itself of the problem of the axiom of choice may have been due to Gödel's having an approach to proving its consistency. (See his introductory note to "The Present Situation" in CW III.)

It can be documented that Gödel obtained the proof of the consistency of CH in June 1937. See Feferman, "Gödel's Life and Work," note s (CW I 36).

[20] Note that in "Remarks before the Princeton Bicentennial Conference," Gödel explores the idea of absolute provability. In this connection it is reasonable to ask whether Gödel is a realist by one criterion suggested by the work of Dummett, according to which realism admits truths that are "recognition-transcendent," that is obtain whether or not it is even in principle possible for humans to know them. In the sphere of mathematics, an obstacle to this view for Gödel is his confidence in

There is another more global and intangible consideration that could lead one to doubt that Gödel's views of the 1930s were the same as those he avowed later. This is the evidence of engagement with the problems of proof theory, in the form in which the subject evolved after the incompleteness theorem. Gödel addresses questions concerning this program in the MAA lecture of 1933 and more thoroughly and deeply in the remarkable lecture given in early 1938 to a circle organized by Edgar Zilsel. This lecture shows that he had already begun to think about a theory of primitive recursive functionals of finite type as something relative to which the consistency of arithmetic could be proved; it is now well known that he obtained this proof in 1941 after coming to the United States. The lecture at Zilsel's also contains a quite remarkable analysis of Gentzen's 1936 consistency proof, including the no-counter-example interpretation obtained later by Kreisel (see his "On the Interpretation of Non-Finitist Proofs"). What he says about the philosophical significance of consistency proofs such as Gentzen's is not far from what was being said about the same time by Bernays and Gentzen, in spite of somewhat polemical remarks about the Hilbert school in this text and in others.[21]

3. Gödel's Avowals of Platonism, 1944–1951

I shall not try to trace the development of Gödel's realism further independently of the notion of mathematical intuition. As I said, it is firmly avowed in RML and further developed in the Princeton Bicentennial remarks, the 1947 version of WCCP, and the Gibbs lecture. It is thus

---

reason; he expresses in places the Hilbertian conviction of the solvability in principle of every mathematical problem. See FMP, pp. 324–325 (on which see note 49 below), cf. also Gödel, "The Modern Development," III 378, 380.

However, the discussion in the Gibbs lecture makes clear that Gödel regards the existence of recognition-transcendent truth as meaningful, since if the mathematical truths that the human mind can know can be generated by a Turing machine, the proposition that this set is consistent would be a mathematical truth that we could not know. And this is presumably what is decisive for Dummettian realism rather than whether recognition-transcendent truths in fact exist, which Gödel was inclined to believe they did not, at least in mathematics.

[21] I owe this observation to Wilfried Sieg. Cf. our introductory note to "Vortrag bei Zilsel" in CW III, at p. 85. [Solomon Feferman has observed that Gödel continued to be engaged with issues connected with the Hilbert program until well after his platonistic views had been publicly presented.]

during the period from 1943 or 1944 through 1951 that it becomes Gödel's public position.[22]

I have discussed elsewhere the position of RML.[23] It is not easy to discern a definite line of argument for realism (which would in turn clarify the position itself); the form of a commentary on Russell works against this. A very familiar argument which is already present in "The Present Situation" (as well as in Bernays, "Sur le platonisme") is that particular principles of analysis and set theory are justified if one assumes a realistic view of the objects of the theory and not otherwise. Gödel applies this point of view particularly in his well-known analysis of Russell's vicious circle principle, where he argues from the fact that "classical mathematics" does not satisfy the vicious circle principle that this is to be considered "rather as a proof that the vicious circle principle is false than that classical mathematics is false" (II 127).

When Gödel says that assuming classes and concepts as "real objects" is "quite as legitimate as the assumption of physical bodies and there is quite as much reason to believe in their existence" (II 128, quoted above), his claim is that classical mathematics is committed to such objects and moreover it must be interpreted so that the objects are independent of our constructions. Gödel reinforces this claim by his analysis of the ramified theory of types in the present paper and by discussions elsewhere in his writings such as the criticism of conventionalism in the Gibbs lecture and "Is Mathematics Syntax of Language?" (actually briefly adumbrated at the end of RML). In a way this is hardly controversial today; an impredicative theory with classical logic is the paradigm of a "platonist" theory. But Gödel's rhetoric has certainly led most readers to think that his reasoning is not just to be reconstructed as an application of a Quinean conception of ontological commitment. Why is this so?

One reason is certainly Gödel's remarks about intuition, of which we are postponing discussion. But that conception plays virtually no role in RML. Another reason more internal to that text is that Gödel makes clear that his realism extends to concepts as well as classes (which in this discussion he does not distinguish from sets). Standard

---

[22] The conversation that was the basis of Russell's remark quoted on p. 153 above would have taken place near the beginning of this period.

[23] In my introductory note in CW II (Essay 5 in this volume); on realism see particularly pp. 108–113.

set theories either quantify only over sets or, if they have quantifiers for (proper) classes, allow a predicative interpretation of class quantification. Thus at most realism about sets seems to be implied by what is common to Gödel and philosophers who have followed Quine. Gödel makes clear that he sees no objection to an impredicative theory of concepts (II 129–130), and the paper contains sketchy ideas for such a theory, which apparently Gödel never worked out in a way that satisfied him. But Gödel does not directly argue for a realism about concepts that would license such a theory; in particular he does not argue that classical mathematics requires such realism.

In what sense does WCCP 1947 offer a further argument for realism?[24] The major philosophical claim of WCCP 1947, that the independence of the continuum hypothesis should in no way imply that it does not have a determinate truth-value, is rather an inference *from* realism. Gödel makes such an inference in saying that if the axioms of set theory "describe some well-determined reality," then "in this reality Cantor's conjecture must be either true or false, and its undecidability from the axioms known today can only mean that these axioms do not contain a complete description of this reality" (II 181). But Gödel then proceeds to give arguments for the conclusion that the continuum problem might be decided. The first is the point going back to the 1933 lecture about the open-endedness of the process of extending the axioms. The second is that large cardinal axioms have consequences even in number theory. Here he concedes that such axioms as can be "set up on the basis of principles known today" (i.e., axioms providing for inaccessible and Mahlo cardinals) do not offer much hope of solving the problem (II 182).[25] The further statement, that axioms of infinity and other kinds of new axioms are possible, was more conjectural, and of course the stronger axioms of infinity that were investigated later (already taken account of to some degree in the corresponding place in WCCP 1964) were shown not to decide the continuum hypothesis

---

[24] I pass over the Princeton Bicentennial remarks, which might, like WCCP 1947, be described as an application of Gödel's point of view to concrete problems. This is not uncharacteristic of Gödel; also in RML he often seems to treat realism as a working hypothesis.

[25] This had been partly shown by Gödel in extending his consistency proof to such axioms; it was subsequently shown that the independence proof also extended, and the consistency and independence of CH were proved even for stronger large cardinal axioms such as Gödel did not have in mind in 1947.

(CH). The third consideration is that a new axiom, even if it cannot be seen to have "intrinsic necessity," might be verified inductively by its fruitfulness in consequences, in particular independently verifiable consequences. It might be added that Gödel's plausibility arguments for the falsity of CH constitute an argument for the suggestion that axioms based on new principles exist, since any such axiom would have to be incompatible with $V = L$.

Another point, which hardly attracts notice today because it seems commonplace, is that the concept of set and the axioms of set theory can be defended against paradox by what we would call the iterative conception of set. In 1947, to say that this conception offers a "satisfactory foundation of Cantor's set theory in its whole original extent" (II 134) was a rather bold statement. Even the point (made in RML, II 134) that axiomatic set theory describes a transfinite iteration of the set-forming operations of the simple theory of types was not a commonplace. Of course in what sense we do have a "satisfactory foundation" was and is debatable. But I think it would now be a non-controversial claim that, granted certain basic ideas (ordinal and power set) in a classical setting, the iterative conception offers an intuitive conception of a universe of sets, which, in Gödel's words, "has never led to any antinomy whatsoever" (WCCP 1947, II 180). I think Gödel wishes to claim more, namely that the axioms follow from the concept of set. That thought is hardly developed in WCCP 1947 and anyway belongs with the conception of mathematical intuition.[26] Overall, the paper was probably meant to offer an indirect argument for realism by applying it to a definite problem and showing that the assumption of realism leads to a fruitful approach to the problem. It is worth noting that he offers arguments for the independence of the continuum hypothesis of which the main ones are plausibility arguments for its *falsity*. An "antirealist" urging upon us the attempt to prove the independence would presumably dwell more on the obstacles to proving it.

The Gibbs lecture, "Some Basic Theorems of the Foundations of Mathematics and Their Implications," seems to complete for Gödel the process of avowing his platonistic position. In some ways, it is the most systematic defense of this position that Gödel gave. At the end it seems

---

[26] In the revised version of 1964, the discussion of the iterative conception of set is somewhat expanded.

to see itself as part of an argument that would "leave the platonistic position as the only one tenable" (III 322–323).[27]

The main difficulty of the Gibbs lecture's defense, however, is not the omission he mentions at the end, of a case against Aristotelian realism and psychologism, but that its central arguments are meant to be independent of one's standpoint in the traditional controversies about foundations; the overall plan of the lecture is to draw implications from the incompleteness theorems. Gödel's main arguments aim to strengthen an important part of his position, which he expresses by saying that mathematics has a "real content."[28] But although this is opposed to the conventionalism that he discerns in the views of the Vienna Circle, and also to many forms of formalism, it is a point that constructivists of the various kinds extant in Gödel's and our own time can concede, as Gödel is well aware. But it is probably a root conviction that Gödel had from very early in his career; it very likely underlies the views that Gödel, in the letters to Wang, says contributed to his early logical work. It would then also constitute part of his reaction to attending sessions of the Vienna Circle before 1930.

4. Mathematical Intuition before 1964

I now turn to the conception of mathematical intuition, beginning with some remarks about its development. I have outlined above the presentation of Gödel's realism in his early philosophical publications RML, WCCP 1947, and the Gibbs lecture. For a reader who knows the 1964 version of WCCP, it is a striking fact about these writings that the word "intuition" occurs in them very little, and no real attempt is made to connect his general views with a conception of mathematical intuition.

In RML the word "intuition" occurs in only three places, none of which gives any evidence that intuition is at the time a fundamental

---

[27] This remark appears to be an expression of a hope that Gödel maintained for many years, that philosophical discussion might achieve "mathematical rigor" and conclusiveness. As he was well aware, his actual philosophical writings, even at their best, did not fulfill this hope, and these remarks are part of an admission that certain parts of the defense of mathematical realism had not been undertaken in the lecture.

[28] This conviction will come up in the discussion of intuition in §§4–5; see also Essays 5 and 6 in this volume.

notion for Gödel himself. The first (II 121) is in quotation marks and refers to Hilbert's ideas. The second is in one of the most often quoted remarks in the paper, in which Russell is credited with "bringing to light the amazing fact that our logical intuitions (i.e., intuitions concerning such notions as truth, concept, being, class, etc.) are self-contradictory" (II 124). Here "intuition" means something like a belief arising from a strong natural inclination, even apparent obviousness. In the following sentence these intuitions are described as "common-sense assumptions of logic." It's not at all clear to what extent "intuition" in this sense is a guide to the truth; it is clearly not an infallible one. In the third place (II 138), Gödel again speaks of "our logical intuitions," evidently referring to the earlier remarks, and it seems clear that he is using the term in the same sense.

One other remark in RML deserves comment. In his discussion of the question whether the axioms of *Principia* are analytic in the sense that they are true "owing to the meaning of the concepts" in them, he sees the difficulty that "we don't perceive the concepts of 'concept' and 'class' with sufficient distinctness, as is shown by the paradoxes" (II 139–140). Since "perception" of concepts is spoken of in unpublished writings of Gödel, this seems to be an allusion to mathematical intuition in a stronger sense. But the remark itself is negative; it's not clear what Gödel would say that is positive about perception of concepts.

The word "intuition" does not occur at all in "Remarks Before the Princeton Bicentennial Conference" and only once in WCCP 1947. Concerning constructivist views, he remarks:

This negative attitude towards Cantor's set theory, however, is by no means a necessary outcome of a closer examination of its foundations, but only the result of certain philosophical conceptions of the nature of mathematics, which admit mathematical objects only to the extent in which they are (or are believed to be) interpretable as acts and constructions of our own mind, or at least completely penetrable by our intuition. (II 180)

Since Gödel does not elaborate on his use of "intuition" at all, one can't on the basis of this text be at all sure what he has in mind. But it appears that intuition as here understood, instead of being a basis for possible knowledge of the strongest mathematical axioms, is restricted in its application, so that the demand that mathematical objects be "completely penetrable by our intuition" is a constraint that strongly limits what objects can be admitted.[29]

The Gibbs lecture is again virtually silent about intuition. I have not found in it a single occurrence of the word "intuition" on its own.[30] But talk of perception where the object is abstract occurs again, this time more positively, but still without elaboration or explanation. Gödel defends the view that mathematical propositions are true by virtue of the meaning of the terms occurring in them.[31] But the terms denote concepts of which he says: "The truth, I believe, is that these concepts form an objective reality of their own, which we cannot create or change, but only perceive and describe" (III 320). At the end, he says of the "Platonistic view":

> Thereby I mean the view that mathematics describes a non-sensual reality, which exists independently both of the acts and the dispositions of the human mind and is only perceived, and probably perceived very incompletely, by the human mind. (III 325)

There is nothing in these early writings to rule out the interpretation that the talk of "perception" of concepts is meant metaphorically. The last quoted statement could come down to the claim that the "non-sensual reality" that mathematics describes is *known* or *understood* very incompletely by the human mind. Thus although there are what might be indications as early as RML of a strong conception of mathematical intuition, in public documents before 1964 they are less than clear and decisive, and Gödel does not begin to offer a defense of it. Nonetheless the allusions to perception of concepts in RML and the Gibbs lecture are very suggestive in the light of his later writings, and it is reasonable to conjecture that although he was not yet ready to

---

[29] The meaning of "intuition" here could agree with that of *Anschauung* in "Über eine bisher noch nicht benützte Erweiterung"; see below. The phrase is replaced in WCCP 1964 by "completely given in mathematical intuition" (II 258); it is hard to be sure whether Gödel saw this as more than a stylistic change. [On this point, van Atten and Kennedy observe ("On the Philosophical Development," p. 468) that the phrase used in 1964 "is completely idiomatic in Husserl's work, while the former is not."]

[30] There are references to intuitionism (e.g., in III 310, n.15), and he does speak (III 319) of the "intuitive meanings" of disjunction and negation.

[31] This is, of course, a sense in which mathematics could be said to be analytic; for further discussion see Essay 6 in this volume.

defend his conception of intuition he already had some such conception in mind.

But of course there is one published writing before 1964 in which a concept of intuition figures more centrally, and that is the philosophical introduction to the *Dialectica* paper "Über eine bisher noch nicht benützte Erweiterung." The German word used is the Kantian term *Anschauung.* I shall not discuss this paper in any detail but only state rather dogmatically that what is at issue are conceptions of intuition and intuitive evidence derived from the Hilbert school. Gödel is concerned with the question of the limits of intuitive evidence, where these limits will clearly be rather narrow. It is contrasted with evidence essentially involving "abstract concepts." Thus the conception of intuition involved is not the strong one, a mark of which is that it yields knowledge of propositions involving abstract concepts in an essential way. There is no doubt that that was Gödel's view of the central concepts of set theory and the axioms involving them. The fact that in the 1972 English version *Anschauung* is translated as "concrete intuition" indicates both that in 1958 he was employing a more limited conception of intuition than that of WCCP 1964 and that it may be a special case of the latter.

There is, however, a source earlier than 1964 for Gödel's thought about mathematical intuition, the drafts of the paper "Is Mathematics Syntax of Language?," which Gödel worked on in response to an invitation from Paul Arthur Schilpp to contribute to *The Philosophy of Rudolf Carnap* but never submitted. Six versions survive in Gödel's *Nachlaß.*

The main purpose of the paper is to argue against the conception of mathematics as syntax that is found in logical positivist writings, especially Carnap's *Logical Syntax of Language.* Gödel had already given a version of his argument in the Gibbs lecture,[32] in a way that does not use the notion of mathematical intuition, and even sketched the ideas in the discussion of analyticity at the end of RML. The basic argument, related to arguments directed at Carnap by Quine, is that in order to establish that interesting mathematical statements are true by virtue of syntactical rules or conventions it is necessary to use the mathematics

---

[32] A large part of it (III 315–319), however, is in a section marked "wegzulassen"; it is possible that this was not included in the lecture as delivered. Cf. editorial note c, p. 315.

itself in its straightforward meaning.[33] In arguing, contrary to the view he is criticizing, that mathematics has a "real content," Gödel is, as I have said, affirming one aspect of his realism. It is, however, only one: The same argument would be open to an intuitionist, and Gödel himself argues that certain fallback positions of his opponent still leave him obliged to concede "real content" to at least finitist mathematics.

The presentation of his argument against Carnap in the Syntax paper does not similarly eschew reference to mathematical intuition, although in the briefer, stripped down presentation of the argument in version V, it does not figure prominently. Before we go into it we should rehearse some elementary distinctions about intuition. In the philosophical tradition, intuition is spoken of both in relation to objects and in relation to propositions, one might say as a propositional attitude. I have used the terms intuition *of* and intuition *that* to mark this distinction. The philosophy of Kant, and the Kantian paradigm generally, gives the basic place to intuition of but certainly allows for intuitive knowledge or evidence that would be a species of intuition that. But talk of intuition in relation to propositions has a further ambiguity, since in propositional attitude uses "intuition" is not always used for a mode of *knowledge*. When a philosopher talks of his or others' intuitions, that usually means what the person concerned takes to be true at the outset of an inquiry, or as a matter of common sense; intuitions in this sense are not knowledge, since they need not be true and can be very fallible guides to the truth. To take another example, the intuitions of a native speaker about when a sentence is grammatical are again not necessarily correct, although in this case they are, in contemporary grammatical theory, taken as very important guides to truth. In contrast, what Descartes called *intuitio* was not genuine unless it was knowledge. Use of "intuition" with this connotation is likely to cause misunderstanding in the circumstances of today; it may even lead a reader to think one has in mind something like intuitions in the senses just mentioned with the extra property of being infallible. It is

---

[33] For discussion see Essay 6 in this volume. However, I barely touch there on the question whether the position Gödel criticizes is what Carnap actually holds. This is questioned by Warren Goldfarb in his introductory note to "Is Mathematics Syntax of Language?," in CW III. Cf. Goldfarb and Ricketts, "Carnap and the Philosophy of Mathematics."

probably best to use the term "intuitive knowledge" when one wants to make clear one is speaking of knowledge.[34]

A difficulty in reading Gödel's writing on mathematical intuition is that he uses the term in both object-relational and propositional attitude senses, and in the latter it is not always clear what epistemic force the term is intended to have. Since, where a strong conception is involved, it is mainly concepts that are the objects of intuition, and Gödel does (as we have already seen) speak of perception of concepts, it might be well in discussing Gödel to use the word "perception" where intuition of is in question, and reserve the term "intuition" for intuition that. I will follow that policy in what follows.[35]

In the Syntax paper Gödel seems to take the propositional sense as primary. I think it is clear that he has first of all in mind what might be called rational evidence, or, more specifically, autonomous mathematical evidence. Thus in stating the view he is criticizing he writes, "Mathematical intuition, for all scientifically relevant purposes . . . can be replaced by conventions about the use of symbols and their application" (version V, III 356). Apart from the conventionalism his argument is directed against, the only alternative to admitting mathematical intuition that Gödel considers is some form of empiricism. Thus the deliverances of mathematical intuition are just those mathematical propositions and inferences that we take to be evident on reflection and do not derive from others, or justify on a posteriori grounds, or explain away by a conventionalist strategy.[36]

It is clear Gödel has primarily in mind mathematical axioms and rules of inference that would be taken as primitive. He does not, however, distinguish mathematics from logic. An example given in a couple of places is modus ponens.[37] In application to logic, what we have presented up to now of Gödel's position does not differ from a quite

---

[34] In the philosophy of mathematics, however, this has the disadvantage that "intuitive knowledge" has a more special sense, for example in "Über eine bisher noch nicht benützte Erweiterung" and its 1972 English version.

[35] [In fact this is Gödel's usage more often than not.]

[36] One might ask, particularly in the light of later writing in the philosophy of mathematics, about the option of not taking the language of mathematics at face value. The only such option considered in Gödel's writings is if-thenism. Apart from other difficulties, in his view the translations have enough mathematical content to raise again the same questions.

[37] Version III, note 30 (III 347); version V, III 359.

widely accepted one, in declining to reduce the evidence of logic either to convention or to other forms of evidence. Such a view is even implied by Quine when he regards the obviousness of certain logical principles as a constraint on acceptable translation, although of course Quine would not agree that this implies an important distinction between logical and empirical principles.

With regard to the epistemic force of Gödel's notion of mathematical intuition, the remarks in the supplement to WCCP 1964 have given rise to some confusion. I think this can be largely cleared up by taking account of the Syntax paper. I think it is clear that for Gödel mathematical intuition is not *ipso facto* knowledge. In a way the existence of mathematical intuition should be non-controversial:

> The existence, as a psychological fact, of an intuition covering the axioms of classical mathematics can hardly be doubted, not even by adherents of the Brouwerian school, except that the latter will explain this psychological fact by the circumstance that we are all subject to the same kind of errors if we are not sufficiently careful in our thinking. (version III, 338 n.8)[38]

In this context, "intuition" has something like the contemporary philosopher's sense, with perhaps more stability and intersubjectivity: Most of us who have studied mathematics find the axioms of classical mathematics intuitively convincing or at least highly plausible. According to Gödel, Brouwer (or for that matter a conventionalist) should grant this much.[39] Elsewhere, where it is clear that he regards mathematical intuition as a *source* of knowledge, it is still clear that possession of intuition isn't already possession of knowledge, for example when he talks of mathematical intuition producing conviction:

> However, mathematical intuition in addition produces the conviction that, if these sentences express observable facts and were obtained by applying mathematics to verified physical laws (or if they express ascertainable mathematical facts), then these facts

---

[38] A parallel passage in version IV is clearer but more controversial in that it introduces the idea of intuition of concepts.

[39] I think Brouwer did grant a good part of what Gödel has in mind here. But to sort this out would be a long story and belong to the discussion of Brouwer rather than Gödel.

will be brought out by observation (or computation). (Version III, III 340)

If the possibility of a disproof of mathematical axioms is disregarded, this is due solely to the convincing power of mathematical intuition. (Version V, III 361)

What he calls the "belief in the correctness of mathematical intuition" (version III, III 341) is not a trivial consequence of acknowledging its existence.

Gödel does (as we shall see) regard mathematical intuition as significantly like perception, but that someone has the intuition that p does not imply p in the way that if he sees that p that implies p. (If someone claims to see that p and p turns out to be false, then he only seemed to see that p.) It is rather more like making a perceptual judgment, which may have a strong presumption of truth but which can in principle be false.

A conclusion I draw from this is that what is at issue between Gödel and his opponents about mathematical intuition is not any basic assertion of its existence, but some questions about its character and especially its ineliminability as an epistemic factor. Gödel attributes to Carnap the view that appeals to mathematical intuition need play no more than an heuristic role in the justification of mathematical claims. Something like this seems also to be true of Quine (although Gödel never comments on Quine's philosophical views).

Even if one grants to mathematical intuition in the sense explained so far a high degree of credence, the question will still arise why it should be called *intuition*. Other philosophers have held that there are non-empirical, non-conventional truths without calling the evidence that pertains to them intuition or using for them a term that could easily be understood as meaning something close to that. A very good example is Frege, who quite on the contrary insists that arithmetical knowledge, because it is a part of logic, does not depend on intuition. For him the term (that is, the German term *Anschauung*) has a roughly Kantian meaning. Gödel himself often speaks of reason in talking of the evident character of mathematical axioms and inferences. This is in agreement with the usage of Frege and others in the rationalist tradition. Yet in speaking of the source of knowledge in these cases as mathematical intuition, without the spatio-temporal connotation of the Kantian tradition, Gödel is not just differing with Frege about terminology.[40]

To analyze the differences between Gödel and Frege would require more exploration of Frege than I can undertake here. But we can see a major difference in the analogy Gödel stresses in places in the Syntax paper between what he calls mathematical intuition and sense-perception. His claims about this analogy are strong but not very much developed. As I remarked earlier, he does not distinguish mathematics from logic. Thus even elementary logic seems to be an application of mathematical intuition:

> The similarity between mathematical intuition and a physical sense is very striking. It is arbitrary to consider "this is red" an immediate datum, but not so to consider the proposition expressing modus ponens or complete induction (or perhaps some simpler propositions from which the latter follows). For the difference, as far as it is relevant here, consists solely in the fact that in the first case a relationship between a concept and a particular object is perceived, while in the second case it is a relationship between concepts. (Version V, III 359)

In this passage and in many others, we find a formulation that is very characteristic of Gödel: In certain cases of rational evidence (of which we can easily grant modus ponens to be one), it is claimed that "perception" of concepts is involved. Indeed, in this passage such perception is even said to be involved in a situation where one recognizes by sense-perception the truth of 'this is red' (with some demonstrative reference or other for 'this'). An inference seems to be made from '*a* perceives that . . . *F* . . .', where '*F*' is a predicate or general term occurring in '. . . *F* . . .', to '*a* perceives (the concept) *F*'. Gödel does not formulate "the proposition expressing modus ponens," but presumably it would involve the concepts of proposition, being of the form 'if *p* then *q*', and implication, so that the claim is that in this case a relation of these concepts is involved. (I am assuming that perceiving a relation

---

[40] Although Kantian intuition plays a role in Gödel's writings, it is not altogether clear whether he accepted some version of the notion or simply explored it as part of a philosophy (that of the Hilbert school) he wished to explore because of its connection with proof theory and constructivity.

between concepts involves perceiving the concepts; I think that can be justified from the texts.[41])

If we ask what the analogy with perception is beyond that of providing an irreducible form of evidence, an appropriate answer is likely to be of the form that certain *objects* are before the mind in a way analogous to that in which physical objects are present in perception. Gödel's answer is "concepts," perhaps concepts of a particular kind. But that in the case of either 'this is red' or elementary logical truths and inferences concepts are present in this way seems to be an assumption, at best part of an explanation of how these things might be evident that is not carried further.[42] This is certainly a point on which Gödel can be criticized.

Gödel actually goes further and sees a close analogy between reason and an "additional sense." After discussing the idea of an additional sense that would show us a "second reality" separated from space-time reality but still describable by laws, Gödel says:

> I even think this comes pretty close to the true state of affairs, except that this additional sense (i.e. reason) is not counted as a sense, because its objects are quite different from those of all other senses. For while with sense perception we know particular objects and their properties and relations, with mathematical reason we perceive the most general (namely the 'formal') concepts and their relations, which are separated from space-time reality in so far as the latter is completely determined by the totality of particularities without any reference to the formal concepts. (Version III, III 354)[43]

In the corresponding passage in version IV, instead of the last sentence above Gödel has: "For while with the latter [the senses] we perceive

---

[41] Cf. the formulation of the same point in version IV, ms. p. 19.

[42] Gödel does in one place ("The Modern Development," III 382, 384) make brief remarks about language learning and suggests that when a child first understands a logical inference this is a step that brings him to a higher state of consciousness. Cf. Essay 6 in this volume, §III.

[43] What does Gödel mean by this last assertion? It seems to say, as Warren Goldfarb remarks in his introductory note to the paper, that "the empirical world is fixed independently of mathematics" (CW III 333). Gödel does not suggest such a view in his discussions of physics, and it is difficult to reconcile with the talk in WCCP 1964 of the "abstract elements contained in our empirical ideas" (II 268).

particular things, with reason we perceive concepts (above all the primitive concepts) and their relations" (Ms. p. 17b). The difference suggests an important uncertainty or change of mind on Gödel's part, as to the exact sphere of reason (which would include mathematical intuition). In IV any perception of concepts seems to be an application of reason, but in III it seems that the concepts involved in 'this is red' belong rather to sense. The passage from V suggests the position of IV but may have been intended to be noncommittal.

It is disappointing that Gödel's logical example is a general principle that involves quantification over propositions or sentences and characteristically logical concepts like implication. He does not, here or so far as I know elsewhere, answer directly the question whether a particular logical truth such as 'it is raining or it is not raining' or a particular inference (say, by modus ponens) is an application of mathematical intuition. This could depend on the question just mentioned, whether any perception of concepts is an application of reason.

Such elementary logical examples would differ from the example that Gödel was most interested in, the axioms of set theory, in that the claim that they are rather directly and immediately evident has a great deal of plausibility. One couldn't argue for them from more theoretical considerations without using inferences or assumptions of much the same kind.

Gödel had another argument for the analogy between reason and perception, based on what in the Gibbs lecture he called the inexhaustibility of mathematics, which he argued for in two ways: from the incompleteness theorem, which implies that a sound formal system for a part of mathematics can always be properly extended, and from the iterative conception of set, where, on his understanding, the conception would always give rise to more sets than a given precise delineation of principles would provide and thus to new evident axioms. Thus there are an unlimited number of independent "perceptions":

> The "inexhaustibility" of mathematics makes the similarity between reason and the senses . . . still closer, because it shows that there exists a practically unlimited number of perceptions also of this "sense." (Syntax, version III, III 353 n.43)[44]

---

[44] Gödel follows this remark by remarks about axioms of infinity in set theory. A parallel remark in IV (ms. p. 19) is followed by an appeal to the second incompleteness theorem.

The concept of set was doubtless for Gödel the most favorable example of "perception" of concepts. (It is also a case where Gödel argued something I have not stressed here, although it is discussed in Essay 6 in this volume, that the propositions known in this way are in a way analytic.) Thus in WCCP 1964, he emphasizes the fact that intuition gives rise to an "open series of extensions" of the axioms (II 268), and of course the incompleteness theorem implies that any such series generated by a recursive rule would be incomplete and would, indeed, suggest a further reflection that would lead to a still stronger extension. Gödel interpreted these considerations by saying that "the mind, in its use, is not static, but constantly developing" ("Some Remarks on the Undecidability Results," II 306). This remark is directed against a mechanist view of mind such as Gödel attributed to Turing. He explicitly offers the generation of new axioms of infinity in set theory as an example.[45] It is interesting that the inexhaustibility of mathematics is used by Gödel both in drawing his analogy between perception and insight into mathematical axioms and in his critical discussion of mechanism.[46] The complex, iterated reflection involved in the uncovering of stronger mathematical axioms and the concepts entering into them strikes me intuitively as very different from perception, and I don't think Gödel has offered more than a rather undeveloped formal analogy. But it is a real problem for what I at the outset called a theory of reason to give a better account.

5. The Position of 1964

Let us now turn to the remarks about mathematical intuition in WCCP 1964. Gödel presents a sketch of his epistemology of mathematics in

[45] One should, however, compare the version of what in CW II is Remark 3 of "Some Remarks" with the version of the same remark published in Wang, *From Mathematics to Philosophy*, p. 325 [and reprinted in CW V 576].

[46] On this subject see the Gibbs lecture (and Boolos's introductory note in CW III), Wang, op. cit., pp. 324–326, and Wang, "On Physicalism and Algorithmism." It would be beyond the scope of this paper to pursue this subject further. But it should be pointed out that what is needed for Gödel's case against mechanism is the inexhaustibility of our potential for acquiring mathematical *knowledge*. He himself makes clear in the Gibbs lecture that that does not follow simply from the mathematical considerations such as the incompleteness theorems. On the other hand it is not intrinsically connected with platonism as opposed to, say, intuitionism.

four paragraphs (II 268–269). Some things that are obscure in the first and third paragraphs should be clearer in the light of our discussion so far.

Gödel begins with a remark that is among the most quoted in all his philosophical writing:

> But, despite their remoteness from sense experience, we do have something like a perception also of the objects of set theory, as is seen from the fact that the axioms force themselves on us as being true. I don't see any reason why we should have less confidence in this kind of perception, i.e. in mathematical intuition, than in sense perception, which induces us to build up physical theories and to expect that future sense perceptions will agree with them, and, moreover, to believe that a question not decidable now has meaning and may be decided in the future. (II 268)

A first problem concerning this passage is how Gödel gets from the axioms "forcing themselves on us as being true," which we might accept as a form of intuition that, to the conclusion that we have something like a perception of the *objects* of set theory, an instance of intuition of. We can see that by "the objects of set theory" Gödel means not just sets but the primitive *concepts* of set theory, "set" itself, membership, what he calls "property of set" (II 260 n.18). And it is clear from the above discussion that he understands rational evidence in general as involving perception of the concepts that are the constituents of the proposition in question. This, I think, is the unstated premise of an inference that at first sight appears to be a *non sequitur*. Although Gödel never so far as I know denies that there is "something like a perception" of *sets,* it isn't on that idea that his conception of our knowledge of axioms of set theory rests.[47] The "new mathematical intuitions leading to a decision

---

[47] Nonetheless there is still a problem, as was pointed out to me by Earl Conee. Perception of "the objects of set theory" does on the face of it include perception of sets, and it is not clear how perception of the concepts explicitly occurring in the axioms of set theory should lead to such perceptions.

Warren Goldfarb has pointed out that in some places in "Syntax" Gödel seems to take "concepts" also to include mathematical objects. (See pp. 332–333 of his introductory note in CW III and the texts cited there; of these the most persuasive to me is version III, note 45.) But even if that were Gödel's general usage, it would not solve this particular problem.

of such problems as Cantor's continuum hypothesis," in particular, would be simply insights into the truth of new axioms that would decide the continuum hypothesis.

In the third paragraph Gödel presents ideas that will be familiar to the reader of the Syntax paper. He is considering a fallback position where intuition concerning the concepts of set theory is not the guide to knowledge that he himself takes it to be. It is still "sufficiently clear to produce the axioms of set theory and an open series of extensions of them," and this "suffices to give meaning to the question of the truth or falsity of propositions like Cantor's continuum hypothesis" (II 268). That reflection on the concepts of set theory gives rise to intuitions of this kind can hardly be doubted if one studies the work of set theorists, although how clear the intuitions are can be questioned, already concerning the axioms of replacement and power set. The meaning that is thus given to CH is that the progress of set theory could give rise to axioms that are supported by the intuitions of set theorists and decide CH. But of course the particular line of inquiry on which Gödel rested his original hopes, large cardinals, proved fruitful in other respects but has not resolved this particular problem. Whether something like what Gödel hoped for is at all likely to happen through the discovery of axioms of another kind deciding CH is so far as I know open, and I would defer to the judgment of experts in any case. But clearly the question can't be suppressed: Couldn't our intuitions concerning sets be conflicting, so that different axioms were discovered that have their own kind of intuitive support but have opposite implications regarding CH?

The spectre of the concept of arbitrary infinite set being a "vague notion" that needs to be "determined in a definite way" by new axioms isn't easily banished, and then one can't rule out the possibility that it might be determined in incompatible ways.

The opening sentence of the paragraph suggests that mathematical intuition might be developed altogether without any commitment as to the extent to which it is a guide to truth:

It can be said that some sets can be identified individually by concepts, and one, ω, is all but explicitly mentioned in the axiom of infinity. Since Gödel would probably have regarded deduction as leading to further or clearer perceptions of concepts, he could very well have thought that individually identifiable sets are "perceived" by way of perception of the concepts that identify them uniquely. This view would have the consequence that natural numbers are also "perceived." So far as I know Gödel nowhere affirms or denies this.

However, the question of the objective existence of the objects of mathematical intuition (which, incidentally, is an exact replica of the question of the objective existence of the outer world) is not decisive for the problem under discussion here. (II 268)

W. W. Tait is led by this remark to compare mathematical intuition to the perceptions of a brain in a vat ("Truth and Proof," p. 82, in note 6 [of reprint]). That would, I think, not square with the view that intuition plays a role in elementary mathematics and logic without which one could not even answer the question whether "intuition" understood more noncommittally is able to decide CH. That Gödel is here making a concession for the sake of argument, and thinking primarily of the intuitions leading to strong axioms of set theory, is suggested by the fact that later in the same paragraph he talks of what "justifies the acceptance of this criterion of truth in set theory."[48] He also, in effect, repeats the point about there being a potentially unlimited number of independent intuitions needed to decide questions not only in set theory but also in number theory.

---

[48] Tait is concerned to argue that mathematical intuition is not "what confers objective validity on our theorems" according to Gödel. I am not sure that Gödel has said enough to make at all clear how he would understand the latter problem. Tait may be denying that according to Gödel mathematical intuition is necessary to us as a ground for asserting mathematical propositions; if so I disagree, and I think the argument of the Gibbs lecture and the Syntax paper would make little sense if the denial is right. The philosophical defense of the objective validity of mathematics is another matter. I agree with Tait that the previous paragraph contains something of what Gödel has to say about that.

Tait seems to reject the interpretation of Gödel as an "arch-Platonist." In part he is rejecting, certainly rightly, the imputation to Gödel of the postulation of a faculty by which we "interact" with mathematical objects. It still seems to me that there is an important sense in which Gödel is a Platonist and Tait is not. Tait's view that "questions about the legitimacy of principles of construction or proof are not . . . questions of fact" and the reasons he gives for this ("Truth and Proof," p. 78) are alien to Gödel (see below). Tait himself sees some disagreement (ibid., pp. 82–83, at end of the note). (Related remarks concerning Tait's remarks about Gödel are made in Yourgrau, Review of Wang, pp. 394–395.)

Note added in proof (December 1994). In correspondence, Tait states that the principal concern of note 3 of his paper just discussed [note 6 of reprint] was to criticize the interpretation expressed by Paul Benacerraf, of Gödel as postulating a faculty with which we "interact" with mathematical objects. I have expressed agreement with Tait's rejection of this interpretation.

[For further discussion of Tait's view of Gödel's realism, taking account of later publications of his, see Essay 12 in this volume.]

I now move back to the second paragraph, possibly the most diffi-cult and obscure passage in Gödel's finished philosophical writing. Only a small part of it is much illuminated by the earlier writings that I have studied. The passage presents new ideas, possibly derived from the study of Husserl that Gödel began in 1959. But it is with Kant and perhaps Leibniz that he seems to make a more direct connection.

Since Gödel is making a comparison between mathematical con-cepts and those referring to physical objects, it may be helpful to recall the most basic elements of Kant's conception of the latter. Knowledge of objects has constituents of two kinds, intuitions and concepts. The former are contributed by the faculty of sensibility (at least at first ap-proximation). But they can't be identified with sensations or sense-data: intuitions are of objects in space and time, and space and time are a priori contributions of the human mind. Intuition gives knowledge its particular reference, but knowledge is in the end propositional, and something must be predicated of objects. Concepts also have both em-pirical and a priori dimensions. Any objective knowledge at least subjects its objects to the categories, a priori contributions of the understanding (the faculty of thought). The categories are "concepts of objects in gen-eral"; referring our knowledge to objects means applying this abstract and a priori system of concepts.

To return to Gödel, after saying that intuition doesn't have to be "conceived of as . . . giving an immediate knowledge of the objects concerned" he says that "as in the case of physical experience we *form* our ideas of these objects on the basis of something else which *is* im-mediately given." It is clear that Gödel intends to say that in the case of physical experience something other than sensations is "immediately given." Here I think he doesn't mean what most analytic philosophers of today (and also, it should be noted, Edmund Husserl) would say, that in some sense real *objects* are immediately given (to the extent that it is appropriate at all to talk of the "given"). The picture resembles Kant's, for whom knowledge of objects has as "components" a priori intuition and concepts. It is, to be sure, un-Kantian to think of pure concepts as given, immediately or otherwise. But Gödel's picture seems clearly to be that our conceptions of physical objects have to be con-structed from elements, call them primitives, that are given, and that some of them (whether or not they are much like Kant's categories) must be abstract and conceptual.

Gödel says, "Evidently the 'given' underlying mathematics is related to the abstract elements contained in our empirical ideas." But the only elaboration of this statement is the remark (note 40) that the concept of set, like Kant's categories, has as its function "'synthesis', i.e. the generating of unities out of manifolds."

Anyway, the general idea is that at the foundation of our conceptions of the physical world *and* of mathematics are certain "abstract elements" which appear to be primitive concepts. So far Gödel is in very rough agreement with Kant. What he mysteriously calls "another kind of relationship between ourselves and reality" (than the causal, manifested in the action of bodies on our sense organs) either consists of, or would account for, the fact that these elements represent reality objectively. They are not "purely subjective, as Kant asserted." Gödel does not offer an interpretation of Kant's transcendental idealism, but it is pretty clear he means to reject it. But in talking of primitive concepts as given and as not being subjective in Kant's sense, whatever that is, Gödel may be following the inspiration of Leibniz.

We should not forget that the concept of intuition is not the basis for Gödel's entire story about mathematical knowledge, since he holds that mathematical axioms can have an a posteriori justification through their consequences. He does not do very much to bring this and the more direct evidence of set-theoretic axioms together; it's as if there were two independent kinds of reasons for which one might accept them. A more holistic view seems to do more justice to the facts and seems even to underlie Gödel's actual argument about the continuum problem. Then one would contrast the more ground-level intuitions (logical inference and elementary arithmetic) with the more theoretical ones. There is a process of mutual adjustment of these. In mathematical practice there are also many "middle-level" intuitions, persuasive propositions about how things should turn out that no one would claim to be evident in themselves or would seriously propose as axioms for a fundamental theory such as set theory.

One can see where Gödel's conception is perhaps eccentric, or at least controversial, by comparing it with another account of rational justification, suggested by John Rawls's views concerning moral and political theories. On this account what would be most properly called "intuitions" are what he calls our "considered judgments" at lower levels, in the moral and political case concrete moral judgments, in particular concerning the justice of social arrangements.[49] Then one constructs theories. Theories may have intrinsic plausibility in their own right and may be defended on theoretical grounds against rivals. But an ineliminable part of their justification is that they yield our considered judgments, or, more likely, a corrected version of them. (If a theory tells us that these judgments are wrong, it explains why they are wrong.) But the process of mutual adjustment of theory and concrete judgment is a dialectical one, which might go through a number of back-and-forth steps. Ideally at least, this process ends in Rawlsian "reflective equilibrium." This view is in two ways more nuanced than Gödel's. First, it allows for a distinction between the kind of intrinsic plausibility possessed by ground-level judgments and that of high-level theoretical principles, and the intrinsic plausibility of the latter is not thought of as analogous to perception. Gödel, in talking about set theory, describes both as instances of intuition and closely analogous to perception. Second, with respect to more theoretical principles, it makes clear that integral to their justification is *both* their intrinsic plausibility and their ability to yield consequences that square with low-level intuition. It may be that that was Gödel's underlying view, but it hardly receives emphasis when he talks about these matters.

If you have begun to think that ideas derived from Rawls offer the kind of theory of reason that the foundations of set theory require, a minimal further look at his writings, not to speak of the extensive controversial literature on them, should disabuse you. At least in later writings than TJ, Rawls makes clear that the procedure of reflective equilibrium should not be expected to yield a unique theory.[50] In his

---

[49] Rawls compares them to the intuitions of a native speaker concerning his language; he sees an analogy between their role and that of speakers' intuitions in a theory of grammar. See *A Theory of Justice* (hereafter TJ), p. 47. In later writings Rawls says that the considered judgments that are relevant are at all levels of generality ("The Independence of Moral Theory," p. 8; *Political Liberalism,* pp. 8, 28). Then the comparison with speakers' intuitions is less apt. The view sketched in the text (which still seems to me a reasonable interpretation of the position of TJ) does not take account of this aspect of Rawls's later view. In particular, I do not go into the question of the status on this [later] view of the distinction between principles and considered judgments (TJ, p. 20).

[50] See "The Independence of Moral Theory," p. 9. In TJ, p. 50, he raises the question whether reflective equilibrium is unique and declines to offer a definite answer. It appears that whether a unique equilibrium is attainable depends on the particular context of application of the method.

later writings, Rawls seems to hold that different "comprehensive doctrines" about morality might be developed so as to achieve reflective equilibrium.[51] The nearest analogue in the philosophy of mathematics would be general philosophical and methodological views about mathematics such as constructivism or some kind or other of platonism.

6. Concluding Remarks

In conclusion, let me return to Gödel's Platonism. I suggested at the beginning that the connection between it and Gödel's conception of mathematical intuition would prove not to be as intrinsic as might appear at first sight. One reason is clear: finitary mathematics, intuitionistic mathematics, and classical mathematics without the characteristic concept formations of set theory (say, what is predicative relative to the natural numbers) each have definite and coherent concepts, and Gödel does not deny intuition concerning these concepts. Indeed, part of his case for the indispensability of mathematical intuition is that attempts to reconstruct mathematics without it require taking some mathematics, at least finitary arithmetic, at face value and therefore appeal to intuition at that level. The position that finitism is the limit of what mathematical intuition underwrites may be blind concerning the obvious, and it closes the door to certain extremely natural forms of reflection (such as whatever convinces us that first-order arithmetic is consistent), but it is not logically incoherent.

A second kind of independence of the two views is that Gödel's epistemology of set theory involves not just recognizing the fact of intuition concerning the concepts and axioms, but giving credence to it. Of course he maintains that that is not just an arbitrary judgment. But he clearly admits that the concepts of higher set theory are not so clear that the claim of intuitions concerning the axioms to yield knowledge is as obvious and unquestionable as Descartes intended *intuitio* to be. Even though there is a difference here between Gödel and someone

---

[51] For an example see *Political Liberalism*, pp. 95–96. Such a possibility is already mentioned in TJ, p. 50. But Rawls also remarks that "the struggle for reflective equilibrium continues indefinitely, in this case as in all others" (*Political Liberalism*, p. 97), which counters the impression he sometimes gives that fully satisfying the demands of reason is a humanly attainable end.

who rejects set theory or who thinks of the axioms either hypothetically or formalistically, the difference can be overestimated.

There is a third point concerning Gödel's Platonism that should be made. Even if we grant Gödel everything he could wish for concerning the clarity of our intuitions concerning the objects of set theory, it is far from clear that he has a case for the transcendental realism concerning these objects that he seems to adhere to, as when he says that the concepts in a mathematical proposition "form an objective reality of their own, which we cannot create or change, but only perceive and describe" (Gibbs Lecture, III 320) and that "the set-theoretical concepts and theorems describe some well-determined reality, in which Cantor's conjecture must be either true or false" (WCCP 1964, II 260). The widespread impression that Gödel is not just affirming CH $\lor$ $\neg$CH, i.e., allowing the application of the law of the excluded middle here, seems to me correct. The view he is expressing is that even if our grasp of the concept of set is not sufficiently clear to decide CH, the concepts themselves form an independent order that, as it were, guides us in developing set theory.[52]

Such a view clearly goes beyond saying that mathematical intuition is intuition concerning *truth*. In Gödel's conception, it is also the unfolding of certain concepts, and tied to a certain kind of development of the concepts. (Intuition concerning inaccessible cardinals requires a prior understanding of lower-level set theory, say ZF.) It is far from clear that it necessarily contains within itself the means of resolving certain disputes. Mathematical intuition itself doesn't tell us that there *must* be a truth of the matter on questions that intuition and other means of arriving at knowledge do not decide. It also does not tell us that given a question such as the continuum problem, it must be possible to develop our intuitions in such a way that we will arrive at principles

---

[52] Gödel's position as expressed here is analogous to what Rawls calls rational intuitionism in moral theory (*Political Liberalism,* pp. 95–96), not surprisingly since Rawls has given Leibniz as an example of a rational intuitionist. (One cannot take for granted that Leibniz was a conceptual realist; see Mates, *The Philosophy of Leibniz,* ch. 10. Although I cannot justify this here, I believe that Leibniz offered a model for Gödel's position.)

Still, in actual argument Gödel sometimes steps back from this position or treats it as a working hypothesis. Although I think it was a conviction of his, I doubt that it is a piece of his philosophy that he claimed to have defended at all adequately.

sufficient for a solution, although Gödel's conviction appears to have been affirmative in both cases.[53]

Gödel would probably argue that unless they reflect an independent reality, we have no explanation of the convergence and the strength of the intuitions we have. It would require a lengthy exploration of foundational issues in set theory to decide whether this reply has any merit. I will only remark that it is prima facie an empirical question whether our intuitions in set theory do or do not have a high degree of convergence and strength, a question to be answered in part by investigating the actual development of set theory. To my not very expert eye, the claim I am here attributing to Gödel has not been at all decisively refuted, but the state of the subject leaves a lot of serious questions, in particular those surrounding the continuum problem which already occupied Gödel.[54]

---

[53] On the second point, see Wang, *From Mathematics to Philosophy* (hereafter FMP), pp. 324–325. (Wang states ["On Physicalism and Algorithmism," p. 119] that the passage cited [from p. 324, last line, through the end of the paragraph] was written by Gödel.) But Gödel nowhere claims that this belief itself is a deliverance of mathematical intuition.

[On Gödel's role in certain parts of FMP, see now §§2.2–2.3 of the introductory note to his correspondence with Wang in CW V.]

[54] Expanded version of the Retiring Presidential Address presented to the annual meeting of the Association for Symbolic Logic at the University of Florida, Gainesville, March 7, 1994. Material from this essay was also presented in lectures to the Fifteenth Annual Wittgenstein Symposium in Kirchberg am Wechsel, Austria, and at the University of Rochester and the University of California, Berkeley.

I wish to thank each of these audiences. I am especially indebted to David Braun, Hans Burkhardt, John Carriero, Erin Kelly, John Rawls, Gila Sher, Wilfried Sieg, Hao Wang, and the referee for their comments and assistance. Wang and my fellow editors of Gödel's works, Solomon Feferman, John Dawson, Warren Goldfarb, Robert Solovay, and Sieg have taught me much of what I know about Gödel's thought.

Unpublished writings of Gödel are quoted by kind permission of the Institute for Advanced Study. The last revisions of the essay were made when the author was a Fellow of the Center for Advanced Study in the Behavioral Sciences, whose support is gratefully acknowledged.

[The section titles were added in the present reprint; in addition, I have moved part of the "note added in proof" (now note 48) to attach it to the text it comments on, so that it is now note 17.]

Since this essay was written, some other writers have expressed their views on a question I discussed, namely the development of Gödel's views, particularly on questions related to "platonism." Before I wrote on the subject, Solomon Feferman and Martin Davis had called attention to passages in Gödel's writings that were difficult to square with Gödel's responses to the questionnaire sent by Burke D. Grandjean, that he had been a "conceptual realist" and a "mathematical realist" since 1925.[1]

I doubt that Gödel meant to say anything so specific as that he already held in 1925 the views that he expressed in the years 1943–1951, which according to my essay was the period in which he avowed his realistic views. It takes time for a philosophical view to develop, and the 19-year-old Gödel, gifted as he undoubtedly was, would have required time in which to develop his philosophy, but in the period from his student days to his emigration to the United States his energy was largely concentrated on logic and mathematics. In his response to Grandjean, Gödel may have tended to read back into his youth his views of much later times. But I don't think that is the whole story; he could have meant primarily that he had a general inclination toward realism about mathematics and toward a realistic stance in issues about universals. Although we do not really know, I think it likely that Gödel had such a general inclination and that this became conscious in his reaction to some of the discussions in the Vienna Circle. That he knew where it would lead, in particular with respect to his view about set theory, is not at all so likely.

---

[1] See *Collected Works* (hereafter CW), vol. 4, pp. 444, 447.

Martin Davis is the writer who has written at greatest length about the problems that must be faced in presenting a coherent view of Gödel's development with respect to issues about realism, "Platonism," or "objectivism."[2] Davis makes some important points, which, however, I see more as amplifying than as correcting the picture presented in my essay. I do have two objections to Davis's discussion, one historical and one more philosophical.

The historical objection is that Davis seems to accept without question the picture Gödel presents in his letters to Wang of 1967–1968 of the attitude toward metamathematics that prevailed when he was doing his early work. The picture as sketched in Gödel's letter of December 7, 1967 represents the view of the Hilbert school as dominant, and it rests on a strongly formalistic interpretation of what that view was. It is true that such an interpretation was widely held then and later, although Paul Bernays repudiated it in a paper of 1930,[3] and that the Hilbert school was dominant in the German-speaking world is no doubt also true. In Britain and America, the problem would rather have been to engage logicians (other than Emil Post) in metamathematical investigations at all.[4] But in Poland matters were quite different, and in 1930 Alfred Tarski published papers on metamathematics using the usual methods of classical mathematics.[5]

Furthermore, Hilbert used the term "metamathematics" to refer to what became proof theory. I do not know what conception he had of model theory, but I don't think it was an explicit and public view of

---

[2] See his review of Dawson and "What Did Gödel Believe?" I tend to prefer the term "realism," although "platonism" (capitalized or not) is more specific to mathematics or at least abstract entities. (On the latter term, see §3 of Essay 12 in this volume.) The term "objectivism" was preferred by Gödel in his conversations with Wang but is not used in his published writings or public lectures, except in his letter to Hao Wang of December 7, 1967, where he speaks of his "objectivistic conception of mathematics"; see *From Mathematics to Philosophy*, p. 9, or CW V 398.

[3] "Die Philosophie der Mathematik und die Hilbertsche Beweistheorie," II §4. The issue came up in the correspondence of Gödel and Bernays; see CW IV 132 and 138.

[4] Of course Gödel's visit to the Institute for Advanced Study in 1933–1934 did something to change that. Davis does cite a "formalist" remark of Church from 1936 (Review of Dawson, p. 121). Church could be an exception, since he spent time in Göttingen as a student.

[5] Remarks to much the same effect as these are made in my introduction to the Gödel-Wang correspondence; see CW V 382–383.

Hilbert and his school that finitary methods had to be used in *all* logical investigations. Bernays did not object to the non-constructive step in Gödel's completeness proof; see his letter to Gödel of November 24, 1930 (CW IV 78).

The more philosophical objection is that in questions about realism, one has to ask: realism about what? The importance, in discussing the question of realism, of distinguishing different domains of discourse has, in our own time, been beaten into the heads of philosophers by Michael Dummett, Hilary Putnam, and Crispin Wright. But distinctions of "platonism" for different domains of mathematics were already made for a mathematical audience in 1935 by Paul Bernays in "Sur le platonisme." Something lacking in Davis's interesting exposition is consideration of the idea that at a certain stage Gödel might have embraced some version of realism about certain levels of mathematics but not about higher levels.

What seems very clearly to have evolved between the early 1930s and the publication of "Russell's Mathematical Logic" and "What Is Cantor's Continuum Problem?" is Gödel's confidence in set theory. Something that is left open by his statements in and about the earlier period is: To what extent did he see impredicativity a problem for classical mathematics, say as formalized in second-order number theory? The context in which he expresses a worry about impredicativity, the lecture of 1933, is motivating the axioms of set theory, indeed axioms that he suggests would go beyond those of ZF. But critiques such as Weyl's (see Essay 2 in this volume) touch the basics of classical analysis. On the other hand, some of his general remarks suggest that he would have seen impredicativity as a problem for classical analysis.

It is noteworthy that just after his notorious remark that the axioms of set theory "necessarily presuppose a kind of Platonism, which cannot satisfy any critical mind, and which does not even produce the conviction that they are consistent" (CW III 50), Gödel says that it is extremely unlikely that they will prove inconsistent and then launches into a discussion of the prospects for proving this consistency by "unobjectionable methods," by some extension of Hilbert's finitism.

There is another relevant point about this part of the 1933 lecture that I failed to note in my essay.[6] That is that Gödel already makes

---

[6] There are remarks in Solomon Feferman's introductory note (CW III 39) which I should have found suggestive but did not.

some of the distinctions that play a role in his discussion in RML of Russell's vicious circle principle. Impredicative definitions are acceptable if one assumes that "the totality of all properties exists somehow independently of our knowledge and our definitions, but there is a vicious circle "if we regard the properties as *generated* by our definitions" (ibid.). Gödel curiously concludes from the fact that an infinite set "can only be given by a characteristic property belonging to its members" (CW III 49) that quantification over all sets of integers presupposes quantification over all properties of integers. He seems not to have regarded the distinction between sets and properties (in his later terminology, concepts) as fundamental in the way he did later.[7]

In the 1967 letter to Wang, Gödel says that he was willing to use the "highly transfinite concept of 'objective mathematical truth'" in the reflections leading to the incompleteness theorem. I don't think we need to suppose that he thought it a quite unproblematic notion, just that he regarded it as a prima facie meaningful notion with which one could reason informally. So much would certainly have been conceded by Bernays and probably by Hilbert, though it is probably incompatible with the strictly formalistic interpretation of Hilbert.

Somewhat speculatively, I would connect Gödel's procedure in this case with his insistence in later writings that mathematics has a "real content" that cannot be explained away, a view he specifically opposed to those he encountered in the Vienna Circle and in Carnap's writings.[8] Since as it stands it would not commit him to any view about the extent of truth in mathematics, there is no clear evidence that he did not already hold it in 1929–1931. It could be part of what he meant by saying that he was a "mathematical realist" already in 1925. In its general form, it falls short of any of the platonisms that Bernays explores, because it does not rule out intuitionism, as Gödel made clear. But Gödel did not give evidence of accepting the constructivist

---

[7] It is possible to hold that properties are "generated by our definitions" while sets are not, so that the properties of integers are an indefinitely extensible totality, while this is not the case for sets of integers. I don't believe that that was ever Gödel's view. But distinguishing the two issues probably did serve to make him more comfortable with impredicativity.

[8] This view or attitude of Gödel is close to what William Tait calls default realism; see *The Provenance of Pure Reason*, especially essays 3 and 4, as well as my discussion in Essay 12 of this volume.

critique of classical methods in general, whatever he may have thought about the specific issue of impredicativity.

In my essay I suggest that what gave Gödel greater confidence in set theory was *mathematical* experience, the success of taking all ordinals as given in working out the model of constructible sets and the definite results he obtained. A hint that this confidence was still incomplete in 1938 is Gödel's remark, cited in §2 above and also by Davis,[9] that $V = L$ "gives a natural completion to the axioms of set theory, insofar as it determines the vague notion of an arbitrary infinite set in a definite way." Juliette Kennedy and Mark van Atten argue that this should not be taken to mean that Gödel took $V = L$ to be true.[10] That would be incompatible with his claim made shortly afterward that $V = L$ is probably absolutely undecidable.

Where Davis is most interesting is in his comments on remarks of Gödel related to this claim. Gödel pointed out, as he already had very briefly in 1931, that the undecidable sentences generated by the incompleteness proof are not absolutely undecidable, but he now suggested that $V = L$ might be. He had already shown that his consistency result extended to inaccessible and Mahlo cardinals, and he conjectured that independence would also be proved.[11]

However, Gödel also mentions a lower-level statement that he conjectures is absolutely undecidable, that every real number is constructible, which is more striking because it belongs to second-order arithmetic. Davis points out that where he first introduced a statement like this, he had wrong its complexity in terms of quantifiers.[12]

In my essay I assumed that what Gödel thought were the available resources for deciding $V = L$ or the continuum hypothesis (CH) were

---

[9] "What Did Gödel Believe?," p. 200.

[10] "Gödel's Modernism," pp. 303–304.

[11] Of course stronger large cardinal axioms refute $V = L$. Dana Scott's result that the existence of a measurable cardinal implies $V \neq L$ appeared only in 1961. Few such cardinals had been proposed by the late 1930s, and they were not well understood.

[12] Review of Dawson, pp. 124–125; "What Did Gödel Believe?," pp. 202–203. The remark about absolute undecidability occurs in "Undecidable Diophantine Propositions," CW III 175; Gödel does not say definitely that "Every real number is constructible" is what he has in mind, but Davis thinks that very likely, in view of the use of that statement for the same purpose in his lecture at Brown University in 1940. The claim of the first text is that there is a Π^1_2 statement that is absolutely undecidable, which would contradict Shoenfield's (later) absoluteness theorem.

the axioms of ZFC plus large cardinal axioms, more or less on the model of strongly inaccessible and Mahlo cardinals. Peter Koellner draws a stronger conclusion, although he considers it a rational reconstruction, in the absence of sufficient textual evidence. He assumes that the only available justifications are "intrinsic justifications," that is spelling out the conception of the universe of sets, and that these "involve spelling out the idea that the universe of sets is 'absolutely infinite', an idea which in turn is made precise by means of reflection principles."[13] He then invokes his own study of higher-order reflection principles to conclude that all such principles, unless they are inconsistent, are compatible with $V = L$. He remarks, "Perhaps intrinsic justifications of a different nature can overcome these limitations and secure axioms that violate $V = L$. This has not happened to date."[14]

What does this information tell us about the question how far Gödel had gone on the path to the platonism he avowed in 1944–1951? It seems clear that his confidence in set theory had increased from where it was in 1933, but in 1938 he was still prepared to describe the notion of arbitrary infinite set as vague. But I don't think anything in the documents from 1938–1940 is incompatible with the claim of RML that sets "can be conceived as real objects" (CW II 128). The notion of a bald man is vague, but that does not lead us to doubt the real existence of bald men.

Still, his remarks of 1938–1940 would be difficult to reconcile with the idea that the axioms of set theory describe a "well-determined reality," which he implicitly endorses in WCCP 1947. That would imply that questions of set theory have determinate answers, whether or not we are able to discover them. At least in the context in which he made this remark, he was entertaining strategies by which CH might be decided. In fact, already in RML he remarked that "it seems likely that for deciding certain questions of abstract set theory and even for certain related questions of the theory of real numbers new axioms based on some hitherto unknown idea will be necessary" (CW II 121). Moreover it is there, as is well known, that Gödel first puts forth the idea

---

[13] "On the Question of Absolute Undecidability," p. 200.

[14] Ibid., p. 201. One might ask: Could not a reflection principle of this kind *imply* $V = L$ and thus decide it? Evidently the results that show that known large cardinals are compatible with both CH and ¬CH, and therefore with $V \neq L$, apply in this case.

that axioms in mathematics may acquire justification from the theory that results from them without being "evident in themselves" (ibid.). Since on the latter point Gödel finds precedent in Russell, one might speculate that it was thinking about Russell that led him to consider the possibility of axioms in mathematics that would have only "extrinsic" justification and to see the possibility that such axioms might decide questions such as those that in 1938–1940 he conjectured were absolutely undecidable.

A rather delicate issue that I did not discuss in my essay and is only treated rather briefly by Davis is Gödel's attitude toward the Hilbert program. The difficulty this issue poses for any view of Gödel's development is that he maintained an interest in proof theory and philosophical issues connected with it until very late in his career, long after he had avowed platonist views. However, one has to reckon with the fact that the research program that was pursued after the war, to which Gödel's *Dialectica* paper was a contribution, was the extended version of Hilbert's program, in which constructive methods more powerful than finitary were used, and proving consistency was less at center stage than it had been before the incompleteness theorems. It should not surprise us that Gödel's relation to the enterprise of constructive proof theory is a rather complicated matter, about which I have little to add to the discussion of Solomon Feferman.[15] Gaisi Takeuti, and following him Feferman, gives an important role in his work and thought to a sort of rivalry with Hilbert, a factor relatively independent of the development of Gödel's philosophical views.

Davis notes that in his informal lecture at Zilsel's in 1938, Gödel shows considerable engagement with the revised Hilbert program.[16] Since the first sketch of his theory of primitive recursive functionals of finite type is given there, and by 1941 it had reached a stage where he could present it in a lecture that was at least semi-public,[17] it is likely

---

[15] "Lieber Herr Bernays! Lieber Herr Gödel! Gödel on Finitism, Constructivity, and Hilbert's Program."

[16] "Vortrag bei Zilsel," CW III 86–113. There is a difference at least in tone between the remarks Davis quotes from Gödel's lecture draft and some rather dismissive remarks about Hilbert's program that he quotes from the lecture draft "The Modern Development" from late 1961 or later (III 378). It should be noted that the latter remarks clearly refer to the original Hilbert program.

[17] "In What Sense Is Intuitionistic Logic Constructive?," CW III 189–200.

that this theory is a product of that engagement with the project of re-
vising the Hilbertian approach to proof theory. But he continued in
later years not only to develop and refine the technical result but to
continue to struggle with the question of its foundational significance.
There is no contradiction in an avowed platonist's having a serious interest in constructive foundations; success in such a program would at
the very least offer a powerful argument that his constructivist opponent should accept. But more important is probably an observation of
Takeuti, that Gödel's canonical work combines the constructive and
the nonconstructive in an unexpected way.[18] He makes this clearest in
a remark on the work on the continuum hypothesis, where Gödel uses
the Russellian ramified hierarchy, which he regards as constructive in a
sense, and yet uses it in "an entirely nonconstructive way" by assuming
ordinals from set theory.[19]

Mark van Atten and Juliette Kennedy's "On the Philosophical Development of Kurt Gödel" is an extended and richly documented discussion about issues concerning the development of Gödel's philosophy. However, their focus is on the dissatisfaction Gödel apparently
felt with the state of his thought in the 1950s and on the motives that
led to his engagement with and advocacy of the philosophy of Husserl.
For that reason, their story begins at a point where Gödel has already
argued in public for Platonic realism and at least begun to develop his
conception of mathematical intuition. It is also concerned with Gödel's
general philosophical outlook and is thus less focused on his philosophy of mathematics. Thus it prompts less comment in this Postscript
than the writings discussed above.

Commenting on the obscure second paragraph of the remarks
about intuition in the supplement to WCCP 1964, I remarked that "the
passage presents new ideas, possibly derived from the study of Husserl
that Gödel began in 1959. But it is with Kant and perhaps Leibniz that
he seems to make a more direct connection" (p. 182). Van Atten and
Kennedy point to a remark about phenomenology in an earlier draft
that was left out in the published version and make the interesting suggestion that "Gödel felt safe enough to make, in print, the negative

---

[18] *Memoirs of a Proof Theorist*, pp. 6–7.
[19] See Gödel's letters to Hao Wang published in *From Mathematics to Philosophy*,
pp. 8–11, esp. p. 10, also CW V 385–389.

point about Kant, but not to make the positive point about Husserl" (op. cit., p. 466), although he was recommending certain of Husserl's writings to visitors to the Institute for Advanced Study. Their conjecture fits well with what we know about Gödel's temperament, but I think it likely that he also had not, to his own satisfaction, sufficiently integrated what he was deriving from Husserl with his earlier ideas.[20]

---

[20] I am grateful to Solomon Feferman for comments on an earlier draft.

PART II

CONTEMPORARIES

8

During the time since World War II, nominalism as a philosophical tendency or research program has been largely identified with what was inaugurated by Nelson Goodman in such works as *The Structure of Appearance*. What was definitive of nominalism for Goodman was the rejection of the assumption of classes in philosophical and logical construction. Quine joined the research program in one well-known joint paper with Goodman, "Steps toward a Constructive Nominalism,"[1] which inaugurated post-war nominalism in the philosophy of mathematics. The opening sentence of that paper, "We do not believe in abstract entities," could serve as the slogan for recent nominalism.

As is well known, if that is what is meant by nominalism with reference to Quine, his nominalism was short-lived, whether one focuses on a philosophical doctrine or on the research program. But the period of his engagement with nominalism in that sense, from toward the end of the 1930s through the 1940s, almost coincides with the period during which he developed his well-known views on ontology and crystallized the views that especially concern me here. So we will not be able to ignore nominalism in the now current sense.

---

[1] The slogan of this essay goes beyond Goodman's position in *Structure*. For him nominalism meant first of all rejecting classes. But he considers two oppositions, nominalism/platonism and particularism/realism. Realism in the latter sense allows qualia that by his own statement are universals. The system of *Structure* is nominalist but realist. However, Goodman certainly held the view that I call Quinean nominalism. Qualia might be described as qualities, but they are not denoted by predicates, or expressed by them in the sense in which attributes might be.

In most of his writings on the subject, what Quine calls nominalism is closer to what Goodman calls particularism, as Quine later commented in "Reply to Nelson Goodman," p. 162.

I will argue that there is another sense in which Quine is a nominalist or at least has a strong nominalist tendency, which appears in his earliest philosophical writings and was to all appearances held for the remainder of his career. It is there on the surface for all to see, but Quine did not use the N-word to describe it. It can be called nominalism about meaning and in particular about predication. It is certainly connected with his questioning of the notion of meaning but appears to have been articulated earlier. I believe that it has merits independently of Quine's more radical views concerning meaning, although more could be said than is said here about possible objections.

I

This section will trace through Quine's early writings the development of a view of predication and of the notions of proposition, property, and attribute that will be called Quinean nominalism. Its avowal is largely complete in "On What There Is" in 1948. Unlike full-blooded nominalism, it is a position that Quine never abandoned. The view can be expressed as two theses:

(1) General terms, predicates, and sentences do not name; in particular they do not name properties, attributes, or (in the case of sentences) propositions.

(2) In addition, understanding and using sentences and predicates does not involve some kind of apprehension of such entities as classes, attributes, concepts, or properties, or (in the case of sentences) propositions or states of affairs.

In Quine's earlier writings, (2) is not really addressed, because of their more or less logical focus. Quine may have held (1) at the time of his earliest writings, but as we shall see he was somewhat slow to avow it clearly.

Reserve toward the notion of proposition is in evidence in his first philosophical publication, "Ontological Remarks on the Propositional Calculus." It is worth noting that Quine was willing to use the adjective "ontological" in the title of this paper and again in one of 1939 to be mentioned shortly. He may already have had the attitude toward the word "ontology" that he expresses in a later discussion of Carnap:

I might say in passing, though it is no substantial point of disagreement, that Carnap does not much like my terminology
here. Now if he had a better use for this fine old word 'ontology',
I should be inclined to cast about for another word for my own
meaning. But the fact is, I believe, that he disapproves of my giving meaning to a word which belongs to traditional metaphysics
and should therefore be meaningless. . . . Meaningless words,
however, are precisely the words which I feel freest to specify
meanings for.[2]

In his "Ontological Remarks" of 1934, Quine argues that understanding propositional logic (or the "theory of deduction," as he also
calls it) as a theory about propositions, so that the letters p, q, . . .
range over propositions, amounts to treating sentences as names of
propositions. He makes it clear that in his view this is a quite optional
way of viewing the theory:

Without altering the theory of deduction internally, we can so
reconstrue it as to sweep away such fictive considerations; we
have merely to interpret the theory as a formal grammar for the
manipulation of sentences, and to abandon the view that sentences are names. Words occurring in a sentence may be regarded severally as denoting things, but the sentence as a whole
is to be taken as a verbal combination which, though presumably conveying some manner of intelligence (I write with delib-
erate vagueness at this point), yet does not have that particular
kind of meaning which consists in denoting or being a name of
something. (p. 267)

Although it is not as explicit as in later writings such as *Methods of
Logic*, Quine's reading of the letters of propositional logic as schematic
letters for sentences is essentially there in 1934, as Quine himself wrote
later.[3] At the end of the paper, Quine does suggest another reading in

---

[2] "On Carnap's Views on Ontology," p. 203.

[3] "Autobiography of W. V. Quine," p. 14. This paper may have been a rather
rapid response to his encounter with Tarski in 1933, when he must have learned
something of Tarski's work on truth, although it was only in 1935 that the *Wahrheitsbegriff* was published in a language Quine could read. Central to both the
non-propositional readings that Quine proposes is the idea that sentences can be

which the letters are true variables ranging over sentences. But that reading does not imply that sentences designate anything or commit him to propositions.

Quine does not address these issues again in print until 1939, as we shall see. But there are indications from unpublished sources that he held thesis (1) about general terms in the mid-1930s. Paolo Mancosu has found in Quine's papers some notes from 1935 in which he at least considers the Millian view that a general term such as 'cat' denotes each cat and not the class of cats or the property of being a cat.[4] A lecture on nominalism given to the Harvard Philosophy Club in October 1937, though in the end critical of nominalism, presupposes the same view. Quine even suggests that although typically nouns denote many objects, the convenience of proper nouns that denote one leads men unconsciously to "force all nouns into the pattern of proper nouns" by inventing a single object (a class or attribute) for them to denote (ms. p. 4).[5]

Central aspects of Quine's ontological views emerge in his lecture "A Logistical Approach to the Ontological Problem," delivered in September 1939 to the International Congress of the Unity of Science in Cambridge, Massachusetts. But this commences a period of Quine's publication that is rather tangled. Quine distributed a six-page abstract of the lecture. This was to be published in *Erkenntnis* but was in fact first published in the first edition of *The Ways of Paradox* in 1966, without any indication that it is an abstract.[6]

At the end of the year Quine published a paper "Designation and Existence," which he describes as constituting "the bulk" of the paper

---

treated as truth vehicles. However, Quine mentions Wittgenstein's *Tractatus* and reads it as identifying propositions with sentences. Although he does not mention it, Carnap's advocacy of the "formal mode of speech" is likely to have been a factor. Thus one can't be certain about Tarski's decisive influence.

[4] Mancosu, "Quine and Tarski on Nominalism," pp. 24–25.

[5] Mancosu also discusses this text; see ibid., pp. 26–27 for a sketch of Quine's criticism there of full-blooded nominalism. The manuscript is in the Quine Papers in Houghton Library, Ms Am 2587 box 40. (This is the same box as Mancosu refers to as MS Storage 299 box 11, no doubt correctly as of when he examined it. The designation and arrangement may well change further as cataloguing proceeds.)

[6] The opening footnote explains the non-appearance in *Erkenntnis*: The publication of the issue was prevented by the German invasion of the Netherlands.

he presented at the Congress.[7] This paper does not appear as such in Quine's collections but is instead combined with other material to form the essay "Logic and the Reification of Universals" in *From a Logical Point of View* (hereafter FLPV). The latter essay, however, mines two additional published papers, "On Universals" and "Semantics and Abstract Objects."

A large part of Quine's view of ontological commitment is already in the abstract, including the slogan "To be is to be the value of a variable." The main idea is simply that whether an expression names something or is "syncategorematic" turns on whether the place where it occurs is subject to existential generalization. In his original example, "Pebbles have roundness," 'roundness' names something just in case the sentence implies '$(\exists x)$(pebbles have x)'. The mark of a name is this connection with quantification. He considers the possibility that the "entities" involved might be fictions and admits this if the quantifiers can be eliminated by suitable definition. That can be done in one case he cites, the addition of "propositional" quantifiers to truth-functional logic, as was done in the inter-war Polish school.

One could certainly infer from the abstract that already in 1939 Quine held that predicates as such do not designate anything. It would be an inference; he does not discuss that case explicitly, even to discuss the relation between "Pebbles are round" and "Pebbles have roundness." But the inference is buttressed by his closing claim that we can obtain a language adequate for science by adding to the first-order language of set theory a number of empirical predicates.

> For this entire language the only ontology required—the only range of values for the variables of quantification—consists of concrete individuals of some sort or other, plus all classes of such

---

[7] Note 1 of the paper makes this remark and also refers to the abstract. The verbal overlap with the abstract is slight, but the slogan "To be is to be a value of a variable" appears in both.

The lecture text is in the Quine Papers (in the box referred to in note 5). It is a 25-page typescript. "Designation and Existence" follows it up to the top of p. 24, with occasional small changes. Quine's statement about the relation of the lecture and published paper is quite accurate. Mancosu, who had earlier examined the typescript, is somewhat misleading in describing "Designation and Existence" as the "first part" of the lecture ("Quine and Tarski on Nominalism," p. 30). The remainder, as Mancosu says, concerns the concrete-abstract distinction; see ibid., p. 31.

entities, plus all classes formed from the thus supplemented totality of entities, and so on.[8]

The inference is further reinforced by "Designation and Existence," where Quine emphasizes that being meaningful is to be distinguished from designating anything: "The understanding of a term does not imply a designatum; it precedes knowledge of whether or not the term has a designatum" (p. 703).

Moreover, Quine consistently uses the notation of first-order logic and regards names as the substituends of variables (p. 707). Regarding abstracta, he is clearly relying on a distinction between general terms and abstract singular terms, even though he does use common nouns as terms for properties in some examples, e.g., *horse* (p. 702), but he regards the statement 'There is such a thing as *horse*' as a singular existence statement. But Quine is not yet as explicit as one would wish on the question whether general terms or predicates designate attributes or possibly classes.[9]

Nonetheless a technical basis for his answering this question in the negative is clearly expressed by 1941. That is his conception of logical schema, in which sentence letters and predicate letters *stand in place* of sentences and predicates but, unlike true variables, do not presuppose that what they stand in place of designates anything. What was implicit in his "Ontological Remarks" is clearly laid out in §5 of "Whitehead and the Rise of Modern Logic," where he describes schematic sentence letters as "dummies" for statements and other letters as dummies for predicates (pp. 18–19).[10] Quine proceeds in essentially the same way in the first edition of *Elementary Logic,* although the terminology used

---

[8] Ibid., p. 201. Curiously, the formulation suggests the "iterative conception of set," which is hardly mentioned in Quine's writing about set theory. [It now seems to me more likely that Quine was thinking of a Russellian simple theory of types.]

[9] However, in addition to the evidence from unpublished sources that Quine held thesis (1) some time before, we note the report of Mancosu that, according to notes by Carnap, in a lecture of December 1940 Quine said that predicates are "syncategorematic, for there are no variables ranging over them." See "Harvard 1940–41," p. 338.

[10] Quine proposes that 'p', 'q', ... and 'φ', 'ψ', ... function in much of their use in *Principia* as schematic letters for sentences and predicates (which he then called matrices) respectively, but this ceases to be so when the latter letters are allowed to occur in quantifiers. He says that this step "embodies the germ of a Platonic ontology of universals. This detail is properly inessential to quantification theory itself, but in some form it is essential to any foundation of classical mathematics" (p. 18).

there did not survive in later writings. At the end of the war, matters are laid out somewhat more systematically in §1 of "On the Logic of Quantification."[11]

On the more philosophical side, it may be that after his return from war service he received a push from Nelson Goodman to be more explicit. He writes that the stimulation for the paper "On Universals" of 1947 "came largely from discussions with him" (p. 74 n.1).

There he says that the platonist (who accepts universals) "is likely to regard the word 'man' as *naming* a universal, the class of men or the property of being a man" (p. 74).[12] He goes on to say that this is not essential to the platonist's position:

> Independently of any question of naming, the platonist feels that our ability to understand general words, and to recognize resemblances between concrete objects, would be inexplicable unless there were universals as objects of apprehension. (Ibid.)

Quine does not directly face at this point the question of the truth of this view; he turns to the question what forms of discourse explicitly presuppose universals and quickly makes the appeal to quantification that is familiar to readers of his better-known writings. But probably he has sketched here a view that he rejected and continued to reject for the rest of his career. It is his first published approach to thesis (2).[13] But we will see that he can't quite get away from the issue of naming.

Obviously, rejection of this "platonist" claim is necessary for any kind of nominalism. In "Steps toward a Constructive Nominalism," immediately after renouncing abstract entities, the authors write:

---

[11] Quine continues to use the term "matrix" and introduces the rather technical use of the term "predicate" that he uses to overcome the difficulties in a correct formulation of the substitution rule for predicate letters; see *Methods of Logic*, §23 and §25 of 1st or 2nd ed. In *Methods*, the term "matrix" has disappeared; its work is done by the now familiar "open sentence."

[12] In writings of this period Quine treats classes as universals, so that they are not limited to properties, attributes, or the like. This usage of Quine, in essentials echoed by Goodman, is probably responsible for the widespread (and in my view deplorable) tendency to view any viewpoint that admits abstract entities as platonistic.

[13] Evidence for my reading of the passage in "On Universals" is that he expresses (2) pretty clearly in §v of the 1946 lecture "Nominalism" discussed in §III below.

We shall not forego all use of predicates and other words that are often taken to name abstract objects. We may still write 'x is a dog,' or 'x is between y and z," for here 'is a dog' and 'is between . . . and' can be construed as syncategorematic: significant in context but naming nothing. (p. 105)

Quine's view that the use of predicates does not unavoidably involve us in commitment to properties or attributes emerges clearly in "On What There Is." He argues against his fictitious McX, who maintains that red houses, red roses, and red sunsets must have something in common, which is what he calls redness. McX is willing to accept the denial that 'red' in such uses *names* redness, and he retreats to the position that it at least has a meaning and that this meaning is a universal. Quine replies:

For McX, this is an unusually penetrating speech; and the only way I know to counter it is by refusing to admit meanings. However, I feel no reluctance toward refusing to admit meanings, for I do not thereby deny that words and statements are meaningful. (p. 11)

Synonymy, which he does not question in this context, would allow meanings in by the back door, as equivalence classes of synonymous expressions. But then they would not be "special and irreducible intermediary entities" that are supposed to have "explanatory value" (p. 12). Quine expresses the same point a little more emphatically in 1951:

A felt need for meant entities may derive from an earlier failure to appreciate that meaning and naming are distinct. Once the theory of meaning is sharply distinguished from the theory of reference, it is a short step to recognizing as the business of the theory of meaning simply the synonymy of expressions, the meaningfulness of expressions, and the analyticity or entailment of statements; meanings themselves, as obscure intermediary entities, may well be abandoned.[14]

---

[14] "Semantics and Abstract Objects," p. 91. As Quine points out (p. 96, n.1), exactly the same remark occurs in "Two Dogmas of Empiricism," p. 22. It is pp. 94–96 of "Semantics and Abstract Objects" that are taken up into "Logic and the Reification of Universals" (FLPV, pp. 114–117).

The point of view where general terms and predicates are explained as true or false of certain objects (so that 'elephant' or 'x is an elephant' is true of each elephant) underlies the treatment of quantificational logic in *Methods of Logic* (see especially §12 and §17).[15] Properties and attributes that might be meant in some sense by such expressions are hardly mentioned, although he does make the following remark in explaining why he uses 'is true of' rather than 'denotes':

> 'Denotes' is so current in the sense of 'designates' or 'names', that its use in connection with the word 'wicked' would cause readers to look beyond the wicked people to some unique entity, a quality of wickedness or a class of the wicked, as named object. The phrase 'is true of' is less open to misunderstanding; clearly 'wicked' is true not of the quality of wickedness, nor of the class of wicked persons, but of each wicked person individually. (p. 65)

The treatment of second-order logic in §38 is also relevant, but I will discuss that topic separately.

It should be clear that at least by 1948 Quine holds (1) and (2). The statement of (2) leaves open the question whether entities of the sort mentioned still have a use in theories of meaning, e.g., for natural language, although it does raise a question about what such theories are meant to do. At any rate, we have in Davidson's conception of a theory of meaning a paradigm of a theory that does not introduce such entities, unless the object language explicitly mentions them. Of course Davidson advances a controversial negative answer to the more particular question (explored by Quine himself) whether they are needed in an account of the meaning of propositional attitudes and similar constructions.

Thesis (1) might be thought to have less impact than is made out here on the ground that predicates might stand to entities such as attributes, properties, or concepts in a semantically significant relation other than naming. If the entities are to belong to the realm of meaning, Quine by the late 1940s regards them as unnecessary. As regards reference, he uses 'designate' and 'denote' rather freely with the same purport as 'name'; it is pretty clear that he does not distinguish different

---

[15] Of the first or second edition.

types of referential relation to objects.[16] This is a ground on which one might object to Quine's view.[17] However, someone objecting to Quine on this ground might well reject thesis (2) as well.

The two theses we have stated are independent of the more radical questioning of the notion of meaning that Quine engaged in. Still, one might well see his embrace of thesis (2) at least as part of a movement of thought that led to his questioning synonymy in "Two Dogmas" and then arguing for the indeterminacy of translation in *Word and Object*, in affect arguing that his earlier demand for a behavioral criterion of synonymy could not be met. There is no clear reason to deny this. I am not sure how far the formal independence of Quinean nominalism from skepticism about the notion of meaning can be carried. The Davidsonian idea of how to construct a theory of meaning suggests that it can be carried rather far.

II

Quine continued to hold for the remainder of his career the views called here Quinean nominalism. One piece of evidence for this, which will not be developed here, is that the revisions of *Methods of Logic* do not show any retreat from them. Something more interesting is to explore Quine's comments about second-order logic. The *locus classicus* of his view on that subject is the section "Set Theory in Sheep's Cloth-

---

[16] See for example the passage from *Methods of Logic* just quoted.

[17] Cf. Tyler Burge, "Predication and Truth," a review essay on Donald Davidson, *Truth and Predication*. In criticizing Davidson, Burge does not directly address Quine. He reads Davidson as holding that no semantic relation of predicates to entities like those we have considered is helpful with the problems about predication that concern him (p. 590). Burge maintains that we can say the predicates *indicate* properties and relations and regards this relation as essentially different from naming (p. 593; cf. p. 598 on Frege). Davidson gives great weight to regress arguments concerning predication that have been discussed since ancient times. Burge argues that carefully distinguishing semantic relations can avoid these regresses while still holding that predicates indicate properties and relations.

Although Burge's argument is not directed against Quine, his discussion brings out a noteworthy fact about Quine: Regress arguments do not play any substantial role in his discussions of predication. (That may help to explain why Davidson does not give Quine more credit for contributing to what he considers to be the solution to the problem of predication.)

ing" in chapter 5 of *Philosophy of Logic*.[18] What he says there has met with a lot of resistance. The criticism of it with greatest impact may be that of George Boolos in 1975.[19] And it is not surprising that Quine's argument has been viewed with disfavor by many in subsequent years, in view of the importance of second-order logic for several programs in the philosophy of mathematics, of which Crispin Wright's neo-Fregean program is the best known.

Second-order logic expands first-order logic by admitting variables that have the form of predicates (of one or more places), which in turn can be bound by quantifiers. From the perspective of contemporary mathematical logicians, the most natural interpretation of the language is that the second-order variables range over subsets of the domain of the first-order variables (or, in the many-place case, over sets of n-tuples of elements of the domain). That is the interpretation Quine focuses on, but he also considers an intensional version where sets are replaced by attributes. Historically, second-order logic has had other interpretations, and such have been considered in our own time. Frege, the inventor of second-order logic, took the second-order variables to range over concepts and relations (functions, possibly many-place, from objects to truth-values), which, since they are unsaturated, could not be objects, as are extensions, the nearest thing to sets that Frege admitted. The Fregean interpretation is still sometimes advanced today. Russell's interpretation in terms of propositional functions is probably closer to the Fregean than to the set-theoretic reading (or to Quine's reading of propositional functions as attributes). With restrictions to insure predicativity, substitutional readings are possible even by Quine's own lights. Boolos's later reading of the language of monadic second-order logic using the English plural has played a prominent role in recent literature.

What is most striking is that Quine's comments in this section are almost entirely focused on the *language* of second-order logic. He begins by considering an interpretation of the predicate letter 'F' of the logic of quantification as a variable ranging over attributes. This might arise through confusion between the schematic role of 'F' as "*standing in place* of an unspecified predicate" and seeing it as naming one (p. 66).

---

[18] The only changes in our section in the second edition are a few changes of wording and notation in the last paragraph (p. 68). But one is substantive; see note 29 below. Citations will work for either edition.
[19] "On Second-Order Logic."

Having got this far, he can round out his confusion by calling *F* an attribute. This attunes his usage to that of the unconfused but prodigal logician who embraces attributes with his eyes open. (Ibid.)

That predicates do not name is something on which he has insisted earlier in the book (pp. 27–28). He is clearly echoing views he had expressed in earlier writings. In *Word and Object,* the distinction of singular and general terms is treated as fundamental, and the simplest predication, symbolized as '*Fa*', combines a singular and a general term:

> The basic combination in which general and singular terms find their contrasting roles is that of predication: 'Mama is a woman', or schematically '*a* is an *F*' where '*a*' represents a singular term and '*F*' a general term. Predication joins a general term and a singular term to form a sentence that is true or false according as the general term is true or false of the object, if any, to which the singular term refers. (p. 96)[20]

A lot of the work of the surrounding text (§§19–20) is devoted to arguing that the singular-general distinction is fundamental for reference in a way which that of nouns, adjectives, and verbs (and even mass and count nouns) is not. The description of predication is in keeping with what is said in the earlier writings canvassed above and with the idea that what is open to quantification are the places occupied by *singular* terms.

To return to *Philosophy of Logic,* he says that "to put the predicate '*F*' in a quantifier is . . . to treat predicate positions suddenly as name positions, and hence to treat predicates as names of entities of some sort" (pp. 66–67). The first candidates for this role are attributes, to which he mentions familiar objections. He then says:

> But I deplore the use of predicate letters as quantified variables, even when the values are sets. Predicates have attributes as their "intensions" or meanings (or would if there were attributes), and they have sets as their extensions; but they are names of neither. Variables eligible for quantification therefore do not belong in predicate positions. They belong in name positions. (p. 67)

---

[20] He remarks later on the page that the notation '*Fa*' has the advantage of being neutral between the cases where '*F*' represents a common noun, a verb, or an adjective.

Quine had said virtually the same thing in 1939:

> It thus appears suitable to describe *names* simply as those constant expressions which replace variables and are replaced by variables according to the usual laws of quantification. Other meaningful expressions (other expressions capable of occurring in statements) are *syncategorematic*.[21]

A little later in *Philosophy of Logic* he writes, "If someone wants to admit sets as values of quantifiable variables, let him write '$x \in y$'; or, if he prefers distinctive variables for sets, '$x \in \alpha$' " (p. 67).

This remark is of interest because the latter is exactly what Quine did in §38 of *Methods of Logic*. One could describe that section as a brief discussion of second-order logic, interpreted as the theory of the subclasses of a given domain, but with 'Fx', 'Gx' with quantifiable 'F' and 'G' replaced by '$x \in \alpha$', '$x \in \beta$', etc. He criticizes the former notation on lines similar to what we have discussed.[22]

Quine's conception of naming and predication could well be thought to have a first-order bias. It is this that Boolos protests against in the early pages of "On Second-Order Logic" (pp. 38–39 of reprint). One might put the matter this way: Why should not the relation of a predicate to its extension, even though not a relation of naming, serve as the basis for regarding predicate places as open to quantification? To be sure, the extension may not always exist, for example when first-order quantifiers do not range over a set. But singular terms may also fail to have objects as their reference.

Quine would probably reply that the meaning would still be expressed more perspicuously in the language he prefers. Boolos offers some other arguments, but I will not pursue them here. My own writings have been critical of many applications of and claims for second-order logic in the philosophy of mathematics but have still not generally followed

---

[21] "A Logistical Approach," p. 199. It is not said there, although it is clearly said later, that the only expressions that are names are singular terms. But that is at least suggested in "Designation and Existence"; see above. Quine seems never to have regarded the distinction between names and other singular terms as fundamental, just as he does not use a distinction between the verbs 'name', 'designate', and 'denote'.

[22] This section is essentially unchanged in the second and third editions of *Methods*, but in the fourth edition the two-sorted language has given way to a one-sorted language. Quine notes how to translate a two-sorted language into a one-sorted one in *Philosophy of Logic*, p. 25.

Quine's linguistic recommendation. The reason is that the ambiguity of the usual second-order language can be fruitful as well as confusing: the language is of interest because of the variety of readings it might have, some of which were noted above. Quine seems to have a positive blind spot for Frege's idea of concepts and functions as "unsaturated." This very likely goes back to his reading of Russell's propositional functions as attributes, already in the essay on Whitehead in 1941. In a late passage he remarks,

> The term 'propositional function' was adapted from Frege, whose functions might indeed be seen as fictitious designata of relative or 'such that' clauses. I say 'fictitious' because a general term does not designate; and this accords with Frege's characterization of his so-called functions as *ungesättigt,* meaning in a way that there were not really any such things.[23]

However, the thesis that second-order logic on the most usual reading is "set theory in sheep's clothing" is supported by other arguments, of which powerful ones have been offered recently by Jouko Väänänen and Peter Koellner.[24]

Two things should be noted that do not arise in the debates about second-order logic known to me. First, Quine's section "Set Theory in Sheep's Clothing" is followed by one entitled "Logic in Wolf's Clothing," which discusses a restricted use of what look like class abstracts without their places being open to quantification, so that the resulting language is a conservative extension of the first-order. Quine had made considerable use of this device in his book *Set Theory and Its Logic.* In fact, the ideas behind it are also expressed in much earlier writings. Thus, at the end of §38 of *Methods,* Quine observes that by itself the Boolean algebra of classes "does no more than reproduce in another form the content of uniform quantification theory" and that it could be formulated using schematic predicate letters.[25] Quine mentions this

---

[23] "The Variable," p. 277. Thanks to Joshua Schwartz for calling this passage to my attention.

[24] Väänänen, "Second-Qrder Logic and the Foundations of Mathematics"; Koellner, "Strong Logics of First and Second Order."

[25] 1st ed., p. 230. By uniform quantification theory he means the fragment of monadic quantification theory that admits only a single variable x and no nesting of quantifiers. He seems to envisage the language of the algebra of classes as having variables, symbols for union, intersection, and complementation, a term for the

idea in "On the Logic of Quantification" and even remarks that we could "develop these portions of logic as a 'virtual' theory of classes and relations which makes no real assumption of such entities."[26] But in fact Quine had already done this in §51 of the book he wrote and published in Portuguese, *O sentido da nova lógica.*[27]

Second, Quine deals with higher-order logic in a rather striking way in "On Universals." After a section on "innocent abstraction" based on identifying indiscernibles (thus closer to the neo-Fregean idea of abstraction than to the virtual abstraction Quine explores elsewhere), Quine discusses binding predicate letters. First he presents an axiomatization of first-order logic. Then he proposes a new rule:

R6. Allow the predicate letters all privileges of the variables '*x*', '*y*', etc.

He then proceeds to derive Russell's paradox (p. 78).[28]

It is easy to see from Quine's text how this comes about. By allowing the predicate letters "all privileges" he means not only that they can be bound by quantifiers, but also that they can be substituted for the individual variables. Then one can derive a version of universal comprehension, expressed in this language as $\exists F \forall H(FH \leftrightarrow GH)$, and he takes **R6** and a rule of substitution to license substitution of $\neg HH$ for G.

Quine has famously said in a number of places that existing set theories result from artificial devices introduced to avoid the paradoxes. It appears that in 1947 at least he thought this even of the distinction of individual and second-order variables in second-order logic.[29]

---

empty set (and by complementation of the "universe"), and identity. The formulae would be identities of terms.

[26] Note 3, *Selected Logical Papers* (hereafter SLP), p. 183. Quine refers to *O sentido* (see below).

[27] [Readers for whom Quine is a figure of another era are reminded that this book is not a translation but was written in Portuguese by Quine.]

[28] The same move is presented in condensed form in "Logic and the Reification of Universals," FLPV, pp. 120–121. I could not make sense of that passage without referring back to its source, "On Universals."

[29] This discussion may be echoed at the end of "Set Theory in Sheep's Clothing" in *Philosophy of Logic* (p. 68 in each edition). He notes (in the *first* edition) that the hypothesis that '*F*' has an extension "falls dangerously out of sight in the so-called higher-order predicate calculus," because it takes the form '$(\exists G)(x)(Gx \equiv Fx)$' "and thus evidently follows from the genuinely logical triviality '$(x)(Fx \equiv Fx)$'." Then he writes, "Set theory's staggering existential assumptions are cunningly hidden now

III

Our study of the emergence of Quinean nominalism shows that it developed in connection with the exploration of nominalism in the more full-blooded sense. That is no doubt the reason why Quine did not disentangle the aspect of nominalism that he retained from what he gave up. Quine's publications during this period, with the one exception of "Steps toward a Constructive Nominalism," emphasize the question what is nominalism? or, looking from the other direction, what locutions or theories commit one to universals or abstract entities. He does not ever say that these categories are coextensive, but he also does not consider possible counterexamples such as "abstract particulars," or as they are nowadays called, tropes. It may seem odd to us to describe classes or sets as universals, but that was in keeping with his conception of classes as extensions of predicates.

Why did Quine pursue the idea of nominalism during this period? We should not be surprised that he had a general interest in the possibility of developing nominalistic views successfully; after all, through almost his whole career he sought ontological economy. In this particular period he certainly had a stimulus in this direction from Tarski. As is well known, Carnap and Tarski were at Harvard in 1940–1941, and there was a long series of discussions, sometimes involving others, in particular Goodman. The discussions are mentioned in autobiographies by both Carnap and Quine.[30] It is well known that Tarski and Quine pressed Carnap about the analytic-synthetic distinction. Carnap, however, makes the following suggestive remark:

> In other problems we came to a closer agreement. I had many private conversations with Tarski and Quine, most of them on the construction of a language of science on a finitistic basis.[31]

---

in the tacit shift from schematic predicate letter to quantifiable set variable." But in the second edition the word "dangerously" is dropped from the first quotation, and the second quotation above is changed to: "There is no actual risk of paradox so long as the ranges of values of 'x' and 'G' are kept apart, but still a fair bit of set theory has slipped in unheralded."

[30] Carnap, "Intellectual Autobiography," pp. 35–36; Quine, *The Time of My Life*, p. 150.

[31] Carnap, op. cit., p. 36.

Carnap's extensive notes on the discussions of that year are studied in Mancosu, "Harvard 1940–41."[32] Clearly "finitistic" in the above remark was meant to include being nominalistic.[33] The views on ontological commitment that Quine had already formed led him to be quite rigorous in understanding this. There seems to have been disagreement between Carnap and Tarski and Quine about finiteness. Quine advocated assuming a definite, though unknown, bound on the number of physical particles in the universe. Carnap and Tarski tended to advocate assumption of finiteness but no definite bound, and at one point Carnap wrote:

> I do not have the intuitive [instinctive] rejection of the notion of possibility as Tarski and Quine do. To me the possibility of always proceeding *(immer Weiterschreitens)* seems the foundation of number theory. Thus, potential but not actual infinity (Tarski and Quine say: they do not understand this distinction).[34]

It may not be easy to reconcile this conception of "the possibility of always proceeding" with strict nominalism; at any rate a reliance on a modal notion was not congenial to Quine at any time.[35] Presumably because of Tarski's and Quine's rejection of it, further discussion of a "finite" arithmetic concentrated on seeing what one could do without the assumption of infinitely many concrete objects, evidently also without the assumption of infinite space, time, or space-time.

Although he seemed not to pursue it until after the war, during a time when he had regular contact with Goodman, Quine seemed to believe at least up to the writing of "Steps toward a Constructive Nominalism" in 1947 that a nominalistic construction of mathematics which would be adequate for science might be possible.

Although Mancosu's study of Carnap's notes confirms the influence of Tarski, one would like also to explain Quine's engagement with nominalism more from his own internal development. One has to be a

---

[32] A full transcription of the notes appears in Gregory Frost-Arnold, "Carnap, Tarski, and Quine's Year Together."

[33] As Carnap stated, op. cit., p. 79.

[34] Mancosu, "Harvard 1940–41," p. 344, his translation. Carnap seems to have been influenced by Hilbert's remarks about finitist arithmetic; this no doubt goes back to the construction of Language I in *Logical Syntax*.

[35] With respect to the project of "Steps," Quine was later quite emphatic on this point; see "Reply to Charles Parsons," p. 397.

little speculative about this. It is clear that views about ontological commitment that Quine developed in the late 1930s made the question of universals (as he then put it) and the issue between nominalism and "platonism" a real one, contrary to the positivist view that it is a metaphysical pseudo-question. And at a general level, his preference as an empiricist would be for nominalism, even apart from his sometimes extreme interest in ontological economy. Another reason was his diagnosis of the paradoxes, according to which the only intuitive or common-sense principle about the existence of classes is the universal comprehension schema that leads quickly to Russell's paradox. He expressed this view in print already in 1941:

> But a striking circumstance is that none of these proposals, type theory included, has an intuitive foundation. Common sense is bankrupt, for it wound up in contradiction. Deprived of his tradition, the logician has had to resort to mythmaking.[36]

We have seen that in 1947 Quine regarded as artificial even the distinction of types embodied in second-order logic. As regards nominalism, on January 5, 1943, Quine wrote to Carnap, reflecting on the discussions of 1940–1941,

> Such an orientation [holding to "the full non-finitistic logic"] seems unsatisfactory as an end-point in philosophical analysis, given the hard-headed, anti-metaphysical philosophical temper which all of us share; and, if this were not enough, evidence against the common-sense admission of universals can be adduced also from the logical paradoxes. So here again we found ourselves envisaging a finitistic constitution system.[37]

On March 11, 1946, Quine gave a lecture, titled "Nominalism," to the Harvard Philosophy Colloquium. The text was rediscovered after his death by Douglas Quine and has been edited for publication by Paolo Mancosu.[38] It summarizes rather nicely the arguments for Quinean nominalism as well as the difficulty for full-blooded nominalism posed

---

[36] "Whitehead and the Rise of Modern Logic," SLP 27. I have not investigated whether Quine expressed this view earlier. For evidence that it was a reason for adopting nominalism, see §x of "Nominalism," discussed below.

[37] Quine and Carnap, *Dear Carnap, Dear Van*, p. 295.

[38] Thanks to Dagfinn Føllesdal for calling this text to my attention and for providing a copy.

by mathematics. The motives he gives for preferring nominalism are "extreme sensationalism" (if the particulars are to be sense-data or the like) or physicalism (if they are to be physical objects, as Quine apparently preferred), the paradoxes, and Gödel's theorem. The relevance of the latter is that since arithmetic is supposed to be a priori, "truth in arithmetic would seem, by normal epistemological standards, to consist of demonstrability" (§x), but no consistent system of arithmetic can make all truths demonstrable.

As for the execution of a nominalist program, the lecture is a progress report on the work of "Steps toward a Constructive Nominalism."[39] He concludes by saying:

> But we can't claim much; the issue of nominalism is still an open problem, though a clearer problem than it used to be to me. I feel sure that nominalism can be executed, but I don't know in which sense.

This is not quite so confident as it seems, since as Mancosu points out, Quine plays here on the ambiguity of 'execute' between the meaning of 'carry out' and the legal meaning of 'put to death'.[40]

Some of the work of "Steps" consists in showing how to express in nominalistic terms some relatively ordinary statements, such as "There are more cats than dogs" and "There are exactly one-third as many Canadians as Mexicans." Heavy use is made of mereology. But the authors then proceed to develop a nominalistic syntax. The construction refrains from assuming any infinity, following Quine's position in the discussions of 1940–1941 of a nominalist/finitist language and in the lecture of 1946.[41] It would seem that this finitism would be a fatal obstacle to the construction even of arithmetic. Goodman and Quine's way around this problem is by resorting to formalism. Inscriptions of formulae apparently expressing statements of classical mathematics can be determined as formulae and their consequence relations to other formulae stated. Goodman and Quine seem to have thought that would be adequate to the role of mathematics in science. The view is

---

[39] There are subtleties ignored here about the relation of the lecture and that paper. See Burgess, "Cats, Dogs, and So On."

[40] "Quine and Tarski," pp. 37–38.

[41] However, the paper does not assume that the number of individuals is finite; it merely refrains from assuming that it is infinite. This was the result of last-minute persuasion by Alonzo Church. See Mancosu, "Quine and Tarski," pp. 40–41.

only sketched at the end of the article; I do not know what they might have done to work it out further.

Quine moves away from nominalism as a doctrine in "On What There Is" and did not pursue the nominalist program further. I don't know of an explicit statement in his writings of why he gave up the position of "Steps" and the hopeful attitude expressed in the 1946 lecture. It appears that he was never fully convinced. Nominalistic syntax is a straightjacket; for example statements of the consistency of theories do not have their usual meaning.[42] Others probably know more about why he rejected the position. These remarks will be limited to what the Quine of a few years later might have said about the motives for nominalism offered in the 1946 lecture. There he talked of giving nominalism priority over mathematics. He may have come to think of that as giving an entry to first philosophy. Physicalism survived in his philosophy, but he soon ceased to think it required an ontology only of physical objects. Using the term 'object' in a general and abstract way, which makes the strictures of nominalism seem artificial, was already part of his practice a few years earlier.[43] In 1947 he was in the process of rejecting the a priori, if he had not done so already, so that the argument from Gödel's theorem would no longer move him. And although he continued to believe (though possibly with decreasing conviction in later years) that solutions to the paradoxes were artificial, he may well have thought that adopting formalism to save nominalism was a far more drastic cure than was needed. Furthermore, the resort to formalism strikes me as an un-Quinean move; this does not seem to be the philosopher for whom, in Burton Dreben's words, "Truth is truth and existence is existence."[44]

What may have been most influential is the abandonment of the analytic-synthetic distinction, and with it the idea of the a priori, and the holistic epistemology sketched in the last section of "Two Dogmas." At that point Quine evidently had the outline of an empiricist

---

[42] Evidently Quine saw this as a problem later and possibly at the time. He writes in his intellectual autobiography, "We settled for a formalistic account of mathematics, but still had the problem of making do with an inscriptional proof theory in a presumably finite universe" ("Autobiography," p. 26).

[43] He explains it briefly to Carnap in a letter of May 7, 1943, *Dear Carnap, Dear Van*, pp. 325–326.

[44] "Quine," p. 87.

epistemology of mathematics that makes it a perfectly meaningful part of science rather than a meaningless calculating device.[45]

The latter motive, and maybe some of the others, can be seen at work in a letter to J. H. Woodger of March 22, 1948, quoted by Mancosu.[46] There he sees the postulation of physical objects (including those of theoretical physics) and the postulation of abstract entities, at least those of mathematics, as more or less on a par. About the artificiality of solutions to the paradoxes he remarks:

> The platonistic acceptance of classes leads to Russell's paradox et al., and so has to be modified with artificial restrictions. But so does the acceptance of a physical ontology, in latter days, lead to strange results: the wave-corpuscle paradox and the indeterminacy.

The viewpoint of the letter is much the same as that of the closing remarks of "On What There Is," which treats the postulation of physical and mathematical objects as parallel, even though from certain points of view they are "myths" (pp. 18–19).[47]

Those remarks seem to signify Quine's abandonment of full-blooded nominalism. An element of instrumentalism did linger in his philosophy of mathematics, in that he continued to think that mathematics gets meaning and truth through its role in the system of science. But this essay has been more concerned to emphasize the survival of the more specific theses that constitute Quinean nominalism.[48]

---

[45] That this factor is given greater weight than in the Oslo lecture is in line with comments in the latter setting by Peter Hylton.

[46] "Quine and Tarski," p. 43.

[47] This passage confirms that by "the indeterminacy" in the letter he alludes to Heisenberg's uncertainty principle.

[48] This essay was presented to the conference Quine at 100 at Harvard University, October 25, 2008. Thanks to Warren Goldfarb for his initiative and work organizing that event and for inviting me to participate. An earlier version was presented at the Quine Centennial Symposium, University of Oslo, June 17, 2008. I am grateful to Dagfinn Føllesdal for the invitation prompting me to write it and to the audience in Oslo, especially Peter Hylton and Thomas Ricketts, for helpful comments. At a late stage I benefited from conversations with Tyler Burge.

9

GENETIC EXPLANATION IN

THE ROOTS OF REFERENCE

Quine's *Roots of Reference*[1] is a puzzling book for a number of reasons. It naturally attracts interest because after his proposal that epistemology should be "naturalized," this work represents an attempt actually to *do* naturalistic epistemology. But the focus of the work on reference means that Quine's attention is drawn away from general epistemological issues.[2] This is not incompatible with its belonging to naturalistic epistemology, but it indicates that the book is concerned with a rather restricted part of what might in principle count as naturalistic epistemology.

It is helpful in understanding Quine's enterprise to observe that the work is an example of "genetic" explanation in philosophy, where some important feature of our thought or knowledge is explained by a hypothetical story about how it came to be as it is. Giving a role to such an explanation is a distinctive feature of Quine's philosophical method. The genetic method is not used by Carnap or other direct antecedents of Quine. Frege, especially, would hardly have approved of it.[3] On the other hand, we can find examples of such accounts in the work of philosophers in other respects very different from Quine, such

---

[1] Page numbers in the text hereafter refer to this work.

[2] It is worth noting that in the frequently cited discussion of Quine's program of naturalistic epistemology by Barry Stroud, the references to *Roots* are all to the first few pages. See *The Significance of Philosophical Scepticism*, ch. 6.

[3] In discussion of this essay at the conference, Professor Lewis Hahn gave examples of the genetic method in pragmatist philosophers. Although Quine, in reply, expressed a sympathetic view of Dewey, James, and Peirce, he did not confirm the suggestion that this feature of his philosophical approach reflected the influence of the pragmatist tradition. A conjecture that might be worth exploring is that pragmatist influence did lead to a tempering in Quine's mind of the extreme antipsychologism of Frege and Carnap.

as Husserl. Although *Roots* is the most developed genetic account in Quine's writings, it is not the first. The clear precedent is the third chapter of *Word and Object*, "The Ontogenesis of Reference," where some of the same issues are discussed. I have not determined where this philosophical method first arises in the Quine corpus, but I think it is in the 1950s.

Genetic explanation in philosophy has puzzled me for a number of years. I have found myself attracted to it, but it is not at all clear to me what its nature is or whether anything can be said that is informative and general about the philosophical motives for engaging in it. It is this rather diffuse puzzlement that motivated the present essay. But my discussion will be limited to some particular points concerning Quine's enterprise in *Roots*. I hope to give some reasons for being dissatisfied with Quine's own description of the nature of his account and with his description of what his 'naturalism' amounts to. But I hope that some positive light will be shed on what Quine is doing in this work.

The essay is in three parts. In the first, I document the change in Quine's method as he proceeds from the earliest stages of language learning through the emergence of full-scale objective reference to his genetic account of set theory. The account loses its behavioristic character and becomes more common-sense and descriptive. In the second part, I pose a puzzle about the possibility of a scientific explanation of reference from Quine's point of view and argue that the answer to this problem implicitly offered by Quine does not comport with the presumption that the account is scientific. The third part concerns the account of set theory. Here Quine implicitly appeals to actual or hypothetical history, as have other genetic accounts in philosophy. I argue that some correction of Quine's story becomes necessary when this historical character is taken into account.

I

Quine conceives his enterprise as a kind of speculative psychology. Speaking of the general program of naturalistic epistemology, he writes:

> For we can fully grant the truth of natural science and still raise the question, within natural science, how it is that man works up his command of that science from the limited impingements that

are available to his sensory surfaces. This is a question of empirical psychology, but it may be pursued at one or more removes from the laboratory, one or another level of speculativity. (4)

Later remarks make clear that the same description is meant to apply to his account of scientific language, which indeed is part of the larger account of the acquisition of science (34–35, 37).[4] When Quine proceeds to set theory, he continues to talk of "psychogenesis" (105, 123).

As Quine goes on, his method gradually changes. The early sections are dominated by behavioristic psychology; the undertaking is, roughly, to give a behavioristic account of the initial stages of language learning.[5] As reference begins to emerge, however, the behavioristic apparatus begins to be supplemented, and in the later parts of the work it disappears altogether. In the end we should not be surprised by this. It is what Quine's indeterminacy thesis would lead us to expect. On the basis of a behavioristic story about how our full language is acquired, we should be able to give a behavioristic account of its meaning, which would give determinacy of translation. But Quine does not say wherein the method continues to be naturalistic. I think it is difficult for Quine to give an account of the method of the whole investigation. The difficulty will perhaps become clearest in §II below.

Quine remarks that as the sentences the language learner uses become more complex and less keyed to observation, "his learning process becomes much harder to picture and to conjecture about" (35–36). He evidently sees a retreat to mentalistic semantics as a temptation, since he then writes:

> It is proverbial, or used to be, that man in his study of nature falls back on the old-time religion to fill in where his scientific explanations leave off. It is at least equally true that man in his study of language falls back on the old-time mentalistic semantics to fill in where his scientific explanations leave off. Mental-

---

[4] Introducing his discussion of relative clauses he says, "Let us proceed, then, with our psychogenetic speculations" (92).

[5] I do not, however, wish to use the term "behaviorism" too strictly. Quine himself points out that his notion of perceptual similarity introduced in §4 is a "theoretical notion" (17); if behaviorism requires that the notions one uses be operationally defined in terms of behavior, Quine's method fails at the outset to be behavioristic. [On this general theme, see now Føllesdal, "Developments in Quine's Behaviorism."]

ism, supernaturalism, and other unwholesome cultures thrive in dark places. (36)

Clearly, however, he is expressing the intention to resist this temptation himself. And indeed, the account as it goes on does not make use of any philosophical theory of ideas or directly presuppose the determinacy of translation. What seems to me to happen is that the picture, more associated with Chomsky than with Quine, of the learner as theorist, is cashed in in terms of conditioning of responses in the account of the earlier phases of language learning but gradually takes on a life of its own. Eventually, the form of explanation used appears to be common-sense psychology, involving attributing propositional attitudes to the learner. I am not entirely sure how the method of the last part should be described; it could be that Quine does not intend to offer an explanation at all rather than something more like description. Thus I do not want to argue that Quine retreats to the old-time religion, but I don't think the account is by his lights strictly naturalistic either.

We can discern some gradual steps away from an account in terms of conditioning of responses. But we will find that the steps taken in part II of the book are quite moderate. But then there is an abrupt transition with part III, where the emergence of reference becomes definitive. I shall therefore have something to say about the question, discussed by previous commentators on *Roots*, whether objective reference is already present in a language of which the methods of part II offer an account.

A step is made in Quine's discussion of the learning of assent and dissent in §12, where the basis of learning to assent to queried observation sentences is "language-dependent" similarity:

The shared feature in which the similarity lies is perhaps an introspective sense of willingness to repeat the heard sentence: a sense of freedom from inhibition. In venturing to speculate thus on inward sense I relax my behaviorism, but not much: I speak only of an incipient drive toward overt behavior. (48)

A similar but countervailing "sense of inhibition" is the basis for the learning of dissent.

Standing sentences such as 'Snow is white' and 'A dog is an animal' pose an obvious problem for an account of learning, since one cannot sort out the occasions of which they are true from those of which they are false, one or the other class being empty. Thus Quine alludes to a "gulf" between occasion sentences and standing sentences and says that beginning to learn the latter puts us "well on the way from observation sentence to scientific theory" (68). Learning of sentences like the above examples occurs by transfer of conditioning from the occasion that prompts assent to the subject term/sentence, so that the child will, on being queried about 'Snow is white', assent as if he had been presented with snow and assented to 'white' (64–65). 'A dog is an animal' is more complex because the subject term is individuative, but the account generalizes. Though this mechanism is behavioristic enough, it brings a sort of original sin into language learning: "confusion of sign and object, use and mention" (68). Quine does not explain how this learning process might be connected with troublesome cases of such confusions. But one would expect it to lead to the child's from time to time saying of the object things appropriate to the word and vice versa. Quine's remark that science is the "redemption" of this "sin" indicates that he thinks that it is only at a much more sophisticated stage, where at least full objective reference is present, that the confusion thus introduced can be sorted out.

In these remarks Quine seems to disclaim one possible answer to the question how an investigation of how science actually came about can replace the epistemologist's attempt to justify science by reasoning from presumed evidence. Quine suggests that his account, being speculative, is rather of how it *might* have come about. The answer I have in mind is that the account represents the process as rational: it tells how a reasonable man, or a reasonable species, would by a long process of learning come out with something close to science as we know it. This conception of the enterprise would probably be circular from Quine's point of view, but in his account of the origin of standing predications he discerns an irrationality at the heart of the process.

In part II of the book Quine gives us an account in terms that involve very moderate departures from behaviorism of a language rich enough so that one might find in it something already counting as objective reference, though falling short of our full apparatus. Indeed, Strawson has urged that at this point objective reference is already present.[6] The language contains individuative terms like 'dog' with associated sameness relations 'same dog as', names such as 'Fido', at-

---

[6] "Reference and Its Roots."

tributive composition that first gives us complex observation sentences like 'yellow paper' but which, once predicational constructions are present, gives complex general terms. The language also contains universal categoricals like 'A dog is an animal'. Some other constructions yielding sentences of the form 'Det N is P' seem not to involve anything essentially new. It seems that once we have assent and dissent we have negation, and since one would assent to 'Some S is P' just when one dissents from 'Every S is not P', we would then have a language that contains all the categoricals of Aristotelian logic. What would make it fall short of full first-order logic is first of all the lack of a supply of general terms that is rich enough to simulate polyadic quantificational logic. Apparently this stage of language also lacks a general predicate or pseudo-predicate 'thing' and 'object' and the nonrelative identity that would go with it.

But it is a question how negation is allowed to occur in this restricted language. The fact that sentence negation is learned at an early stage might lead one to think that the terms entering into early universal categoricals will also have negations. Such negation, which I will call term negation, Quine sees as rather a by-product of the 'such that' idiom which appears much later (§25, see below), so that it appears in his story only when the learning of the apparatus of quantification is nearly complete.

To obtain the Aristotelian categoricals, negation of general terms needs to be admitted only in predicate position. To distinguish it from unrestricted term negation, I will use the designation 'copulative negation'. A restriction to copulative negation might be defended on common-sense grounds, such as that if α is a term that we have learned by confrontation with instances of it, we know what we are talking about when we say 'Every α is a β' but not when we say 'Every non-α is a β'. Also, because 'non-α' will in general not have been projected by the child's elders, he will not be in a position to see how a generalization about non-α's can be true.

The universal categorical construction together with copulative and sentence negation yields the Aristotelian categoricals. With unrestricted term negation as well as term conjunction, given that we have truth-functional composition of sentences, we have a language with the expressive power of what Quine in *Methods of Logic* called uniform quantification, with only monadic predicates and no nesting of quantifiers. I see no convincing reason why objective reference should not be attributed to the latter language, in accord with Strawson's view, and it is not clear that the difference between that language and the Aristotelian should mark the difference between the presence and absence of objective reference.

An argument to the contrary was offered by Gilbert Harman.[7] Harman describes as emerging at 'the stage at which truth functions enter into the attributive and universal categorical constructions' a language that is roughly equivalent to the uniform quantificational language. He argues that the language contains only finitely many atomic predicates and can therefore express only a finite number of nonequivalent sentences. The latter would be a convincing reason. But it seems reasonable to suppose that the child will have acquired a demonstrative construction and thus would be able on an indefinite variety of occasions to make statements of the form 'This is an α' or 'This α is a β'. Therefore, contrary to Harman's claim, it is not like the case of a finite listed ontology, which Quine says is 'no ontology'.[8]

Although Strawson seems to me right as against Harman if we assume, as both of them seem to do, that the kind of learning Quine describes in part II of *Roots* gives rise to a language with term negation, I see no principled reason why term negation or even copulative negation should be admitted at this stage, or why it should not.[9]

Essential to the expressive power of the full quantificational language is that a sentence containing a singular term even in embedded or multiple occurrences can be treated as a predicate true or false of what the singular term designates, applied to that object. The most widely used device to this effect in ordinary language is the relative clause, which then allows the formation of general terms. Thus, Quine points out, from 'I bought Fido from a man that found him' we have the relative clause 'that I bought from a man that found him', which allows formation of the general term 'dog that I bought from a man that found him'.

Quine bypasses the problem of the learning of the relative clause construction in its full generality, whose syntactic and logical compli-

---

[7] Review of *Roots*, pp. 395–396.

[8] *Theories and Things*, p. 7.

[9] Gareth Evans also relies on the presence of term negation in the language in one part of his argument against Quine's *gavagai* argument for the indeterminacy of translation. See "Identity and Predication," esp. pp. 350–352.

cations are well known, by concentrating on the 'such that' idiom characteristic of mathematicians' and logicians' English. From a sentence '. . . *a* . . .' we have a transformation leading to a sentence

$$a \text{ is an } x \text{ such that } \ldots x \ldots$$

or, more idiomatically,

$$a \text{ is such that } \ldots \text{he/she/it.} \ldots$$

This transformation is also available with other subject phrases. Thus

$$\ldots \text{every } \alpha \ldots$$

can be rendered as

$$\text{Every } \alpha \text{ is such that } \ldots \text{it} \ldots$$

though in both cases we have to be sure that cross-references are not disturbed.

By itself this transformation yields no increase in expressive power. But it creates general terms that can then be put into other places, in particular into categoricals, yielding sentences such as 'Everything that we salvaged from the wreck is in the shed' (93–94). Quine suggests that the 'such that' idiom would be learned by an equivalence transformation: "The mechanism of learning an equivalence transformation seems simple: the learner is merely brought to see, by abundant examples, the interchangeability of certain constructions" (94). Interchangeabilty *salva* what? Presumably assent and dissent on the part of others and eventually the learner himself.

But then this usage is extended by analogy. Clauses created by 'such that' are analogous to general terms, hence they can be used in positions appropriate to general terms, in particular in subject position in categoricals. It is not clear to the child or to us how far one can go on the basis of this analogy, as Quine illustrates by his amusing deduction of how the child might go "down the garden path and into the very jaws of Russell's paradox" (96). This threatened catastrophe, however, involves an extension of the 'such that' device to a substitution operator where the variable can take other than singular term places.

From his view that the 'such that' construction, which introduces the variable by a substitution transformation, contains the essentials, Quine concludes that variables of quantification are substitutional in their origin. But the universal categorical, the prototype of quantification, is not substitutional. When these two elements are put together, that is when clauses formed by 'such that' are introduced into categoricals, the variables go objectual. "It goes objectual with a vengeance" (100). This is, Quine writes, "an irreducible leap in language learning" (99).

As I remarked earlier, in Quine's account term negation arises with the 'such that' idiom. We can then quantify over a range that is not explicitly limited, since 'Everything is α' can be rendered as 'Everything that is not an α is an α'. This artificial derivation conceals a problem.

Mastery of unrestricted quantification, at least together with non-relative identity, can be described as possession of a highly general concept of object, which might be described by Kant's phrase 'the concept of an object in general'. The fact that *any* sentence of the form '. . . *a* . . .' is allowed to enter into the such–that transformation and then yield a general term that can be the subject term of a universal categorical is what enables us to generate unrestricted quantification. The same would be accomplished more simply just by treating 'thing' or 'object' as a general term by itself. But the former route is a more plausible one, because it provides by the way a point of origin for the word 'thing' or 'object' itself.

Quine regards this as an irreducible leap in language learning and forswears an account of its mechanism. And in most of the further discussion of the development of reference to abstract objects and set theory, Quine similarly describes steps of extensions by analogy of what one is prepared to do. In one case this is described as "learning various linguistic constructions by language-dependent similarities" (115), but there is no real attempt to show that similarity works in the way it does in the early part of the book.

There are fairly obvious reasons why this attempt is not made. The learner is now dealing with extensions of the language beyond what has assent and dissent conditions that he has any hope of mastering by the methods used at the earlier stages. He is on the true terrain of theory. The outside world does not offer a straightforward control on what he comes up with. He will, of course, find his elders assenting to many sentences of language thus extended and will be prompted by this to assent to them himself. There is some kind of control, because he will be conversing with others and adjusting his usage to that of others. At this level, however, in some ways others will adjust their usage to his, for example if the conversation concerns matters where there are facts that he knows and they do not. At the most advanced level of all, where our learner becomes a true theoretical innovator, there will be no comparable usage of his elders for him to adjust to and no previously given conditions determining assent or dissent.

II

A puzzle of a more philosophical kind concerns the very idea of the enterprise in which Quine is engaged. This is what the doctrine of the inscrutability of reference allows in the way of scientific investigation of the human referential apparatus. Put baldly and probably too crudely, if there is no fact of the matter as to what the expressions of a language refer to, how can there be a scientific explanation of their referring to what they do? What is it, in other words, that a naturalistic or scientific account of reference is intended to explain?

One way of looking at the puzzle is the following. Suppose we have given an explanation of the genesis of reference, or of reference to objects of certain given kinds such as bodies, events, numbers, and classes. Then the explanation seems to be threatened because the language that is the vehicle of this reference can be reinterpreted so as to change the ontology, while leaving the underlying hypotheses about verbal behavior undisturbed. An example would be a reinterpretation by proxy functions, which Quine uses to argue for the inscrutability of reference.[10] In this kind of case, where the reinterpretation interprets the language as talking about a structure isomorphic to what it was previously interpreted as talking about, the explanation will avoid the difficulty if it is couched in an invariant way, so that it is robust under isomorphisms of the structure that provides the domain of quantifiers, the extension of predicates, and the denotations of singular terms.

Do Quine's explanations have this kind of robustness? At the higher reaches, where one is dealing with the language of mathematics and theoretical science, one might well expect it. The matter is less clear for the beginning stages. Here we attribute our own reference to such objects as bodies to the child, on the ground that this makes psychological sense: the regularities of behavior on the basis of which we translate an alien language, or on the basis of which the child learns a first language, are first of all regularities involving bodies, because the simi-

---

[10] *Theories and Things,* p. 19. Quine's view might be called the 'structuralist view of objects in general'.

larity of different manifestations of a body is of the sort that "we are innately disposed to appreciate":

> Bodies, for the common man, are basically what there are; and even for the esoteric ontologist bodies are the point of departure. Man is a body-minded animal, among body-minded animals. Man and other animals are body-minded by natural selection; for body-mindedness has evident survival value in town and jungle. (54)

We can reinterpret our own language so that predicates true of bodies are instead true of, say, quadruples of real numbers. Leaving our own language alone, can we interpret the child along such lines and still preserve the psychological explanation? Putting quadruples of real numbers into the child's ontology is already a questionable step; given that the child has not acquired that much mathematics, it introduces gratuitous complexity into the theory by which we interpret him. Moreover, there is a question how we can then subsequently explain his acquisition of the mathematical concepts.

However this may be, Quine implicitly responds to our puzzle in a quite different way, in remarks he makes about the relevance or lack of it of the indeterminacy of translation to his enterprise. He writes:

> I shall speculate on the steps by which the child might progress from that primitive stage until we are satisfied by his easy communication with us that he has mastered our apparatus of reference. Thus I shall be concerned with our language, not with translation. . . . I spoke of translation from the foreigner's language just to make one point clear: to show that reference involves more than the simple ability to acknowledge a presence. Now that this point is made, I can forget about foreign languages and translation. (83)

Thus the investigation is, in Gilbert Harman's terms, an immanent one;[11] it is concerned with our own language. The problem, then, is how someone can come to refer in the way that *we* do. If our child uses the word 'rabbit' we take him to be referring to rabbits and we undertake to explain how it could have come about that he *could* be referring to rabbits.

Even if his explanations do have the kind of robustness under reinterpretation by proxy functions that the first answer to the puzzle

---

[11] "Immanent and Transcendent Approaches to the Theory of Meaning."

requires, it seems that Quine would have to appeal to something like this second answer when one considers that the language of the person whose capacity for reference is to be explained is subject to the deeper indeterminacy of translation.

I do see the following difficulty with this: The account is supposed to be at least potentially a scientific account, when filled out with more evidence and presumably more precise hypotheses. But then it is somewhat odd that its field of application is defined as it were in relation to ourselves. The theory is about speakers of *our* language. Of course it isn't really intended to be just about the speakers of our language. Quine would not, I think, think it just irrelevant if a speaker of Chinese considered some of these questions and came to the conclusion that some of the things that are said in the explanation are just wrong in application to speakers of Chinese. It's supposed to derive its generality of application in a way that any semantical investigation has to derive that generality: we apply it to the speaker of Chinese by using a manual of translation of the Chinese speaker's language into our own. But then it's not quite true that the indeterminacy of translation doesn't have any bearing on the matter. The indeterminacy of translation doesn't enter in for the investigation itself, but it does enter in, it seems to me, for the investigation to have a general application to human language, and it does seem to me that a scientific project should have an application, a field of application which is more a natural kind than the speakers with whom I or W. V. Quine can communicate in a fluent way. It isn't in the end limited in that way, but the way in which it gets beyond that limit is by using manuals of translation, with the resulting indeterminacy.

Professor Harman, I think, takes a similar view of semantics in general and compares a semantics that begins immanently and then gets its general application by translation to *Verstehen* and similar ideas about the methods of the human sciences. It seems to me that this is something about which a naturalistic epistemologist ought to be uncomfortable. That Quine's conception of a theory of language is really consistent with naturalism seems to me in question.

III

Concerning set theory, it's pretty clear that the first problem a genetic account faces is how the subject itself gets off the ground. Although

Quine continues to describe his concern as psychogenesis, the setting is no longer really that of a child learning from his elders or students learning from teachers. To be sure, some of the process by which the subject originated might be repeated in the learning of later generations. If by means of certain analogies the discoverer of some new set-theoretic concepts extends what he was able to do from where it stood before, then once this is part of the established corpus of set theory, it may be by getting students to see the same analogies that a teacher conveys it.

I don't take this observation to be a criticism of Quine; I think the genesis of the subject was part of his concern. But it does have a bearing on the question what empirical data we could now or eventually appeal to in checking our speculations, because it puts at our disposal an obvious array of empirical data, namely from the history of the subject.[12]

Quine builds into his genetic account of set theory two distinctive and controversial features of his philosophy of set theory, which in my view (developed elsewhere[13]) are linked with each other: the conception of sets as extensions, that is objects associated with *predicates* such that if two predicates are coextensive, then the objects thus associated with them are identical, and the view that the intuitively most natural set theory would be based on the (inconsistent) universal comprehension schema. In these respects Quine's view is Fregean.[14] Actual set theories are based on artificial devices to block the paradoxes while preserving as much as possible of this naive theory. Both views appear at the outset of the account in *Roots*, as is evident in §27, where Quine begins with the concept of attribute (object associated with a predicate with some intensional identity criterion or other). Criticizing the view advanced by his critics that "even before the paradoxes it was not usual to suppose there was a set, or class, for every membership condition," he writes:

---

[12] From a very different point of view, Edmund Husserl also proposed to illuminate science by a genetic investigation. In his conception, in contrast to Quine's, it is through and through historical, although of a quite different character from actual work in the history of science. See especially "Vom Ursprung der Geometrie."

[13] "Quine on the Philosophy of Mathematics," §VI.

[14] Frege argued that the notion of extension is the notion that contemporaries who talked of sets or systems were somewhat confusedly aiming at. See part I of my "Some Remarks on Frege's Conception of Extension." Frege's own interest in these concepts, however, was for the foundations of arithmetic and analysis, and he did not have a great interest in set theory as a branch of mathematics in its own right or as a tool for attacking other mathematical problems.

What is myopic about the view, in any event, is that it looks back only to the first systematic use in mathematics of the word 'set' or 'Menge', as if this were uncaused. For surely it is traditional to talk as if everything we say about an object assigned an attribute. It is evident nowadays, further, that this attitude toward attributes is involved in paradoxes just like those of set theory. And it should be evident that classes, or sets, are wanted simply as the extensional distillates of attributes. It is implausible that Cantor or anyone else would narrow this universe of classes for other than sophisticated reasons, either nominalistic scruples or fear of paradox. (102–103)

Both views have been criticized by a number of writers, in particular by me.[15] These criticisms are relevant to the question of the genesis of set theory, since a different conception of what a set is, and of the intuitive basis of set theory and the force of paradoxes, will call for a different genetic explanation. At this point the history of set theory is relevant. It is appealed to in criticisms of the two theses of Quine mentioned above. It also suggests a different genesis for the concept of set from that given in *Roots*.

To turn to the latter point, one finds in the history of discussion of the concept of set in addition to the notion of extension two other rough intuitive ideas about what a set is, which I will call the notion of collection and the notion of plurality. Thought of as a collection, a set is something that is constituted by objects that are its elements, perhaps even formed from them by some kind of activity, so that if one wanted to create a kind of genetic story about the conception of collection, one would start, as some writers on the subject do, with the activity of collecting, but obviously sets as collections have to be an extreme generalization of that. For the notion of sets as pluralities, one might appeal to the plural form in a language like English. Examples of pluralities would be the people in this room, the critics who, in a famous sentence, are said to admire only one another, the finite sets of natural numbers, and so on.[16]

---

[15] Ibid. The relevance of competing conceptions of set, in particular that of collection discussed below in the text, is further defended in essays 8 and 10 of *Mathematics in Philosophy*. [See now also *Mathematical Thought and Its Objects*, ch. 4.]

[16] In the history of the foundations of set theory, the notion of plurality occurs most clearly in Russell's *Principles of Mathematics*. Russell's concept of a "class as

One might expect Quine to reject all three of these conceptions because of his adherence to a structuralist view of mathematical objects. His doctrine of ontological relativity implies that different types of mathematical object are "known only by their laws." Unlike some who have written about mathematical objects in this vein, he makes clear that sets are no exception: "Sets in turn are known only by their laws, the laws of set theory."[17] But it seems that Quine can hold this view and still hold to the conception of sets as extensions, because extensions are themselves hardly characterized in more than structural terms.

A virtue of the view that "sets are known only by their laws" is that one can then see the genesis of the concept of set as involving different ideas. Thus I argued earlier (see notes 13 and 15) that intuitions coming from the notion of collection were important in Cantor's own conception of set and play a role in motivating the axioms of set theory today, even when they are purged of ideas that cannot be taken literally, such as that of sets being formed from their elements by some mental or other activity. It is clear from Cantor's explanations of the concept of set that he did not think of sets primarily as extensions, although it would also be too much to say that he thinks of them primarily as collections. The collection idea does seem dominant in the most widely cited one, from 1895:

> By a 'set' we understand any collection (Zusammenfassung) M into a whole of determinate, well-distinguished objects of our intuition or our thought (which will be called the 'elements' of M).[18]

---

many" is the notion of plurality, that of a "class as one" a Cantorian notion of set. A contemporary version of the kind of explanation of the notion hinted at in the text can be derived from George Boolos, "Nominalist Platonism." Boolos, however, uses it to interpret second-order logic and the notion of class in set theory rather than that of set. Moreover, he would deny that second-order logic so interpreted commits one to pluralities as entities. I criticize the latter view in "The Structuralist View of Mathematical Objects," §6; see also Michael Resnik, "Second-Order Logic Still Wild."

[17] *Ontological Relativity and Other Essays*, p. 44.

[18] *Gesammelte Abhandlungen*, p. 282. (Translations are my own unless otherwise indicated.) Elsewhere Cantor writes, "Every set of well-distinguished things can be viewed as a *unitary thing for itself,* in which those things are parts *(Bestandteile)* or consititutive elements" (ibid., p. 379, from 1887).

Richard Dedekind seems more explicit in thinking of what he calls "systems" as what I have called collections; see *Was sind und was sollen die Zahlen?*, p. 1.

However, in the earlier characterization of a set as "every Many, which can be thought of as One"[19] something like the notion of plurality is present, and it occurs again in what may be Cantor's most mature statement, in the correspondence with Dedekind in 1899. There Cantor begins with the idea of 'multiplicity' *(Vielheit)*. A set is a multiplicity whose elements "can be thought without contradiction as 'being together', so that they can be gathered together into 'one thing'."[20] In this they are contrasted with "absolutely infinite" or "inconsistent" multiplicities, of which Cantor offers all ordinals as an example. It is plausible that these formulations are a late modification responding to paradoxes. But they share with the 1883 conception starting with a notion of plurality, and then taking a set to be a plurality that can be thought of as a unity, i.e., a single object. Pluralities are thus not necessarily objects in the full sense, in analogy with Frege's notion of function. Cantor's explanations of the concept of set encourage the idea that all three of the above rough notions play a role in its genesis.

To see the contemporary use of the idea of collection in discussion of the elements of set theory, consider a genetic story that is part of textbook set theory, the so-called iterative conception of set. As commonly presented, it supposes that sets are formed from their elements. One starts with a situation where there are no sets, and in all possible ways forms sets from objects that are already given. One then arrives at sets of individuals. One then goes through the process again, forming sets from what one has at that stage and then keeps on repeating the process. At any given stage one can form sets from objects that are available at that stage. The process can be iterated into the transfinite, with no limit on how far it can be iterated.[21]

An important precursor of set theory is Bernard Bolzano. He gives explanations of the concept of set that are rather clear examples of the notion of collection. See *Wissenschaftslehre*, §§82–83, and *Paradoxien des Unendlichen*, §3. In Bolzano, the connection between the idea of set and traditional notions of whole and part (to which the idea of extension is foreign) is very clear. He uses the term *Teil* (part) for what we would call the elements of a set; nonetheless his explanation distinguishes the set concept from that of a mereological sum.

[19] Op. cit., p. 204 (from 1883).

[20] Cantor, *Gesammelte Abhandlungen*, p. 443; translation from van Heijenoort, *From Frege to Gödel*, p. 114.

[21] A good example of an exposition of the iterative conception based on the idea of sets as formed from their elements is Shoenfield, "Axioms of Set Theory."

It is crucial that a set can only be formed from objects that are already available. That rules out self-membership, and together with the principle that it is always possible to form sets from any available objects, it rules out a set of all sets and thus Russell's paradox. The principle involved, that the elements of a set are prior to the set, is a natural interpretation of the idea of collection but does not require that one take the conception of sets as formed from their elements at all literally. Within set theory, of course, it is reflected by the idea that the membership relation is well-founded, and thus by the Axiom of Foundation, but more fundamentally by the fact that the existential axioms of set theory are all true in an inner model satisfying Foundation.

Quine would, I think, have two objections to appealing to the iterative conception as sketched above as an argument for the claim that the concept of set derives genetically at least in part from the concept of collection. First, the iterative conception is historically a late development, certainly "wisdom after paradox." This is undoubtedly true. One can see it developing in papers by Mirimanoff, von Neumann, and Zermelo;[22] the first had paradoxes very much in mind; the last two were concerned with the interpretation of axiomatic set theory, itself a post-paradox development and motivated in part by the problem of the paradoxes. Moreover, the concept of collection does not figure at all prominently in their papers. It is rather a feature of more recent expositions of the iterative conception.

The second objection that Quine would make is that the concept of collection as it occurs in the expositions of the iterative conception that I have in mind is not very clear. In fact, this version of the iterative conception has well-known difficulties. Quine himself comments little on the notion of collection. Here, however, there is a point of historical interest. Nelson Goodman is well known for his defense of nominalism and for taking the rejection of classes to be a key element of nominalism. Goodman appeals repeatedly to what is essentially the idea of collection. For example he writes, "Nominalism bars the composition of different entities out of the same elements. . . . Platonism, using full set theory, admits a vast infinity of different entities made up of the

---

[22] Dimitri Mirimanoff, "Les antinomies de Russell et de Burali-Forti"; J. von Neumann, "Über eine Widerspruchsfreiheitsfrage in der axiomatischen Mengenlehre"; Ernst Zermelo, "Über Grenzzahlen und Mengenbereiche."

same atoms."[23] The classes Goodman rejects are collections; curiously, he does not in his writings address the notion of extension. Quine, replying to Goodman, takes "composition" to be a spatial metaphor and not very helpful.[24] Although he has not emphasized the point in his writings, Quine agrees with Goodman in rejecting collections. Goodman, on the other hand, has never said why he cannot follow Quine in accepting extensions. Perhaps he agrees that it imparts arbitrariness to the principles of set theory and does not want to use a theory of this character in logical construction.

Quine is justified in rejecting the idea that the concept of collection can by itself serve as the basic notion of set theory. Certain intuitions going with the notion of collection are appealed to in motivating the axioms of set theory, but so also are intuitions associated with the notion of extension, for example for the axiom schema of separation. By itself, neither conception is capable of yielding a convincing system of set existence principles that yields set theory as it has actually developed. What does more justice both to the history of the subject and our contemporary understanding of it is a genetic picture in which the concept of set develops from all three of the above rough notions. There is a parallel between this story about the concept of set and Quine's story about quantification, where there is the substitutional origin of one aspect of it and the quite essentially non-substitutional origin of another aspect of it from the universal categorical.

Though I think it true as far as it goes, this picture does not do entire justice either to the facts or to Quine's more underlying philosophy. So far, we have concentrated too much on conceptions of what a set is; Quine's structuralism should remind us that it is in the last analysis set *theory* more than the rough intuitive notions of set I have been discussing that is primary. In Quine's genetic story his second distinctive thesis, that devices for avoiding the paradoxes are unavoidably artificial because the intuitively most natural set theory would be the one based on the inconsistent universal comprehension schema, does not figure very prominently although expressed here and there (103 [quoted above], 121–122). This is unsurprising because the theory in which the story eventuates is the simple theory of types, and since he

---

[23] "Nominalisms," p. 160. Goodman's fuller, more definitive exposition of these ideas is "A World of Individuals."

[24] "Reply to Nelson Goodman," p. 162.

seems to think of cumulative types as at least as natural as Russell's own disjoint types, the story could readily be carried further to lead at least to Zermelo set theory, that is the original part of axiomatic set theory as we know it today. In fact, elsewhere Quine says as much.[25] In this respect Quine is guided more by the actual shape of set theory and less by his own distinctive theses.

A structuralist conception of set theory plays a role in the history of the subject and therefore deserves a place in a genetic account. It seems to arise in the wake of the paradoxes; it can be discerned in the writings of Ernst Zermelo, the founder of axiomatic set theory. In introducing his original paper on his axiomatization, Zermelo takes the "naive" concept of set to be given by Cantor's 1895 'definition'. Although I have claimed that this is a formulation of the concept of collection rather than that of extension, Zermelo does interpret it to license unrestricted comprehension and therefore to lead to paradoxes.[26] The procedure Zermelo then adopts in effect disclaims the project of offering an alternative account of what a set is, beyond saying that sets and their elements are a domain of objects satisfying the axioms of set theory.

> Set theory is concerned with a *domain* B of individuals, which we shall call simply *objects* and among which are the *sets*. . . . We say of an object that it "exists" if it belongs to the domain B. . . . Certain *fundamental relations* of the form $a \in b$ obtain between the objects of the domain B. If for two objects a and b the relation $\in$ holds, we say "a is an element of the set b". . . . An object b may be called a *set* if and—with a single exception (Axiom II [the null set])—only if it contains another object, a, as an element.[27]

---

[25] "Reply to Joseph S. Ullian," p. 590. Note that Quine there describes his own systems NF and ML as "a more radical and artificial departure."

[26] "Untersuchungen über die Grundlagen der Mengenlehre I," p. 261; trans. from van Heijenoort, op. cit., p. 200.

This interpretation was widely held; for example it is expressed in the standard works of A. A. Fraenkel. His *Einleitung in die Mengenlehre,* a very influential textbook of set theory in the 1920s which deals extensively with foundational questions, begins the subject with the same Cantorian explanation (2nd ed., p. 1). When he comes to discuss the paradoxes and the axiomatization of set theory, he again takes Cantor's explanation to imply unrestricted comprehension (p. 157).

[27] "Untersuchungen über die Grundlagen der Mengenlehre I," p. 262, trans. p. 201. A later, more radical structuralist understanding of set theory is Skolem's

No doubt this procedure came naturally as an application of Hilbert's influential outlook on the axiomatic method. Where Zermelo's procedure differs from Quine's at least as the latter describes it is that what Zermelo thought himself responsible to in constructing his axiom system was the body of set theory as it had developed in the work of Cantor and others; just after his remarks on Cantor's 'definition' he writes:

> Under these circumstances there is at this point nothing left for us to do but to proceed in the opposite direction and, *starting from set theory as it is historically given,* to seek out the principles required for establishing this mathematical discipline.[28]

Quine takes axiomatizations of set theory rather to be attempting, necessarily unsuccessfully, to capture the Fregean conception of extension.

Although Zermelo's interpretation of Cantor is a natural one, not clearly contradicted in explanations of the concept of set in his published writings, it has been convincingly argued by Michael Hallett that Cantor's understanding of the set concept, in contrast to Frege's conception of extension, never incorporated an unrestricted comprehension principle.[29] One can discern some heuristic role of the idea of collection. But much more fundamental was the distinction Cantor made early on between the transfinite and the Absolute or absolutely infinite, which colored the whole development of his theory of transfinite numbers. This was Cantor's modification of traditional views of infinity in philosophy and theology. The traditional conception of the infinite as innumerable and beyond the comprehension of our minds Cantor took to apply to the Absolute; the transfinite, on the other hand, while it is genuinely ("actually") infinite, can be treated mathematically. Sets have cardinal numbers and are "increasable"; that is, they are related to larger sets. Absolute infinity is beyond mathematical treatment. Already in the early 1880s, Cantor recognized that the totality of transfinite cardinal and ordinal numbers are absolutely infinite, that is do not

---

conception of "the axiomatic way of founding set theory," more radical because Skolem insists on a first-order interpretation of separation and replacement and presses its consequences.

[28] Ibid., p. 261 (trans. p. 200); emphasis mine. A similar methodology underlay the later addition to the established axioms of the axiom of replacement by Fraenkel, Skolem, and von Neumann. See for example Michael Hallett, *Cantorian Set Theory and Limitation of Size,* §8.2.

[29] *Cantorian Set Theory,* ch. 1.

themselves have cardinals or ordinals. Where his thought at that time differs from that of 1899, it is that the earlier way of thinking hardly allows for "all ordinals" to be regarded as a mathematical concept at all, while in 1899 Cantor attempts to say how the ordinals are like a set, while yet not constituting a set.[30] It would be wrong to say that the paradoxes are no problem for Cantor's set theory. But as Cantor developed it the paradoxes are relevant only at the margins, and he had already sketched well in advance of paradoxes some crucial distinctions that could be applied in understanding them and in axiomatizing set theory so as to avoid them, though Cantor's own contribution in 1899 was not a self-conscious application of the axiomatic method (in contrast to Zermelo's in 1908) and was certainly too sketchy to meet as it stood the standards of that method. But retrospect certainly justifies Cantor in not seeing paradoxes as catastrophic for set theory. Kurt Gödel seems to me right in remarking that what we would call the iterative conception of set offers "a satisfactory foundation of Cantor's set theory in its whole original extent and meaning."[31] What the paradoxes required of Cantor's set theory was not trimming of the theory itself but clarification of its underlying principles.

Although Cantor himself viewed the concept of set in a more ontological way, the structuralist view fits well with certain aspects of Cantor's thought. The constraint on set existence was not so much some interpretation of his explanations of the set concept as the coherence Cantor expected between the concept of set and the concept of number, as extended to include transfinite cardinals and ordinals. Though he did not formulate the matter in this way, this amounted to seeing sets and transfinite numbers as related to each other in such a way as to satisfy a certain theory. Where Quine goes wrong, it seems to me, is in seeing this theory as vainly trying to capture Frege's concept of extension, rather than growing, as Cantor's set theory did, by proceeding

---

[30] Thus Hallett sees Cantor in 1899 as giving more mathematical content to his earlier conception of the absolute, which he had then rather tended to exclude from mathematics (ibid., pp. 165–166).

[31] "What Is Cantor's Continuum Problem?" (1964 version), p. 258. Gödel's description of this conception does not commit him to the idea of sets as formed from their elements, although his remark later that the function of the concept of set is, like that of Kant's categories, "'synthesis', i.e. the generating of unities out of manifolds" (p. 268 n.40) suggests that he held a probably more sophisticated version of this conception.

from arithmetic to analysis, from traditional analysis to a theory of arbitrary sets and functions on the real and complex numbers, and from there to higher infinities, so that a general account of sets and transfinite numbers would capture this hierarchical structure.

In his genetic story, Quine attempts to explain the genesis of two things at once: the concept of attribute, as the paradigm "common sense" abstract objects, and set theory. He sees this as requiring only the simplifying assumption that the objects one is talking about are individuated extensionally. This is a consequence of the conception of sets as extensions. In this connection, it is worth pointing out that Frege was not the only earlier writer to have a sets-as-extensions conception. The same view underlies the criticisms of set theory by a number of French writers after the turn of the century, who objected to arguments (such as Zermelo's proof of the well-ordering theorem) that assumed the existence of sets that may not be definable. Their implicit idea was that a set is the extension of a predicate. The natural demand that the predicate should be antecedently understood leads to the rejection of impredicativity, as Poincaré saw clearly.[32] Quine himself sees proceeding in this way to be the natural one in introducing reference to abstract objects by first nominalizing predicates; it is then possible to take the quantification involved substitutionally. From there, however, Quine proceeds into set theory, where the requirements of the theory lead to admitting impredicativity and thus allowing class quantifiers to go objectual (112).

I would see the genesis of set theory, and the genesis of "linguistic" abstract entities, as two different though of course not unrelated problems. Common-sense notions of attribute and class do not require impredicativity; where the latter comes in is in the use in mathematics of the notions of set and function. Here the "leap" involved in introducing impredicativity doesn't have to be tied with a conception of these entities as extensions; indeed, a genetic story could adopt Bernays's description of quantification over all subsets of a given set, or over all functions from one set into another, as "quasi-combinatorial."[33] In this description, the analogical extension involved is less arbitrary than it appears to be in Quine's story. There is, however, a definite kinship of

---

[32] I have used similar considerations to argue that sets, as conceived in set theory, are not reducible to extensions, in "Sets and Classes," pp. 216–218.

[33] "Sur le platonisme," p. 54; trans. pp. 259–260.

general epistemological outlook on set theory between Bernays and Quine, in that both of them see set theory as arising by extending more elementary theories by means of analogies.[34] This outlook seems to me to have considerable merits. Although in some respects I have found Quine's account remote from actual set theory, this aspect of it mirrors part of the thinking of actual set theorists.[35]

---

[34] In commenting on my talk at the conference, Quine said that in the part of *Roots* concerned with set theory he was speculating on "how we as a species could have got on to talking about properties" and that he was not concerned with "sophisticated set theory." This is hard to reconcile with the text of *Roots*, where considerations about mathematics are introduced very quickly. Although he still seems to hold to the view, contrary to what I express in the text, that a theory of properties of the sort to be explained will be impredicative, his remarks can perhaps be taken as conceding that the problem of the genesis of talking of properties or attributes, and the problem of the genesis of set theory, are two distinct problems.

On the general issues, cf. also Penelope Maddy, "Believing the Axioms."

[35] [This essay was written for and first presented at the conference on Quine's philosophy at Washington University, St. Louis, in April 1988.]

I am indebted to Sidney Morgenbesser for once again helping me to clarify my ideas about Quine's philosophy. In the final preparation of the essay, I have benefited greatly from Professor Quine's comments at the St. Louis conference. Gilbert Harman's lecture was especially helpful with §II.

10

HAO WANG AS PHILOSOPHER

AND INTERPRETER OF GÖDEL

In this essay I attempt to convey an idea of Hao Wang's style as a philosopher and to identify some of his contributions to philosophy. Wang was a prolific writer, and the body of text that should be considered in such a task is rather large, even if one separates off his work in mathematical logic, much of which had a philosophical motivation and some of which, such as his work on predicativity, contributed to the philosophy of mathematics. In a short essay one has to be selective. I will concentrate on his book *From Mathematics to Philosophy*,[1] since he considered it his principal statement, at least in the philosophy of mathematics. It also has the advantage of reflecting his remarkable relationship with Kurt Gödel (which began with correspondence in 1967[2]) while belonging to a project that was well under way when his extended conversations with Gödel took place. Although Gödel's influence is visible, and in some places he is documenting Gödel's views by arrangement with their author, the main purpose of the book is to expound Wang's philosophy. I will also discuss the Wang-Gödel relation and Wang's interpretations of Gödel's thought and the role of his work as a source. That discussion will necessarily focus on the two books on Gödel written after Gödel's death, *Reflections on Kurt Gödel* and *A Logical Journey*.

---

[1] This work is referred to as FMP and cited merely by page number.

[2] Wang in fact first wrote to Gödel in 1949, and they had a few isolated meetings before 1967. (See *A Logical Journey* (cited hereafter as LJ), pp. 133–134.) But the closer relationship originated with an inquiry of Wang with Gödel in September 1967 about the relation of his completeness theorem to Skolem's work. Gödel's reply is the first of the two letters published in FMP, pp. 8–11. [For the letters as originally received by Wang, see Gödel, *Collected Works* (hereafter CW), vol. V, pp. 396–399, 403–405.]

1. Style, Convictions, and Method

Wang's writings pose difficulties for someone who wishes to sort out his philosophical views and contributions, because there is something in his style that makes them elusive. FMP, like other writings of Wang, devotes a lot of space to exposition of relevant logic and sometimes mathematics, and of the work and views of others. Sometimes the purpose of the latter is to set the views in question against some of his own (as with Carnap, pp. 381–384); in other cases the view presented seems to be just an exhibit of a view on problems of the general sort considered (as with Aristotle on logic, pp. 131–142). The presence of expository sections might just make Wang's own philosophizing a little harder to find, but it is not the most serious difficulty his reader faces. That comes from a typical way Wang adopts of discussing a philosophical issue: to raise questions and to mention a number of considerations and views but with a certain distance from all of them. This makes some of his discussions very frustrating. An example is his discussion (FMP, ch. VIII) of necessity, apriority, and the analytic-synthetic distinction. Wang is sensitive to the various considerations on both sides of the controversy about the latter distinction and considers a greater variety of examples than most writers on the subject. But something is lacking, perhaps a theoretical commitment of Wang's own, that would make this collection of expositions and considerations into an *argument,* even to make a definite critical point in a controversy structured by the views of others.[3]

This manner of treating an issue seems to reflect a difference in philosophical aspiration both from older systematic philosophy and from most analytic philosophy. In the preface to FMP Wang writes:

> This book certainly makes no claim to a philosophical theory or a system of philosophy. In fact, for those who are convinced that philosophy should yield a theory, they may find here merely data for philosophy. However, I believe, in spite of my reservations about the possibility of philosophy as a rigorous science, that philosophy can be relevant, serious, and stable. Philosophy should try to achieve some reasonable overview. There is more philosophical value in placing things in their right perspective than in solving specific problems. (p. x)

---

[3] But see the remarks in §4 on Wang's discussion of "analytic empiricism."

Nonetheless one can identify certain convictions with which Wang undertook the discussions in FMP and other works. He is very explicit about one aspect of his general point of view, which he calls "substantial factualism."[4] This is that philosophy should respect existing knowledge, which has "overwhelming importance" for philosophy. "We know more about what we know than how we know what we know" (p. 1). Wang has primarily in mind mathematical and scientific knowledge. Thus he will have no patience with a proposed "first philosophy" that implies that what is accepted as knowledge in the scientific fields themselves does not pass muster on epistemological or metaphysical grounds, so that the sciences have to be revised or reinterpreted in some fundamental way. He would argue that no philosophical argument for modifying some principle that is well established in mathematical and scientific practice could possibly be as well-grounded as the practice itself.[5]

Factualism as thus stated should remind us of views often called naturalism. In rejecting first philosophy, Wang is in agreement with W. V. Quine, as he seems to recognize (p. 3), and yet his discussions of Quine's philosophy emphasize their disagreements.[6] Wang's factualism differs from a version of naturalism like Quine's in two respects. First, natural science has no especially privileged role in the knowledge that is to be respected. "We are also interested in less exact knowledge and less clearly separated out gross facts" (p. 2). Even in the exact sphere, Wang's method gives to mathematics an autonomy that Quine's empiricism tends to undermine. Second, in keeping with the remark from the preface quoted above, Wang has in mind an essentially descriptive method. Quine's project of a naturalistic epistemology that would construct a comprehensive *theory* to explain how the human species constructs science given the stimulations individuals are subjected to is quite alien to Wang. In one place where Wang criticizes epistemology, his target is not only "foundationalism"; he says his

---

[4] Wang reports (LJ, p. 144) that it was Gödel who suggested the adjective "substantial," after Wang had used "structural" in an earlier draft.

[5] All views of this kind have to recognize the fact that scientific practice itself undergoes changes, sometimes involving rejection of previously held principles. There is a fine line between altering a principle for reasons internal to science and doing so because of a prior philosophy.

[6] See especially *Beyond Analytic Philosophy* (cited hereafter as BAP), "Two Commandments," and "Quine's Logical Ideas."

point of view "implies a dissatisfaction with epistemology as it is commonly pursued on the ground that it is too abstract and too detached from actual knowledge" (p. 19). That comment could as well have been aimed at Quine's project of naturalistic epistemology as at "traditional" epistemology.[7] Wang proposes to replace epistemology with "epistemography which, roughly speaking, is supposed to treat of actual knowledge as phenomenology proposes to deal with actual phenomena" (ibid.). But he does not make that a formal program; I'm not sure that the term "epistemography" even appears again in his writings. One way of realizing such an aspiration is by concrete, historical studies, and there is some of this in Wang's writing.[8]

There are, I think, convictions related to his "factualism" that were at work in Wang's work from early on. One I find difficult to describe in the form of a thesis; one might call it a "continental" approach to the foundations of mathematics, where both logicism after Frege and Russell and the Vienna Circle's view of mathematics and logic have less prominence, and problems arising from the rise of set theory and infinitary methods in mathematics, and their working out in intuitionism and the Hilbert school, have more. In an early short critical essay on Nelson Goodman's nominalism he wrote:

> ... there is ... ground to suppose that Quine's general criterion of using the values of variables to decide the "ontological commitment" of a theory is not as fruitful as, for instance, the more traditional ways of distinguishing systems according to whether they admit of infinitely many things, or whether impredicative definitions are allowed, and so on.[9]

Wang's sense of what is important in foundations probably reflects the influence of Paul Bernays, under whose auspices he spent much of the academic year 1950–1951 in Zürich. Wang's survey paper "Eighty Years of Foundational Studies" is revealing. There he classifies positions in the foundations of mathematics according to a scheme he attributes to Bernays (and which indeed can be extracted from "Sur le

---

[7] In a brief comment on Quine's project (BAP, p. 170), Wang expresses a similar philosophical reservation but also asks whether the time is ripe for such a program to achieve *scientific* results.

[8] Good examples are "The Axiomatization of Arithmetic" and the chapter on Russell's logic in FMP.

[9] "What Is an Individual?," pp. 416–417 of reprint.

platonisme"), so that the article ascends through strict finitism (which he prefers to call "anthropologism"), finitism, intuitionism, predicativism, and platonism. He shared the view already expressed by Bernays that one does not need to make a choice among these viewpoints and that a major task of foundational research is to formulate them precisely and analyze their relations. His own work on predicativity, most of which was done before 1958, was in that spirit. Although he was familiar with the Hilbert school's work in proof theory and already in the mid-1950s collaborated with Georg Kreisel, his own work on problems about the relative logical strength of axiom systems made more use of notions of translation and relative interpretation than of proof-theoretic reductions.[10]

Another conviction one can attribute to the early Wang, more tenuously connected with factualism, is of the theoretical importance of computers. Interest in computability would have come naturally to any young logician of the time; recursion theory and decision problems were at the center of interest. Wang concerned himself with actual computers and spent some time working for computer firms. His work in automatic theorem proving is well known. A whole section of the collection *A Survey of Mathematical Logic* consists of papers that would now be classified as belonging to computer science.[11] Issues about computational feasibility had some concrete reality for him.

---

[10] Here I might make a comment about my own brief experience as Wang's student. In the fall of 1955, as a first-year graduate student at Harvard, I took a seminar with him on the foundations of mathematics. I had begun the previous spring to study intuitionism, but without much context in foundational research. Wang supplied some of the context. In particular he lectured on the consistency proof of Wilhelm Ackermann, "Zur Widerspruchsfreiheit." I knew of the existence of some of Kreisel's work (at least "On the Interpretation of Non-Finitist Proofs") but was, before Wang's instruction, unequipped to understand it. In the spring semester, in a reading course, he guided me through Kleene's *Introduction to Metamathematics.* Unfortunately for me he left Harvard at the end of that semester, but his instruction was decisive in guiding me toward proof theory and giving me a sense of its importance. [Further to this subject see note 2 to the preface of Parsons and Link, *Hao Wang.*]

Wang's seminar was memorable for another reason. Its students included two noteworthy undergraduates, David Mumford and Richard Friedberg. Friedberg gave a presentation on problems about degrees growing out of Post's work; I recall his mentioning Post's problem and perhaps indicating something of an approach to it. It was just a few weeks after the seminar ended that he obtained his solution.

[11] See also *Computation, Logic, Philosophy,* which has some overlap with the earlier collection but also contains additional papers even from before 1962.

One can see Wang's familiarity with computers at work in the essay "Process and Existence in Mathematics," partly incorporated into chapter VII of FMP. This essay is one of the most attractive examples of Wang's style. Many of the issues taken up seem to arise from Wittgenstein's *Remarks on the Foundations of Mathematics,* but Wittgenstein's name is not mentioned; in particular he does not venture to argue for or against Wittgenstein's general point of view as he might interpret it. But the notion of perspicuous proof, the question whether a mathematical statement changes its meaning when a proof of it is found, the question whether contradictions in a formalization are a serious matter for mathematical practice and applications, and a Wittgensteinian line of criticism of logicist reductions of statements about numbers are all to be found in Wang's essay. But it could only have been written by a logician with experience with computers; for example it is by a comparison with what "mechanical mathematics" might produce that Wang discusses the theme of perspicuity. Computers and Wittgenstein combine to enable Wang to present problems about logic in a more concrete way than is typical in logical literature, then or now. But except perhaps for the criticism of Frege and Russell on '7 + 5 = 12', one always wishes for the argument to be pressed further.

In the remainder of the essay I shall discuss three themes in Wang's philosophical writing where it seems to me he makes undoubted contributions that go beyond the limits of his descriptive method. The first is the concept of set, the subject of chapter VI of FMP. The second is the range of questions concerning minds and machines, the subject of chapter X and returned to in writings on Gödel and the late article "On Physicalism and Algorithmism." The third is the discussion of "analytic empiricism," a term which he uses to describe and criticize the positions of Rudolf Carnap and W. V. Quine. This only becomes explicit in BAP and "Two Commandments." Each of these discussions reflects the influence of Gödel, but it is only of the second that one could plausibly say that Wang's contribution consists mostly in the exposition and analysis of Gödel's ideas.

2. The Concept of Set

Chapter vi of FMP is one of the finest examples of Wang's descriptive method. It combines discussion of the question how an intuitive concept of set motivates the accepted axioms of set theory with a wide-ranging exploration of issues about set theory and its history, such as the question of the status of the continuum hypothesis (CH) after Cohen's independence proof. As an overview and as a criticism of some initially plausible ideas,[12] it deserves to be the first piece of writing that anyone turns to once he is ready to seek a sophisticated philosophical understanding of the subject. Not of course the last; in particular, even given the state of research when it was published, one might wish for more discussion of large cardinal axioms,[13] and some of the historical picture would be altered by the later work of Gregory Moore and Michael Hallett.

The inconclusiveness that a reader often complains of in Wang's philosophical writing seems in the case of some of the issues discussed here to be quite appropriate to the actual state of knowledge and to respond to the fact that philosophical reflection by itself is not likely to make a more conclusive position possible, not only on a specific issue such as whether CH has a definite truth-value but even on a more philosophically formulated question such as whether "set-theoretical concepts and theorems describe some well-determined reality" (199). (Incidentally, Wang is here suspending judgment about a claim of Gödel.)

The most distinctive aspect of the chapter, in my view, is his presentation of the iterative conception of set and "intuitive" justifications of axioms of set theory. It is very natural to think of the iterative conception of set in a genetic way: Sets are formed from their elements in successive stages; since sets consist of "already" given objects, the elements of a set must be available (if sets, already formed) at the stage at which the set is formed. Wang's treatment is the most philosophically developed presentation of this idea. A notion of 'multitude' (Cantor's *Vielheit* or *Vieles*) is treated as primitive; in practice we could cash this in by plural or second-order quantification. Exactly how this is to be understood, and what it commits us to, is a problem for Wang's

---

[12] For example, that independence from ZF itself establishes that a set-theoretic proposition lacks a truth-value (pp. 194–196). This discussion has a curious omission, of the obvious point that if ZF is consistent, then on the view in question the statement of its consistency lacks a truth-value. Possibly Wang thought the holder of this view might bite this particular bullet; more likely the view is meant to apply only to propositions that, in their intuitive content, essentially involve set-theoretic concepts, so that the theorems of a progression of theories generated by adding arithmetized consistency statements would be conceded to have a truth-value.

[13] Wang himself seems to have come to this conclusion; see "Large Sets."

account, but not more so than for others, and for most purposes we can regard the term 'multitude' as a place-holder for any one of a number of conceptions.

Wang says, "We can form a set from a multitude only in case the range of variability of this multitude is in some sense intuitive" (p. 182). One way in which this condition is satisfied is if we have what Wang calls an intuitive concept that "enables us to overview (or look through or run through or collect together), in an *idealized* sense, all the objects in the multitude which make up the extension of the concept" (ibid.). Thus he entertains an idealized concept of an infinite intuition, apparently intuition of objects. An application of this idea is his justification of the axiom of separation, stated "If a multitude A is included in a set x, then A is a set" (p. 184):

> Since x is a given set, we can run through all members of x, and, therefore, we can do so with arbitrary omissions. In particular, we can in an idealized sense check against A and delete only those members of x which are not in A. In this way, we obtain an overview of all the objects in A and recognize A as a set. (Ibid.)

Some years ago I argued that the attempt to use intuitiveness or intuitability as a criterion for a multitude to be a set is not successful. The idealization that he admits is involved in his concept of intuition cuts it too much loose from intuition as a human cognitive faculty.[14] For example, the set x can be very large, so that "running through" its elements would require something more even than immortality: a structure to play the role of time that can be of as large a cardinality as we like. Moreover, it seems we need to be omniscient with regard to A, in order to omit just those elements of x that are not in A. It is not obvious that intuitiveness is doing any work that is not already done by the basic idea that sets are formed from given objects.

Two considerations raise doubts as to whether my earlier discussion captured Wang's underlying intention. First, Wang lists five principles suggested by Gödel "which have actually been used for setting up axioms." The first is "Existence of sets representing intuitive ranges of variability, i.e. multitudes which, in some sense, can be 'overviewed'" (p. 189). This suggests that Gödel gave some level of endorsement to

---

[14] "What Is the Iterative Conception of Set?," pp. 275–279, a paper first presented in a symposium with Wang. (His paper is "Large Sets.")

Wang's conception. But from remarks quoted in LJ it appears that Gödel's understanding of "overview" included some of what I was criticizing. Gödel is reported to say, "The idealized time concept in the concept of overview has something to do with Kantian intuition" (in remark 7.1.17, in ch. 7). In remark 7.1.18 he speaks of infinite intuition, and of "the process of selecting integers as given in intuition." Speaking of idealizations, he says, "What this idealization . . . means is that we conceive and realize the possibility of a mind which can do it" (7.1.19). This mind would of course not be the human mind but an idealization of it.[15]

Second, Wang evidently saw some justice in my criticisms, at least of some of his formulations (see "Large Sets," p. 327) but still holds that the just quoted principle "is sufficient to yield enough of set theory as a foundation of classical mathematics and has in fact been applied . . . to justify all the axioms of ZF (ibid., p. 313).

Wang's use of the term "intuition" in chapter VI of FMP is confusing. I don't think it is entirely consistent or fits well either the Kantian paradigm or the common conception of intuition as a more or less reliable inclination to believe. The way the term "intuitive range of variability" is used also departs from Gödel's use of "intuition" in his own writings, although it could be an extension of it rather than inconsistent with it.

One possibly promising line of attack is to think of what Wang calls "overview" as conceptual. In going over some of the same ground in note 4 of "Large Sets," Wang encourages this reading; for example he seems to identify being capable of being overviewed with possessing unity. A reason for being confident that the natural numbers are a set is that the concept of them as what is obtained by iterating the successor operation beginning with 0 gives us, not only a clear concept of natural number, but some sort of clarity about the *extension* of "natural number," what will count as a natural number. It is this that makes the natural numbers a "many that can be thought of as one."[16] For

---

[15] Note, however, that the "*idealized* intuitive evidence" of remark 7.1.15 is what would be evident to an "idealized *finitary* mathematician" (emphases added). But also in that remark Gödel seems to envisage carrying the idealization further. It should be pointed out that this remark is not from Wang's conversations, but is quoted from Toledo, *Tableau Systems*, p. 10.

[16] Cantor in 1883 famously characterizes a set as "jedes Viele, welches sich als Eines denken läßt" (*Gesammelte Abhandlungen*, p. 204).

well-known reasons we cannot obtain that kind of clarity about all sets or all ordinals. The difficult cases are the situations envisaged in the power set axiom and the axiom of replacement. In the footnote mentioned, Wang discusses the set of natural numbers in a way consistent with this approach and then reformulates his treatment of the power set axiom as follows:

> Given a set b, it seems possible to think of all possible ways of deleting certain elements from b; certainly each result of deletion remains a set. The assumption of the formation of the power set of b then says that all these results taken together again make up a set. ("Large Sets," p. 327)

The difference between thinking of all possible ways of deleting elements from a given set and thinking of all possible ways of generating sets in the iterative conception is real, but not so tangible as one would like, as Wang concedes by talking of the *assumption* of the formation of the power set. There is a way of looking at the matter that Wang does not use, although other writers on the subject do.[17] In keeping with Wang's own idea that it is the "maximum" iterative conception that is being developed, any multitude of given objects can constitute a set. Suppose now that a set x is formed at stage α. Then, since its elements must have been given or available at that stage, any subset of x could have been formed at stage α. But we would like to say that *all* the subsets of x are formed at stage α, so that $\wp(x)$ can be formed at stage $\alpha + 1$. This amounts to assuming that if a set could have been formed at α, then it *is* formed at α. This is a sort of principle of plenitude; it could be regarded as part of the maximality of the maximum iterative conception. But it is certainly not self-evident.[18]

---

[17] For example Boolos, "The Iterative Conception," p. 21; cf. "What Is the Iterative Conception?," p. 274.

[18] Cf. my "Structuralism and the Concept of Set," pp. 86–87. But there I ask why we should accept the appeal to plenitude in the case of subsets of a set but not in the case of sets in general or ordinals. This question is thoroughly confused. If by "sets in general" is meant all sets, then there is no stage at which they *could* have been formed, and hence no application for plenitude; similarly for ordinals. If one means *any* set, then in the context of the iterative conception, the version of plenitude leading to the power set axiom already implies that when a set could have been formed (i.e., when its elements are available) it is formed. [Cf. now *Mathematical Thought and Its Objects*, pp. 135–136.]

Wang's discussion of the axioms has the signal merit that he works out what one might be committed to by taking seriously the idea that sets are *formed* in successive stages. Thereby difficulties are brought to light that stimulate one to attempt a formulation of the iterative conception that does not require the metaphor of successive formation to be taken literally. The task is not easy. An attempt of my own (in "What Is the Iterative Conception?") relied on modality in a way that others might not accept and might be objected to on other grounds. Moreover, much of Wang's discussion, for example of the axiom of replacement, can be reformulated so as not to rely on highly idealized intuitability.

I have left out of this discussion the a posteriori aspect of the justification of the axioms of set theory. Following Gödel, Wang does not neglect it, although in the case of the axioms of ZFC itself, I think he gives it less weight than I would.

3. Minds and Machines

Clearly Wang was prepared by experience to engage himself seriously with Gödel's thought about the concept of computability and about the question whether the human power of mathematical thought surpasses that of machines. Chapter X of FMP contains the first informed presentation of Gödel's views on the subject, and Wang returned to the question several times later.

Most of the chapter is a judicious survey of issues about physicalism, mechanism, computer simulation as a tool of psychological research, and artificial intelligence. The general tone is rather skeptical of the claims both of the mechanist and the anti-mechanist sides of the debates on these subjects. Section 6 turns to arguments in this area based on the existence of recursively unsolvable problems or on the incompleteness theorems; the difficulties of establishing conclusions about human powers in mathematics by means of these theorems are brought out, but as they had already been in the debate prompted by J. R. Lucas's claim that the second incompleteness theorem shows that no Turing machine can model human mathematical competence.[19]

Only in the last section of the chapter do Gödel's views enter, and Wang confines himself to reporting. On the basis of the then unpublished 1951 Gibbs Lecture, the "two most interesting rigorously proved

[19] See his "Minds, Machines, and Gödel."

results about minds and machines" are said to be (1) that the human mind is incapable of formulating (or mechanizing) all its mathematical intuitions; if it has formulated some of them, this fact yields new ones, e.g., the consistency of the formalism. (2) "Either the human mind surpasses all machines (to be more precise: it can decide more number theoretical questions than any machine) or else there exist number theoretical questions undecidable for the human mind" (324).[20] The disjunction (2) is now well known. A statement follows of Gödel's reasons for rejecting the second disjunct, expressing a point of view that Wang calls "rationalistic optimism."[21] As an expression of the view that "attempted proofs of the equivalence of mind and machines are fallacious" there follows a one-paragraph essay by Gödel criticizing Turing (325–326).[22] Wang reports Gödel's view that the argument attributed to Turing would be valid if one added the premises (3) "there is no mind separate from matter" and (4) "the brain functions basically like a digital computer" (326).[23] He also reports Gödel's conviction that (3) will be disproved, as will mechanism in biology generally.

Wang dropped a kind of bombshell by this reporting of Gödel's views, with very little explanation and only the background of his own discussion of the issues. But he did not leave the matter there. It is taken up in *Reflections*, pp. 196–198, but only to give a little more explanation of Gödel's theses. The questions are pursued in much greater depth in "On Physicalism and Algorithmism."[24] Much of this

---

[20] Cf. CW III, pp. 307–308, 310.

[21] Wang subsequently stated ("On Physicalism and Algorithmism," p. 119) that the paragraph consisting of this statement (324–325) was written by Gödel. He also described the formulations of (1) and (2) as "published with Gödel's approval" (ibid., p. 118 n.12). [It is now known from correspondence that this section and some other places in the book were put into final form and sent to the publisher by Gödel while Wang was in China. See Gödel's letters to Ted Honderich and §§2.2–2.3 of my introductory note to the correspondence with Wang, in CW V.]

[22] This essay is an alternate version of the third of three "Remarks on the Undecidability Results" in CW II. Wang states (op. cit., p. 123), presumably on Gödel's authority, that the version he published is a revision of what subsequently appeared in CW. [The version published by Wang is reprinted in CW V, appendix B.]

[23] These theses are numbered (1) and (2) in FMP; I have renumbered them to avoid confusion with the theses of Gödel already numbered (1) and (2).

[24] The ground is gone over once again in ch. 6 of LJ. There the term "algorithmism" is replaced by "computabilism," which I find less euphonious. In general I find the treatment of LJ less attractive and have therefore focused my discussion on the paper. In the book chapter, concern to document the conversations with Gödel

essay consists of commentary on Gödel, either on the views just mentioned reported in FMP or on other remarks made in their conversations or in other documents such as the 1956 letter to John von Neumann containing a conjecture implying P = NP.[25]

By "algorithmism" about a range of processes Wang means the thesis that they can be captured or adequately modeled by an algorithm (and so simulated by a Turing machine). Physicalism and algorithmism about mental processes have to be distinguished, because one can't take for granted that *physical* processes can be so modeled.

General physical grounds will support algorithmism about the *brain* (some version of Gödel's (4)) only if physical processes are in some appropriate sense computable. Wang thought that it was such grounds that one should look for in order to decide whether to accept (4).[26] Gödel thought it practically certain that physical laws, in their observable consequences, have a finite limit of precision. Wang concludes that numbers obtained by observations can be approximated as well as makes observational sense by computable numbers, and therefore the best approach to the question of algorithmism about the physical is to ask whether physical theories yield computable predictions on the basis of computable initial data. Wang was skeptical about the physical relevance of the well-known negative results of Pour-El and Richards and mentions a conjecture of Wayne Myrvold that "noncomputable consequences cannot be generated from computable initial data within quantum mechanics" ("On Physicalism," p. 111).[27] Gödel is reported to have said that physicalism amounts to algorithmism, but Wang expresses doubts because the conclusion is based on arguments that tend to show that computable theories will agree with observation, but that is not the only requirement on a physical theory. Although he is not able to formulate the point in a way he finds clear, he is also given pause by the nonlocality of quantum mechanics.

---

may have got in the way of Wang's presentation of his own analysis. The comparison illustrates how little justice we do to Wang's engagement with Gödel's thought if we put too much emphasis on his role as a documenter and reporter.

[25] The letter is now published, with an English translation, in CW V 372–377.

[26] Wang evidently regards it as most prudent to consider the brain as a physical system, leaving open the question whether characteristic properties of the mind can be attributed to the brain.

[27] Cf. Myrvold, "Computability in Quantum Mechanics," which contains results and discussion relevant to this question, not limited to quantum mechanics.

Thus it is not so clear one way or the other whether physical processes are algorithmic. Wang does not go far into the question whether mental processes, considered apart from any physicalist or anti-physicalist thesis, are algorithmic, although he gives a useful commentary on Gödel's remarks on Turing's alleged argument for mechanism about the mind (see above).[28] The question about the mental has been much discussed by philosophers of mind, often in a highly polemical way. Wang may have thought it not fruitful to engage himself directly in that debate. He follows Gödel in confining himself to what can be said on the basis of rather abstract considerations and the structure of mathematics. There the upshot of his remarks is that it is very difficult to make Gödel's considerations about the inexhaustibility of mathematics into a convincing case for anti-mechanism.

Considered as a whole, Wang's discussions of these issues consist of a mixture of his characteristic descriptive method with commentary on Gödel's rather cryptically expressed views. The latter does lead him into an analytical investigation of his own, particularly into whether processes in nature are in some sense computable, and if so what that sense is. Philosophers of mind, when they discuss the question of mechanism, tend to take the "machine" side of the equation for granted. The value of Wang's exploration of Gödel's thoughts is in bringing out that one cannot do that.

4. Wang on "Analytic Empiricism"

Underlying FMP is undoubtedly a rejection of empiricism as unable to give an adequate account of mathematical knowledge. This had long been recognized as a problem for empiricism. The Vienna Circle thought it saw the way to a solution, based on the reduction of mathematics to logic and Wittgenstein's conception of the propositions of logic as tautologies, as "necessarily true" because they say nothing. The most sophisticated philosophy of logic and mathematics of the Vienna Circle is that of Carnap. Quine rejected Carnap's views on these matters but maintained his own version of empiricism. In "Two Commandments," Wang undertakes to describe a position common to both

---

[28] Wang also raises the corresponding question about the evolutionary process in biology but does not pursue it far, although his quotations from Edelman, *Bright Air, Brilliant Fire* are very provocative.

and to criticize it as not giving an adequate account of mathematics. A more general treatment of Carnap and Quine, in the context of a development beginning with Russell, is given in *BAP*.

Wang states the "two commandments of analytic empiricism" as follows ("Two Commandments," p. 451):

(a) Empiricism is the whole of philosophy, and there can be nothing (fundamental) that can be properly called conceptual experience or conceptual intuition.

(b) Logic is all-important for philosophy, but analyticity (even necessity) can only mean truth by convention.

(b) implies that the Vienna Circle's solution to the problem posed by mathematics for empiricism will be to view it as true by convention. Wang attributes this view to Carnap and criticizes it on lines close to those of Gödel's remarks on the subject in the Gibbs Lecture and "Is Mathematics Syntax of Language?" Apart from sorting out some different elements in the position being criticized and considering options as to changing one or another of them, Wang's main addition to Gödel's discussion is to include the views of Quine, on whose philosophical writings Gödel nowhere comments (even in the remarks from conversations in LJ). Wang brought out the interest of Gödel's views for the Carnap-Quine debate, by pointing to the fact that the stronger sense of analyticity that Gödel appeals to allows the thesis that mathematics is analytic to be separated from the claim that mathematics is true by convention or by virtue of linguistic usage, so that Gödel's view is a third option. I argue elsewhere that Gödel, in his reliance on his notion of concept, does not really have an answer to the deeper Quinian criticism of the ideas about meaning that underlie the analytic-synthetic distinction.[29] But that is not the end of the matter.

Concerning Carnap, Wang assumes, with Gödel, that Carnap aspires to answer a somewhat traditionally posed question about the nature of mathematical truth. His discussion is vulnerable to the objection to Gödel posed by Warren Goldfarb,[30] who points out reasons

---

[29] See Essay 6 in this volume. This essay was influenced and in some respects inspired by "Two Commandments."

[30] See Goldfarb's introductory note to "Syntax," CW III 329–330. Moreover, Howard Stein has called my attention to late texts of Carnap (in particular, "Replies," pp. 978–999) that raise a question whether Carnap at that time altogether

for questioning that assumption. It is hard to be comparably clear about how Wang intends to criticize Quine on these issues. In "Two Commandments" he makes rather generalized complaints, for example against Quine's tendency to obliterate distinctions. The extended discussion of Quine's philosophy in *BAP* is not very helpful; it is rather rambling and does not really engage Quine's arguments.[31]

I think one has to reconstruct Wang's reasons for the claim that Quine's version of empiricism does not give a more satisfactory account of mathematics than that given by logical positivism. An indication of the nature of the disagreement is the difference already noted between Wang's factualism and Quine's naturalism. Wang grants an autonomy to mathematics that Quine seems not to; mathematical practice is answerable largely to internal considerations and at most secondarily to its application in science. By contrast, for Quine mathematics forms a whole of knowledge with science, which is then answerable to observation. At times he views mathematics instrumentally, as serving the purposes of empirical science. This leads Quine to some reserve toward higher set theory, since one can have a more economical scientific theory without it.

Wang's basic objection, it seems to me, is that Quine does not have a descriptively adequate account of mathematics, because he simply does not deal with such matters as the more direct considerations motivating the axioms of set theory, the phenomena described by Gödel as the inexhaustibility of mathematics, and the essential uniqueness of certain results of analysis of mathematical concepts such as that of natural number (Dedekind) or mechanical computation procedure (Turing). Thus he writes:

> In giving up the first commandment of analytic empiricism, one is in a position to view the wealth of the less concrete mathematical facts and intuitions as a welcome source of material to enrich philosophy, instead of an irritating mystery to be explained away. ("Two Commandments," p. 459)

---

denied anything "that can be properly called conceptual experience or conceptual intuition," as (a) of the text has it.

[31] The reader gets the impression that some rather basic features of Quine's outlook repel Wang, much as he admires Quine's intellectual virtuosity and persistence. The rather limited focus of the present essay means that it does not attempt to do justice to BAP.

His primary argument against Quine, then, would consist of descriptions and analyses of various kinds, such as those discussed above concerning the concept of set and the axioms of set theory.

Wang goes further in describing mathematics as "conceptual knowledge," where he evidently means more than to give a label to descriptive differences between the mathematician's means of acquiring and justifying claims to knowledge and the empirical scientist's. He is evidently sympathetic to Gödel's way of regarding mathematics as analytic and thus true *by virtue* of the relations of the concepts expressed in mathematical statements. If that is actually Wang's view, it is then troubling that one does not find in his writings any real response to the very powerful objections made by Quine against that view or the availability of a notion of concept that would underwrite it.

Closer examination shows, however, that Wang does not commit himself to Gödel's view. In a later paper, Wang returns to the theme of conceptual knowledge. He states what he calls the Thesis of Conceptualism: "Given our mathematical experience, the hypothesis stating that concepts give shape to the subject-matter (or universe of discourse) of mathematics is the most natural and philosophically, the most economical."[32] Wang's discussion of this thesis is too brief to give us a very clear idea of its meaning. One theme that emerges is the importance of a general overview or understanding of mathematical situations and also of organizing mathematical knowledge in terms of certain central concepts; he cites Bourbaki as a way of arguing for some version of the thesis, presumably because of the idea underlying their treatise of organizing mathematics around certain fundamental structures. Wang recognizes that he has not given an explanation of the notion of concept.[33] But even though this discussion is undeveloped, it should be clear that he is less committed to the philosophical notion of conceptual truth, of which analyticity as conceived by Gödel is an instance, than appears at first sight. In this passage, what Wang contrasts

---

[32] "To and from Philosophy," p. 263. Wang is at pains to distinguish this thesis from realism, which is stated as a separate thesis about which he is more reserved though not unsympathetic.

[33] Ibid., p. 264. In 1959 Gödel made a similar admission in a letter to Paul Arthur Schilpp explaining why he had not finished "Syntax," quoted at p. 143 above from CW V 244. Sidney Morgenbesser informs me that in discussions he and Wang had in the last years of Wang's life the notion of concept figured prominently, but it appears that Wang did not arrive at a settled position.

conceptual knowledge with is not empirical knowledge but technical knowledge or skill.

5. Wang as Source for and Interpreter of Gödel's Thought

The relation between Wang and Gödel that developed after 1967 was a curious one. They had extensive conversations about philosophy and the foundations of mathematics, but that is what one might expect given their intellectual interests. But early on Wang took on the role of a documenter of Gödel's thought and of some aspects of his life and career. As already noted above, some places in FMP are presentations of views of Gödel that he had not previously published, by a very specific arrangement, which included actual writing explicitly announced as such (8–11, 326), but also places where Wang's reporting of Gödel's views was examined and approved by him, even short passages that report Gödel's views in the third person but were actually written by him. The article "Some Facts about Kurt Gödel" was prepared in 1976 on the basis of statements by Gödel; it is a presentation approved by him of some facts about Gödel's life and work, to be published after his death.[34] Beyond these arrangements, Wang took extensive notes on their conversations and mined them for his writings after Gödel's death, especially of course LJ.

This unusual relationship led Wang to become a scholar of Gödel already during Gödel's lifetime, and he was then one of the first scholars to work with Gödel's posthumous papers after they were catalogued by John Dawson.[35] Although he did not participate in the editing of the *Collected Works*,[36] his knowledge of Gödel was of great assistance to the editors, in particular the present writer.

---

[34] John Dawson remarks, "It is doubtful that Gödel ever checked Wang's article in detail, for had he done so he would surely have corrected a number of factual errors in it" (*Logical Dilemmas*, p. 250). In conversation Dawson has added that the copy of the manuscript in Gödel's papers contains no markings by Gödel, evidence that he did not examine the paper closely. [This is not quite accurate. See CW V 391–392, in my introductory note to the correspondence with Wang.]

[35] Dawson completed this work in 1984. I recall Wang's telling me before the publication of *Reflections* of some things he had found in these papers.

[36] Wang declined an invitation to write the introductory note to the Gibbs Lecture. The editors did use a draft translation by Wang and Eckehart Köhler of "The Modern Development" but revised it before publication.

The interest in Gödel's biography, his intellectual development, and the totality of his thought emerge in nearly all Wang's writings on Gödel, in particular in the two books. Wang undertook to see Gödel whole, to a greater degree than he did with other figures in whom he took an intense interest, such as Russell, Wittgenstein, and Quine. In particular, although he says that Gödel did not work out in writing the systematic philosophy to which he aspired, he clearly took this aspiration seriously and reports views of Gödel that are not very closely connected with his published work and are not developed in writing known to us.

Wang brought out the fact that Gödel turned his energy principally to philosophy from about 1943 on; the most important technical work he did after that, the construction of his cosmological models, originated in the context of philosophical reflection about time. Even so the philosophical work that Gödel published or left in a state so that it could be published in CW III is almost entirely in the philosophy of mathematics or of relativity theory. Wang's reports of his conversations with Gödel, as well as other available documents, indicate strongly that this was where Gödel's views were most worked out. Nonetheless Gödel seems to have aspired to construct a more comprehensive philosophy and had some views about what its main outlines would be. Wang quotes the following remarks:

0.2.1 My theory is a monadology with a central monad [namely, God]. It is like the monadology by Leibniz in its general structure.

0.2.2 My theory is rationalistic, idealistic, optimistic, and theological.

Much of what we know about these convictions of Gödel comes from Wang, although for Gödel's rationalism we have considerable documentation from his writings on the philosophy of mathematics, and his theological views and optimism are expressed in letters to his mother. These remarks raise some obvious questions: How much did Gödel work out a monadology with God as central monad, and how did he face the problems, quite evident in Leibniz and his successors, of how to interpret the physical world on such a scheme? How much did Gödel develop a rationalistic view beyond the philosophy of mathematics, for example in philosophy of physics or moral philosophy? What were Gödel's views about the problems of philosophical theology? What did Gödel mean by describing his view as "idealistic"?

On these questions Wang gives valuable testimony, but still leaves much open. The theological side of Gödel's thought, in particular his rationalistic optimism, was not congenial to Wang.[37] The reader of "Ontological Proof" and the related notes published as an appendix in CW III would like first of all to know how much the Leibnizian framework of the proof was something Gödel was committed to, and to what extent the remarks in the notes were considered opinions of Gödel. Perhaps because Wang was not deeply interested in philosophical theology, and because he had not seen these texts at the time of his exchanges with Gödel, he does not press these questions.

Where Wang is of greatest value as a source for Gödel's thought is in the philosophy of mathematics and related areas; we have already seen some of this in the examples of §§2–4. More is to be found in LJ than I have mined here, especially on logic and mathematics (chapters 7–8) and on general views of philosophy (chapters 5 and 9). However, this way of classifying things does not do Wang justice: because he was concerned to see Gödel whole and to document his life as well as his thought, Wang's writing on Gödel presents a whole intellectual personality, with a great deal of information that others thinking about various aspects of Gödel can use.

It may be the concern to see Gödel whole that leads Wang to deemphasize a feature of Gödel's intellectual style that he is certainly aware of, the distinction in his philosophical work between his own more or less private opinions and what he was prepared to defend in writing destined for a public (whether or not he actually published it).[38] There is also very little comment in Wang's writing on what seems to me a

---

[37] This is, in my view, a place where Wang's Chinese cultural background shows itself. A study of Wang's thought by someone conversant with China is much to be desired.

[38] A distinction of this kind was first urged in Feferman, "Kurt Gödel: Conviction and Caution."

Howard Stein reminded me of the tension between remarks 0.2.1–2 quoted above and Oskar Morgenstern's statement in his diary in 1970 that Gödel hesitated to publish his ontological proof for fear it would be thought "that he actually believes in God, whereas he is only engaged in a logical investigation." (Quoted from the introductory note to "Ontological Proof," CW III 388.) That is surely an instance of the phenomenon mentioned in the text. [See now the comment on the matter in my "Gödel and Philosophical Idealism," p. 186.]

serious weakness in Gödel's engagement with philosophy in the latter part of his career: his largely self-imposed isolation after 1940 from the philosophical community around him and probably very limited knowledge of what went on in philosophy after the war. As remarked in §4, I have not found in Gödel's known writings or in Wang's extensive reports a single reference to the philosophical views of Quine. This might surprise us in view of Wang's own concern with Quine's views; perhaps his strong disagreements with Quine made him reluctant to bring his views up with Gödel. Gödel evidently read some of the posthumous publications of Wittgenstein, and discussed Wittgenstein with Wang, but there is no evidence that Gödel followed any of the extensive discussion of Wittgenstein and development of Wittgensteinian themes by other philosophers. At one point Gödel mentions favorably some conservative figures of the older generation of American philosophers (LJ, pp. 141, 145), but it does not appear that he took up any intellectual contact with them.[39]

I suspect that this policy of Gödel did not much bother Wang because part of his kinship with Gödel was that both felt themselves outsiders in the contemporary philosophical world.[40] Wang does mention a number of times some of the features of Gödel's temperament that led to this isolation: his shunning of controversy, his tendency to overestimate the hostility that his own views would encounter, his concern for his health. There is a change in Wang's attitude between *Reflections* and LJ that he does not remark on: in the earlier work, Wang in recounting Gödel's life seems to think that Gödel arranged it in an optimal way for theoretical work. This theme disappears in the later work; it is quite possible that Wang changed his view.

Given the importance of Wang's writing as a source, it is natural to ask how accurate he is as a reporter.[41] It would be difficult to answer this question with any definiteness, because for so much no other

---

[39] This evidence makes it hard to make judgments about what Gödel read in postwar philosophy. In spite of the silence about Quine's philosophy, I find it impossible to believe that Gödel did not read at least some of Quine's philosophical writing. (Quine's logical work is mentioned in several places.) There is some discussion of Quine in FMP, a draft of which Gödel discussed at length with Wang. The only philosopher of a younger generation than Wang himself mentioned in the conversations as Wang reports them is Saul Kripke, and only very briefly (LJ, p. 138).

[40] Cf. LJ, p. 132. Wang, however, certainly did not isolate himself in the same way.

[41] I am indebted to John Dawson and Warren Goldfarb for pressing this question.

comparable source exists. What is reported in FMP is undoubtedly what Gödel wanted to have published and was gone over carefully by him, so that it has an authority almost equal to that of Gödel's own writings. What Wang published after Gödel's death is another matter. In particular, he himself says in several places in LJ that the remarks from the conversations are reconstructions based on notes (said on p. 130 to be "very incomplete"), and he evidently experienced considerable difficulty in the task, although in his description of the difficulty he does not separate the problem of reconstructing what Gödel said from the problem of interpreting it and placing it in perspective. A student of Gödel owes Wang immense gratitude for his work and will for some aspects of Gödel's thought have to rely on Wang's testimony, but he will nonetheless have to exercise caution in using it.

6. Conclusion

In his discussions of Carnap and Quine (especially in *BAP*), Wang expresses general dissatisfaction with the philosophy of his own time. That is not the only context in which he does so. Neither time nor space would permit a full discussion of his reasons. By way of conclusion I want to mention a respect in which Wang's philosophical aspirations differed importantly from those of most of his philosophical contemporaries, of different philosophical tendencies and differing levels of excellence. The major philosophers of the past have left us systematic constructions that are or at least aspire to be rather tightly organized logically, and many of the problems of interpreting them derive from the difficulty of maintaining such a consistent structure while holding a comprehensive range of views. Systematic constructions have been attempted by some more conservative figures of recent times as well as some of the important "continental" philosophers, in particular Husserl. But they have also been aimed at by some figures in the analytic tradition whom Wang knew well, such as Quine and Dummett. Those who have not sought such systematicity have generally approached problems in a piecemeal manner or relied heavily on a technical apparatus.

As I see it, Wang's aspiration was to be synoptic, in a way that the more specialized analytic philosophers do not rise to, but without being systematic in this sense, which I think he saw as incompatible with descriptive accuracy and a kind of concreteness that he sought. Moreover, part of what he sought to be faithful to was philosophical thought itself as it had developed, so that he was led into an eclectic tendency.

All but the last of these aspirations might be thought to be shared with a figure who greatly interested Wang but whom I have mentioned only in passing so far: Ludwig Wittgenstein.[42] Yet Wang was no Wittgensteinian. I am probably not the person to explain why, although clearly there were deep differences in their approaches to mathematics and logic.[43]

In Gödel, Wang encountered someone who shared many of his views and aspirations but who did have the aspiration to system that he himself regarded skeptically. I think this was the most significant philosophical difference between them. Because of its rather general nature, it is most often manifested by Wang's reporting and working out rather definite claims of Gödel about which he himself suspends judgment. On some matters, such as the diagnosis of his problems with analytic empiricism, Gödel's influence made it possible for Wang to express his own views in a more definite way; one might say that the ideal of systematicity imposed a discipline on Wang's own philosophizing even if he did not in general share it. I would conjecture that the same was true of Wang's treatment of the concept of set, although in that case it is difficult to pin down Gödel's influence very exactly.

Wang has written that Gödel did not believe he had fulfilled his own aspirations in systematic philosophy.[44] Very often he also gives the impression that he did not find what he himself was seeking in philosophical work. Both Gödel and Wang struggled with the tension inherent in the enterprise of general philosophizing taking off from a background of much more specialized research. I would not claim for Wang that he resolved this tension satisfactorily, and often, even when

---

[42] Juliet Floyd informs me that from 1990 to 1994 she had regular meetings with Wang, mostly devoted to discussion of Wittgenstein. That shows his continuing strong interest in Wittgenstein into his last years. I believe that at one time he planned a book that would juxtapose Gödel and Wittgenstein; the article "To and from Philosophy" may have resulted from that plan. [See now Floyd, "Wang and Wittgenstein."]

[43] In a note to me of January 3, 1997, Burton Dreben writes, "Wang was constantly deeply attracted to Wittgenstein, especially to Wittgenstein's criticisms of so much of contemporary philosophy that Wang despised. But Wang couldn't accept the full force, the dialectical nature, of Wittgenstein's criticisms because it would cut too deeply into Wang himself."

[44] *Reflections*, pp. 46, 192, 221.

he is most instructive, what one learns from him is less tangible than what one derives from the arguments of more typical philosophers, even when one disagrees. I hope I have made clear, however, that Wang's "logical journey" gave rise to philosophical work that is of general interest and to contributions that are important from standpoints quite different from his own.[45]

---

[45] This essay is a revised and expanded version of "Hao Wang as Philosopher," in Petr Hájek (ed.), *Gödel '96: Logical Foundations of Mathematics, Computer Science, and Physics*, Lecture Notes in Logic 6 (Springer-Verlag, 1996). Material from that paper is reprinted by permission of Springer-Verlag, the editors of Lecture Notes in Logic, and Professor Hájek. A version closer to the present one was presented to a symposium in memory of Wang at Boston University in December 1996. I wish to thank John Dawson, Burton Dreben, and Howard Stein for comments and suggestions.

11

The origin of this essay lies in a puzzlement I have felt about how Putnam understands Quine's views on existence and ontology. I have found it difficult to put my finger on what his quarrel with Quine about this has been. But one could also see it as a fragment of a commentary on the following recent remark of Putnam. Speaking of the lectures "Ethics without Ontology" published in the book of that name, Putnam writes that they

> ... provided me with an opportunity to formulate and present in public something that I realized I had long wanted to say, namely that the renewed (and continuing) respectability of Ontology (the capital letter here is intentional!) following the publication of W. V. Quine's "On What There Is" at the midpoint of the last century had disastrous consequences for just about every part of analytic philosophy.[1]

It will be pretty clear from what follows in the lectures themselves that Putnam is largely concerned with a rather broad philosophical tendency that followed Quine's essay and not directly with Quine's views themselves. He is not very explicit about the extent to which he holds Quine responsible for what he deplores. In various writings Putnam has written quite a lot about Quine. He has not, however, written a lot on Quine on ontology. What I discuss will be based on a relatively small number of texts. I hope they represent his thought well enough for this exercise to be worthwhile. Although Quine's views and Putnam's view

---

[1] *Ethics without Ontology,* p. 2. "On What There Is" is cited as OWTI. I remind the reader that papers of Putnam and Quine reprinted in their collections are cited according to the collections, in the latest edition where applicable.

of them will be prominent in what follows, I hope to offer more than just a commentary on the Putnam-Quine relation.

I might frame an issue rather crudely by distinguishing between what could be called a modest and extravagant sense of ontology and concern with it. Quine's discussion of ontological commitment belongs in the first instance to the modest side. Ontological commitment as Quine understands it is a property of theories. The question of the ontological commitment of a theory can be discussed in abstraction not only from the question whether the theory is true or to be accepted, but also from the question how much it takes in, how comprehensive it is. Extravagant views in ontology are in all the examples I can think of views about what there is. Quine does not commit himself to such views in OWTI, and to the extent that he gives indications of his views there, they tend to be either deflationary, on the matter of universals and possibilia, or tentative, as on the matter of physicalism vs. phenomenalism.[2]

Putnam's concern with ontology in his later writings seems to be with views that can be rather tendentiously classified as extravagant. To the extent that he is quarreling with the Quine of OWTI, it is with the idea that one can have an ontology at all, not a view of ontological commitment but a view of what there really is. Does such an aspiration commit one to an extravagant ontology or at least the program of seeking one? I think this raises questions of ontological and other forms of relativity, about which both Quine and Putnam have had things to say.

I

In the case of Quine, OWTI seems to license a concern with ontology, although apparently not with extravagant ontology. Sometimes Quine does flirt with ontological extravagance, although usually in a negative rather than a positive sense, as when he suggests much later that physical objects might be eliminated from our theory in favor of sets (thus producing economy, since on his view we need sets anyway).[3] The views about reference that he advances in "Ontological Relativity" and later do imply that even given a manual of translation that determines

---

[2] However, in later writings of Quine phenomenalism drops out as an option.
[3] See *Theories and Things*, pp. 17–18. Cf. "Whither Physical Objects?"

the truth-values of the sentences of a theory, it is not settled what objects are referred to. But both his views on assessing ontological commitment and his own firm theoretical preferences mean that the options will be different domains of objects for a first-order theory.

Putnam's early writings indicate an inclination to downplay the significance of ontology. He mentions equivalent descriptions in physics, and it is likely that that phenomenon made an impression on him at the very beginning of his career. About mathematics, he has always tended to treat ontology as secondary. Thus in his important early paper, "Mathematics without Foundations," he writes, "In my view the chief characteristic of mathematical propositions is the wide variety of equivalent formulations that they possess" (p. 45).[4] Then he singles out two ways of formulating mathematical statements that he regards as equivalent in roughly the same sense, Mathematics as Modal Logic and Mathematics as Set Theory. The second is just the familiar framing of classical mathematics in axiomatic set theory, ZFC or some extension by additional axioms. The first is illustrated by translating a number-theoretic statement as a modal statement asserting that, necessarily, if the axioms of arithmetic hold, then the statement holds.[5] In MWF he proposes a more complicated method of translating statements of set theory into a modal language.[6] In both cases Putnam avoids explicit statements of the existence of the relevant mathematical objects, but the method does presuppose the *possible* existence of objects that stand in relations corresponding to the structure of the mathematical domain in question. Unlike the ontological relativity discussed by Quine just a little later, this way of formulating mathematical theories changes the logical form in such a way that however one describes the change, it's not just a change in what objects are referred to.

Putnam also differs from most writers influenced by his construction, who took it as a way of eliminating or at least greatly limiting reference to mathematical objects. Taken in that way it is subject to

---

[4] This paper will be cited as MWF.

[5] For first-order arithmetic, the antecedent would have to be the conjunction of some finite subset of the axioms, since first-order arithmetic is not finitely axiomatizable.

[6] Putnam never published the technical details of this translation, and his readers have had some difficulty working out his sketch. However, constructions serving the same end have been worked out by others, most extensively by Geoffrey Hellman in *Mathematics without Numbers*.

some criticisms.[7] Putnam, however, viewed the more conventional set-theoretic formulation of mathematics and the modal-logical formulation as equivalent descriptions, each corresponding to different pictures in something like Wittgenstein's sense. Each would shed light in its own way on the concepts involved. For example, the set-theoretic formulation serves to make clear at least the logic of the modalities, while the modal formulation brings out what is limited in the sense in which mathematics postulates objects (MWF 49). Putnam does not say in so many words that the two formulations differ in their ontology, but this is clearly implied by some of his statements (e.g., ibid., p. 48). I think he wants to conclude that its ontological commitment, whether in the precise sense derived from Quine or some other, is not an essential or intrinsic feature of a mathematical theory. Rather than accepting or rejecting the "mathematical-objects picture" outright, his aim is to relativize it.

One of the features of the "metaphysical realism" that Putnam began to attack in the late 1970s is that according to it what objects there are and their individuation are fixed by "reality" independently of our theories and their formulation.[8] As regards mathematical objects, Putnam did not hold this view even in his early essays. To that extent, he was never a metaphysical realist about mathematics. However, he seems to have thought that some degree of "ontological relativity" was compatible with metaphysical realism.[9] In MWF, however, he advances equivalent descriptions in which the objects of one do not arise in the other even as constructions. That appears to be a more radical departure from metaphysical realism as he formulated it later. So far as I know, however, he does not address the question of the compatibility of the view of MWF with metaphysical realism. It may be that the ontological side of the rejection of metaphysical realism in the late 1970s

---

[7] Putnam himself deflates the claim made by some writers that such a modal translation avoids the epistemological problem widely thought to exist for reference to mathematical objects. Whether in the case of higher set theory the possibility statements referred to above can be cashed in by objects that are in any plausible sense concrete or nominalistically admissible is questioned in §6 of my "The Structuralist View of Mathematical Objects." That point affects remarks Putnam makes but, it seems to me, not the issues discussed in the text.

[8] See in particular "Realism and Reason."

[9] Ibid., p. 132.

represents an extension to reference in general of a view that he already held about mathematical reference.

II

I now turn to an early text where Putnam does engage to some extent with views of Quine on ontology, the little book *Philosophy of Logic*.[10] But I confess I also have the aim of reminding us of something about Quine, which is easily forgotten especially if one concentrates on the metaphysical flights of some philosophers influenced by him. I hesitate a little in using this work for the purpose, for two reasons. First, it was evidently intended as an introductory book and so might simplify issues more than he would in another context. Second, in much of it he adopts nominalism as his target. In fact the book is probably now best known for its presentation of an "indispensability argument" for admitting mathematical objects. And in that place he makes clear his *agreement* with Quine. If by nominalism one means the rejection of abstract objects, the refusal to quantify over them, then of course Quine was no nominalist. Even his experimental interest in seeing how far one could go on a nominalist basis, reflected in his well-known paper with Nelson Goodman, was temporary.[11]

Nonetheless I think there is a nominalist tendency in Quine, not about mathematical objects but about meaning. It is certainly connected with his skepticism about the notion of meaning but appears to have been formed earlier. I would also argue that it has merits independent of that skepticism.

What I have in mind is Quine's view of predication and the notions of proposition, property, and attribute. His reserve toward these notions is one of his earliest philosophical stances. As regards propositions, it is expressed in "Ontological Remarks on the Propositional Calculus," his first philosophical publication. It is worth noting that Quine was willing to use the adjective "ontological" in the title of this paper and again in one of 1939 that I will mention. He may already

---

[10] Citations are of the reprint in *Mathematics, Matter, and Method,* 2nd ed., with the original pages in brackets.

[11] Goodman and Quine, "Steps toward a Constructive Nominalism." Note that Quine did not reprint the paper in any of his own collections, although Goodman reprinted it in *Problems and Projects.*

have had the attitude toward the word "ontology" that he expresses in a later discussion of Carnap:

> I might say in passing, though it is no substantial point of disagreement, that Carnap does not much like my terminology here. Now if he had a better use for this fine old word 'ontology', I should be inclined to cast about for another word for my own meaning. But the fact is, I believe, that he disapproves of my giving meaning to a word which belongs to traditional metaphysics and should therefore be meaningless. . . . Meaningless words, however, are precisely the words which I feel freest to specify meanings for.[12]

In the paper of 1934 Quine argues that understanding propositional logic (or the "theory of deduction," as he also calls it) as a theory about propositions, so that the letters p, q, ... range over propositions, amounts to treating sentences as names of propositions. He makes it clear that in his view this is a quite optional way of viewing the theory:

> Without altering the theory of deduction internally, we can so reconstrue it as to sweep away such fictive considerations; we have merely to interpret the theory as a formal grammar for the manipulation of sentences, and to abandon the view that sentences are names. Words occurring in a sentence may be regarded severally as denoting things, but the sentence as a whole is to be taken as a verbal combination which, though presumably conveying some manner of intelligence (I write with deliberate vagueness at this point), yet does not have that particular kind of meaning which consists in denoting or being a name of something.[13]

Although it is not as explicit as in later writings such as *Methods of Logic,* Quine's reading of the letters of propositional logic as schematic letters for sentences is essentially there in 1934.[14] At the end of the paper, Quine does suggest another reading in which the letters are true

---

[12] "On Carnap's Views on Ontology," p. 203.
[13] *The Ways of Paradox,* p. 267.
[14] As he wrote later in "Autobiography of W. V. Quine," p. 14.

variables ranging over sentences.[15] But that reading also doesn't imply that sentences designate anything or commit him to propositions.

Another central element of Quine's nominalistic side emerges in a paper written and presented in 1939, "A Logistical Approach to the Ontological Problem."[16] A large part of Quine's view of ontological commitment is already there, including the slogan "To be is to be the value of a variable." The main idea is simply that whether an expression names something or is "syncategorematic" turns on whether the place where it occurs is subject to existential generalization. In his original example, "Pebbles have roundness," 'roundness' names something just in case the sentence implies '$(\exists x)$(pebbles have x)'. The mark of a name is this connection with quantification. He considers the possibility that the "entities" involved might be fictions and admits this if the quantifiers can be eliminated by suitable definitions, as could be done in one case he cites, the addition of "propositional" quantifiers to truth-functional logic, as was done in the inter-war Polish school.

One could certainly infer from this paper that already in 1939 Quine held that predicates as such do not designate anything. It would be an inference; he does not discuss that case explicitly, even to discuss the relation between "Pebbles are round" and "Pebbles have roundness." But the inference is buttressed by his closing claim that we can obtain a language adequate for science by adding to the first-order language of set theory a number of empirical predicates.

> For this entire language the only ontology required—the only range of values for the variables of quantification—consists of concrete individuals of some sort or other, plus all classes of such entities, plus all classes formed from the thus supplemented totality of entities, and so on. (Ibid., p. 201)

---

[15] I conjecture that this paper represents a rather rapid response to his encounter with Tarski in 1933, when he must have learned something of Tarski's work on truth, although it was only in 1935 that the *Wahrheitsbegriff* was published in a language Quine could read. Central to both the non-propositional readings that Quine proposes is the idea that sentences can be treated as truth vehicles. However, Quine mentions Wittgenstein's *Tractatus* and reads it as identifying propositions with sentences, so that one can't be certain about Tarski's decisive influence.

[16] First published in *The Ways of Paradox*, 1st ed., pp. 64–69. (On the complicated history of what Quine presented at the 1939 conference, see "Quine's Nominalism" [Essay 8 of this volume], pp. 202–203.)

Quine's view that the use of predicates does not unavoidably involve us in commitment to properties or attributes emerges in OWTI. He argues against his fictitious McX, who maintains that red houses, red roses, and red sunsets must have something in common, which is what he calls redness. McX is willing to accept the denial that 'red' in such uses *names* redness, and he retreats to the position that it at least has a meaning and that this meaning is a universal. Quine replies:

> For McX, this is an unusually penetrating speech; and the only way I know to counter it is by refusing to admit meanings. However, I feel no reluctance toward refusing to admit meanings, for I do not thereby deny that words and statements are meaningful. (p. 11)

Synonymy, which he does not question in this context, would allow meanings in by the back door, as equivalence classes of synonymous expressions. But then they would not be "special and irreducible intermediary entities" that are supposed to have "explanatory value" (p. 12).

The point of view where general terms and predicates are explained as true or false of certain objects (so that 'elephant' or 'x is an elephant' is true of each elephant) underlies the treatment of quantificational logic in *Methods of Logic* (see especially §§12 and 17).[17] Properties and attributes that might be meant in some sense by such expressions are hardly mentioned, although he does make the following remark in explaining why he uses 'is true of' rather than 'denotes':

> 'Denotes' is so current in the sense of 'designates' or 'names', that its use in connection with the word 'wicked' would cause readers to look beyond the wicked people to some unique entity, a quality of wickedness or a class of the wicked, as named object. The phrase 'is true of' is less open to misunderstanding; clearly 'wicked' is true not of the quality of wickedness, nor of the class of wicked persons, but of each wicked person individually. (p. 65)

I shall call the collection of views mentioned here Quinean nominalism.[18]

---

[17] Of the first or second edition (1959).

[18] For more on this theme see "Quine's Nominalism" (Essay 8 in this volume).

Let us now turn to Putnam's *Philosophy of Logic*. Early in the book Putnam focuses on a simple syllogism:

All *S* are *M*. All *M* are *P*. Therefore all *S* are *P*.

He considers the contemporary textbook view to be that this syllogism is valid if it is so "no matter what *classes* may be assigned to *S*, *M*, and *P*" (p. 324 [4], emphasis in the text). He probably has in mind standard model-theoretic definitions of validity and consequence, although he doesn't put it that way. In fact he focuses on an "object language" formulation:

(A) For all classes *S*, *M*, *P*: if all *S* are *M* and all *M* are *P*, then all *S* are *P*.

Shortly after, at the beginning of the next chapter, Putnam introduces nominalism. As I have said, his target is full-blooded nominalism rather than Quinean nominalism. However, about the above syllogism, what he offers as the nominalist's substitute for the formulation in terms of classes is a formulation close to one that Quine would also use:

The following turns into a true *sentence* no matter what *words* or *phrases* of the appropriate kind one may substitute for the letters *S*, *M*, *P*: 'if all *S* are *M* and all *M* are *P*, then all *S* are *P*'. (p. 325 [9–10], emphasis in the text)

However, it doesn't seem to be Quine's use of it that Putnam has in mind. He worries about what is meant by the appropriate kind of word, and writes that "what is meant is all *possible* words or phrases of some kind or other, and that *possible words and phrases* are no more 'concrete' than classes are" (p. 326 [10], emphasis in the text).

This example shows that it is not so easy to tell where and to what extent Putnam dissents from Quinean nominalism. Quine's point of view, as we noted, depends on treating sentences as truth bearers, and at one point Putnam seems to deny this (p. 327 [13]), but a little later he considers arguments on the issue and ends with an inconclusive result.

A more definite disagreement arises when the immediate issue is the scope of logical truth. He mentions that according to Quine quantification theory does not really assert (A) but rather something without the class quantifiers, in which *S*, *M*, and *P* are schematic letters for predicates. Since by "quantification theory" Quine surely means first-order logic, Putnam could hardly quarrel with the first part of Quine's claim: (A) is no theorem of first-order logic, and after all first-order logic is complete. But he rejects the view that such instances of the schema as

> (9) If all crows are black and all black things absorb light, then all crows absorb light.

are truths of logic, while evidently they are paradigm truths of logic according to Quine. Putnam writes:

> In my view, logic, as such, does *not* tell us that (9) is true; to know that (9) is true I have to use my knowledge of the logical principle (A), *plus* my knowledge of the fact that the predicates 'x is a crow', 'x is black' and 'x absorbs light' are each true of just the things in a certain class, namely the class of crows, (respectively) the class of black things, (respectively) the class of things which absorb light. (p. 335 [29])

This argument clashes pretty directly with Quinean nominalism, but I find it pretty dubious. Putnam seems to be saying that to know (9) we have to infer it with the help of (A) and comprehension principles such as

$$(\forall y)(y \in \{x: x \text{ is a crow}\} \leftrightarrow y \text{ is a crow}).$$

That would be a more complicated quantificational argument than the syllogism (and perhaps conditionalization) involved in arriving at (9) without any appeal to classes. There is a kind of circularity: in order to arrive at a certain first-order logical conclusion I have to appeal to a more complex, second-order piece of reasoning. I doubt that this argument really expresses Putnam's considered view.

I suspect that the more important point from Putnam's point of view is the observation that he goes on to make, first, that there is an idealization involved in treating the predicates contained in (9) as well-defined, and second, that "knowing a good bit about both language and the world" is involved.

> That 'x is a crow' is a pretty well-defined predicate, 'x is beautiful' is pretty ill-defined, and 'x is a snark' is meaningless, is not logical knowledge, whatever kind of knowledge it may be. (Ibid.)

However, I don't think this point locates the disagreement with Quine. It would lead to the conclusion that there is not an autonomous body of logical knowledge, knowledge that one can arrive at without at any point making commitments that are not logical. But I doubt that Quine thinks otherwise; his holism would suggest a similar conclusion.

Putnam also remarks that insofar as the logician working with first-order logic makes assertions, they are assertions of such properties as validity:

> ... when a logician builds a system which contains such theorems as (5′), *what does he mean to be asserting?* . . . The simple fact is that the great majority of logicians would understand the intention to be this: the theorems of the system are intended to be valid formulas. Implicitly, if not explicitly, the logician is concerned to make assertions of the form 'such-and-such is valid', that is, assertions of the kind (A). Thus even first-order logic would normally be understood as a "metatheory." (p. 336 [31])

Now notions of validity and consequence that involve quantification over classes or sets are entirely standard in logic.[19] One would have to be a nominalist in the general sense, not just the Quinean sense, to refuse to appeal to them. Still, this remark enables us to identify the disagreement between Putnam and Quine. It is in part about the demarcation of logic (e.g., whether it is reasonable to regard second-order logic as logic), but I think it is really about the status or interest of different kinds of statement a logician (or even someone reasoning logically) might make.

Let us consider three cases, (9), by Quine's lights a logical truth, and two formulations of the general principle that it instantiates:

(A′) For any classes α, β, γ, assigned to the letters S, M, P as their extensions, the conditional 'If all S are M and all M are P, then all S are P' is true.

(B) No matter what *terms* or *predicates* are instantiated for S, M, and P, 'If all S are M and all M are P, then all S are P' is true.

Here terms and predicates are to be understood as linguistic items. Quine, of course, has no quarrel with either of these formulations. I think Putnam's reservations about (B) would disappear if one thinks of it as presupposing a definite language, definite enough so that anything

---

[19] I don't know whether Putnam speaks of classes rather than sets to accommodate the case where predicates involved might not have sets as their extensions.

that counts as a term will be true or false of any element of the domain.

Putnam seems to be arguing in *Philosophy of Logic* that (A′) is the proper formulation of the general logical principle and that (B) is in some way derivative. Quine does not directly engage this issue, but the procedure of his own *Philosophy of Logic* (written about the same time as Putnam's) suggests a certain primacy to (B), since it is what is arrived at after his discussions of truth-bearers, grammar, and truth.[20] (B) has a well-known problem, in that it only accords with intuitions concerning logical truth if the language is sufficiently rich. And (A′) has the advantage that it need not refer to a particular language, particularly if it is emended in the natural way so as also to generalize over the domain of quantification. In the end Quine seems to give some fundamental role to both, as well as to the syntactic characterization by a proof procedure.

We are left with less clarity concerning Putnam's attitude toward Quinean nominalism than we would wish for.

III

Putnam does discuss logical relations with an eye to ontological issues in *Ethics without Ontology*. He seems to admit that it is natural to use the language of objects in discussing cases of logical validity. However,

> ... few philosophers today think it right to take talk of "describing relations between statements" with inflationary metaphysical earnestness, that is, to think we are literally *describing* a certain sort of *relation* between certain *intangible objects* when we point out the validity of a simple logical inference in this way. (p. 56)

He then points out that in mathematical logic the terms of logical relations are typically sentences or formulae. However, he seems to regard the notions used in mathematical logic as ersatz: "The standard way of treating "validity" in mathematical logic is a way of *doing without* the notion, not a way of analyzing it" (p. 57). His meaning might be better expressed by saying that it is a way of giving and operating with a mathematical model of these notions.

---

[20] Quine, *Philosophy of Logic*.

Putnam goes on to take up, with a reference to *Methods of Logic,* the question of definitions of validity like (B) above. He is critical of such a definition, partly on grounds familiar from *Philosophy of Logic.* But unlike there, however, he does not seem to be breaking a lance for an alternative like (A'), although one of his objections, that the definition of 'tautology' in *Methods* "fails to capture what may be called the *universality* of logical truth" (p. 57), may point in that direction, since he interprets Quine as intending to apply his definition to a particular formalized language. As an interpretation of *Methods* this does not strike me as obvious, since so much of the book is devoted to analyzing inferences in natural language.

It's not so clear what the specific criticisms of Quine have to do with the main theme Putnam is pursuing in the lecture. That one shouldn't "take talk of 'describing relations between statements' with inflationary metaphysical earnestness" is a recommendation that Quine and many other logicians agree with. The relative roles of statements of principles of logic involving linguistic items, classes, and propositions or attributes still give rise to delicate questions. Are they in the end not ontological?

I'm a little troubled by the position Putnam takes here because he seems to want to give up what I myself see as one of the insights achieved by more formal analytic philosophy, in particular Quine's writing on quantification and ontology. One might put it by saying that Quine offered a way of making reasonably definite and tractable questions about when we are making reference to objects and committing ourselves to their existence. The effect of this in his own work (already anticipated in many respects by Frege and Carnap) was to forestall the kind of metaphysical inflation that Putnam is most concerned to combat.

IV

Putnam suggests that the passages we have just discussed are a non-technical restatement of parts of his extended paper "Was Wittgenstein *Really* an Anti-realist about Mathematics?"[21] There we do find a more direct criticism of Quine's views on existence and ontology, with the focus more on mathematics. Early in this discussion he makes a

---

[21] Obviously the essay is only incidentally concerned with Quine.

very strong assertion: "Wittgenstein believes—and I think he is right—that it does not make the slightest sense to think that in pure mathematics we are talking about objects" (p. 366).

I don't want to enter into Wittgenstein's views. My own interpretation of Quine has something in common with what Putnam proposes on the next page, that he "has interpreted the statement that in mathematics we are talking about objects in a way that totally robs it of its supposed metaphysical significance." I'm not sure what "metaphysical significance" is, but I do think that in talking about ontological questions about mathematics Quine is in the sphere of modest rather than extravagant ontology and that that is true of most of his writing about ontology generally. That is not always the case, and this bothers Putnam. To state in simple terms what bothers him, it is that Quine holds that what there really is is determined in the end by science, but science rather narrowly conceived. Scientific truth does not uniquely determine ontology (as we have noted), but it does greatly constrain it. And in later writings Quine is not sure to what extent scientific truth is itself uniquely determined.

But I want to return to modest ontology. Putnam refers in a footnote to remarks of his own on continually extending the sense of the word 'object', but he seems curiously to resist the very well-entrenched extension involved in understanding first-order logic and its application to mathematical theories. His Wittgenstein sees a difference between the use of quantifiers in talking of ordinary objects and talking of objects in pure mathematics as "just too great" (ibid.). Putnam writes that Paul Benacerraf's "Mathematical Truth"

> ... clearly assumes that if we take the quantifiers in classical mathematics to be "objectual" quantifications in Quine's sense, then we are committed to regarding them as ranging over objects in a sense perfectly analogous to the sense in which quantifiers over the books in my office range over objects. (p. 367)

Whether or not Benacerraf regards the two senses as "perfectly analogous," he makes himself vulnerable by arguing that to understand knowledge of objects, the best available theory (as of 1973) is a causal theory. That is obviously a specific epistemological assumption, in no way part of Quine's view of what ontological commitment involves, and even for the Benacerraf of 1973 a separate thesis. Many later writers who have argued that Benacerraf has raised a serious problem for mathematical knowledge have not adopted this epistemological thesis but have sought an alternative that would do the same critical work.[22]

The analogy that exists between the use of quantifiers over numbers (natural, real, and complex) and over the books in Hilary Putnam's office is first of all formal: reasoning about both can be modeled by first-order logic. I would guess that children begin to use the determiners of natural language that are modeled as quantifiers with respect to numbers fairly early. I have argued elsewhere that it is logic that tells us what an object is, or better, perhaps, what the concept of object is.[23] That makes the concept of object a formal concept. It certainly doesn't give encouragement to the idea that Putnam has so often objected to since the late 1970s, that Reality (presumably including "mathematical reality") consists of a fixed totality of objects with determinate properties.

It is also true that the articulation of such a concept of object is a modern development. It is at most hinted at by Kant, even though the notion of object is a central part of his conceptual apparatus. But Kant's developed conception is that of an object of experience, which will precisely not be an abstract object according to a common way of demarcating abstract objects, which I myself favor, since a Kantian object of experience is located in space and time and enters into causal relations. Probably the first articulation of something close to the formal conception of object was by Frege, although there are probably other nineteenth-century sources.

This, I think, offers the way to look at Putnam's suggestion in a footnote that "of course if one wants to simply extend the notion of object by speaking of 'mathematical objects' in connection with mathematical propositions, that is something different" (p. 366 n.23).[24] This extension was not just a late invention of philosophers, but a

---

[22] Most prominently Hartry Field, *Realism, Mathematics, and Modality*, introduction and the title essay.

[23] "Objects and Logic" or ch. 1 of *Mathematical Thought and Its Objects*.

[24] Putnam goes on to say that "then the application of 'intangible' makes no sense, since the tangible/intangible distinction goes with the ordinary, unextended use." I am not persuaded. Why should that not have been extended as well? What is Putnam to make of "intangible drilling costs" in the Federal income tax code? There is even a Wittgensteinian reply suggested by statements in writings of William Tait: to say that numbers are "intangible" is a grammatical remark, roughly that it does not make sense to say that we touch them.

gradual development of which the important background was the nineteenth-century transformation of mathematics.

Putnam raises a substantial issue in noting that the language of mathematical logic might be viewed as "the skeleton of an *ideal* language," a view that he says drives Quine's philosophy, or one can regard it as "simply a useful canon of rules of inference, and of formalization as simply a technique of idealization that facilitates the statement of these rules, their representation as (recursive) calculating procedures, and so on" (p. 368).

If we mean by the former outlook that the language is supposed to be language as it ought to be, that represents the world as it is while the language we ordinarily use (even in mathematics) is only a rough and ready approximation, then it is hard not to discern metaphysical excess. It's a delicate question of Quine interpretation to what extent Quine's view of the language of science leads him into such excess. Although Putnam is suggesting that it does, I don't think that he means to retract here his earlier statement that Quine is not a metaphysical realist.[25] In view of his thesis of the inscrutability of reference, Quine could hardly be a metaphysical realist in Putnam's sense, even though Putnam is critical of that thesis.

But seeing mathematical logic as working with a "language" that is an idealized model of actual language does not mean that it is not as workable a model for many purposes as models in science generally are. The application of logic that Putnam might be inclined to question is in developing theories of meaning and reference in natural languages. But in practice he does not question it in application to mathematics, for example that Gödel's incompleteness theorem tells us something about what certain means of proof cannot achieve. Another example would be independence proofs in set theory. Consider the statement "All Σ_2^1 sets of real numbers are Lebesgue measurable." This is one of a number of statements that the classical descriptive set theorists of the early twentieth century tried to prove. They failed. Gödel already showed that this particular statement is refutable if one assumes $V = L$, so that it could not be proved in ZFC. A forcing model showed in the 1960s that it could not be refuted in ZFC. It follows that the methods the classical theorists used could not have decided this question. But

---

[25] *Realism and Reason*, p. 223 n.; cf. also *The Many Faces of Realism*, p. 31, and *Words and Life*, p. 362.

that follows only if the representation of their methods of proof in the formal theory ZFC did not lose anything essential about what could be proved and what not.

I doubt that Putnam means to dispute any of this. But he does seem to resist applying the same point of view to ontology.

V

My own puzzlement about Putnam's view of Quine on ontology began with hearing him say in seminars that according to Quine, "exists" is *univocal*.[26] This was something he wished to deny. My first reaction was that this could not be a fundamental question for Quine, because it would make it a substantial issue whether the difference in the use of 'exist' in "Tables and chairs exist" and "Complex numbers exist" is a difference of meaning. I don't think he would have been much moved by the comment that these statements represent a different *use* of 'exist'; too many possible distinctions could be intended in talking of different uses. He would, however, observe that the difference between the basis on which we judge that tables and chairs exist and the basis on which we judge that complex numbers exist is better described as resting on a difference between tables and chairs and complex numbers, rather than on a difference in what we mean by 'exist'. Could we use 'exist' in the tables-and-chairs sense and consider "Complex numbers exist" so understood? Maybe; it seems that the answer would be that they don't, so that that might capture the nominalist's intention. But then it seems there would be no argument about nominalism; it

---

[26] Putnam does say this in print in a note to his first Dewey Lecture; see *The Threefold Cord*, p. 179 n.12. (I had not noticed this before the Dublin conference.) Putnam seems to recognize there that the question of the univocality of 'exist' should not be a fundamental one for Quine, but he says that Quine's practice with regard to issues of ontology is inconsistent with this. The explanation I would offer is that according to Quine, questions of ontological commitment properly arise only for statements and theories regimented in first-order logic. I think he takes that to imply not that the words 'object' and 'exist' as we inherit them are univocal in a way in which other philosophically significant words are not, but rather that the schematic and formal concepts of object and existence embodied in first-order logic suffice for his philosophical purposes, and possibly for any serious discussion of ontology. Like Putnam, I have argued that Quine's framework is too narrow, but Putnam's objection is much more far-reaching.

would just turn on an ambiguity. Leaving Quine aside, that is not what Putnam thinks.

In the paper discussed in the last section, Putnam discusses two more substantially different uses of the existential quantifier, as substitutional and as meaning $\neg\forall\neg$ in the fragment of intuitionistic logic that obeys the classical laws. I wouldn't myself object to saying that these represent a different meaning of the quantifier. Quine considers the substitutional quantifier as offering only an ersatz concept of object and existence. I proposed otherwise some years ago. I don't think the issue is of great moment.[27] If one enters into the intuitionistic way of thinking, which of course Quine does not, one may be tempted to say that it is also not important whether one says that the intuitionistic $\neg\forall\neg$ expresses a different meaning of the existential quantifier from the classical logician's. For lower levels of mathematics it is probably not. For higher set theory, matters are more complicated than I can go into here.

The remark about Benacerraf that we quoted above suggests that it is the objectual interpretation, as opposed to the substitutional, that incorporates allegedly unique meanings of 'object' and 'exist'. Quine says that in the case of a language like that of arithmetic, where for each element of the domain there is a term designating it, there is no difference between the two interpretations. That suggests that he doesn't assign any special metaphysical force to the objectual reading. The substitutional reading is unavailable in cases where the language's provision of names is inadequate, most prominently with the variety of very local objects that we encounter in perception and, at the other end as it were, where uncountable totalities are involved.

VI

Here again, I don't think I have succeeded in putting my finger on what troubles Putnam. Although I am not sure I can attribute this view to Quine, I would suggest that what the univocality claim amounts to is that there is a common formal structure, first-order logic, and that in fact theories can be formulated, admittedly in the idealized way that

---

[27] However, the substitutional interpretation is a sort of first approximation to the understanding of quantifiers in constructive type theories such as Per Martin-Löf's; see my "Intuition in Constructive Mathematics."

the application of symbolic logic implies, so as to maintain this structure and still be quite comprehensive, though not perfectly so. However, at least one type of gap in comprehensiveness is corrected by talking of an expanded domain of objects, still within this framework.

This is in keeping with the view of the concept of object as a formal concept, briefly set forth in §IV and developed in more detail in my own writings (see note 25). Invoking it brings us to one of the deeper disagreements between Putnam and Quine, about the inscrutability or indeterminacy of reference. In a late essay Putnam writes,

> And I still see ontological relativity as a refutation of any philosophical position that leads to it. For what sense can we make of the idea that the world consists of objects any one of which is a quark in one admissible model, the Eiffel tower in a second admissible model, but is no more intrinsically one of these than any other? Surely the very notion of "object" crumbles if we accept this. The reason that Quine does not see that it does is, I think, his belief that the laws of quantification theory by themselves give enough content to the notion of an object to render notions like "object" and "ontology" usable in metaphysics.[28]

In a footnote Putnam says that he argues against "this view" in the first lecture of The Many Faces of Realism (see note 27). By "this view" he seems to refer to the belief he attributes to Quine. I am not sure what argument from that lecture he has in mind. It may be the introduction in the last part the lecture of the idea of conceptual relativity, which is deployed again in the second lecture.[29]

The main theme of these lectures is criticism of the view that we can make at a global level a distinction between discourse and theory that represents the world "as it is in itself" and other discourse that is in one way or another second-class. The example with which he begins is Wilfrid Sellars's contrast between the manifest image of the world and the scientific image. Much of Putnam's argument in this lecture does not engage the issue of a general concept of object at all, although he might very well say that a metaphysical realist view implies that there

---

[28] "Realism without Absolutes," pp. 280–281.
[29] See also "Truth and Convention."

is a single "objective" conception of object.[30] Since he doesn't regard Quine as a metaphysical realist, that would not help with my present puzzlement.

Where he does touch on these issues is in introducing the idea of conceptual relativity, which he illustrates by an example of a simple world with three atomic individuals and the contrast between two ways of conceiving it, a "Carnapian" way in which the three individuals are all there is in that world, and a "Polish logician's" way in which there are also mereological sums of these objects.[31]

Evidently this example is meant to offer a very simple model of more serious cases where different conceptual schemes talk of objects in very different ways, with far-reaching consequences as to what, according to the scheme, there is. Putnam writes:

> And so it is no accident that metaphysical realism cannot really recognize the phenomenon of conceptual relativity—for that phenomenon turns on the fact that *the logical primitives themselves, and in particular the notions of object and existence, have a multitude of different uses rather than one absolute 'meaning'.*[32]

What is right about the second of the ways we considered of reconciling the two versions or 'worlds'—reinterpreting the existential quantifier—is that the notions of 'object' and 'existence' are not treated as sacrosanct, as having just one possible use. It is very important to recognize that the existential quantifier itself can be used in different ways—ways consonant with the rules of formal logic. What would be wrong . . . would be to accept this

---

[30] There is a connection between metaphysical realism and the recently much discussed question about quantification over absolutely everything, whether it makes sense, or at least makes sense in the way that quantification ordinarily does. I have argued that the affirmative view implies a form of metaphysical realism; see "The Problem of Absolute Universality." That the converse does not hold, that someone holding the negative view need not deny metaphysical realism, is argued in Michael Glanzberg, "Quantification and Realism."

My claim would not have been news to Putnam, who, in the context of explaining his rejection of metaphysical realism, writes of the notion of object as "inherently extendable" and glosses conceptual relativity as "the idea that there are many usable extensions of the notion of object, and many alternative ways of describing objects" ("The Question of Realism," p. 305).

[31] The same example is used in "Truth and Convention."

[32] *The Many Faces of Realism*, p. 19.

idea, and then go on to single out *one* use as the only metaphysically serious one.[33]

It is not clear to me whether Putnam means in these passages (or in the larger text of which it is a part) to offer an argument against the understanding of the concept of object as a formal concept. In the situation of the simple model, singling out one of the uses as "the only metaphysically serious one" would necessarily involve putting more into that use than the basic formal concept, which would only decide between the two ways of describing the toy world if we beg the question by assuming that the formal logic underlying the object concept includes mereology. That would not fit Quine's understanding, in view of his privileging of first-order logic. As for my own view, I don't privilege any particular logic, but I also haven't taken a position on whether changing the logic (e.g., adding modality, or going intuitionistic) changes the concepts of object. I do think that certain steps, in particular using modal logic or the most plausible version of the Meinongian idea of nonexistent objects, does introduce ambiguities about *existence* that are glossed over by Quine's view.[34]

Examples of this kind make plausible the view that a formal concept of object is a sort of skeleton, which is amplified in actual applications, possibly by a more determinate choice of logic, but more fundamentally by the flesh that is put on such a skeleton by various instances of objective reference, particularly the most basic such as that to bodies and other objects of our immediate environment.

Some of Putnam's more everyday examples are probably meant to suggest that even a more liberal version of the skeleton, such as I have sketched here, will not be adequate to the data. I will defer this question for the moment. Already relevant to the univocality issue is that the more liberal attitude toward logic than Quine's introduces what I have described as ambiguities.

The remark at the end of the above quotation from "Realism without Absolutes" suggests a different objection to Quine: that a concept of object along roughly these lines cannot suffice for *metaphysics*. The reply that the formal concept gives only the most general concept of

---

[33] Ibid., p. 35.

[34] On modality and its relation to mathematical existence, see some of the essays in *Mathematics in Philosophy*; on nonexistent objects, see §6 of *Mathematical Thought and Its Objects*.

object and that in particular cases (e.g., common sense or fundamental physics) something more needs to be added would, if I am reading Putnam right, be seen as conceding the criticism rather than answering it. Although the reply is one that I might make, I don't think that Quine would make it, given his thesis of the inscrutability of reference. Roughly, the possibilities of reinterpretation that imply that there is no fact of the matter about which objects individual predicates of a theory are true of also affect the predicates that would lay out the "something more" that would distinguish other fundamental kinds of objects.[35]

Already in *The Many Faces of Realism,* and even more in later writings, Putnam engages in a defense of common sense. Of course that separates him from Quine in many ways. I am in general sympathetic to this project and to the complaint that *perception* does not get its due in Quine's account of reference, probably not even in his genetic accounts.[36] But my main concern is with the question whether Putnam has ground for objecting to the formal concept of object, or at least objecting to reading discourse in which objective reference occurs in such a way as would fit it.

I don't know of a place where he focuses on the issue in this way. However, he appears to raise doubts about it in the first Dewey Lecture. He gives a variety of examples: the sky, mirror images, intentional objects that don't exist, a lamp in his house whose shade falls off whenever it is moved (so that its parts do not in general move together), and particles in quantum mechanics.[37] The context is the question whether the questionable objects should be recognized as objects by a realistic *theory.* That makes it unclear whether the examples are meant to show a fundamental inadequacy in the formal concept of object.

However, we can consider the examples without answering this question about Putnam's intentions. The most challenging may be that of quantum particles, and here I don't have the knowledge to give an answer.[38] Intentional objects, in my view, are just objects in some other

---

[35] An instructive treatment of the issues is Peter Hylton, "Quine on Reference and Ontology." Putnam is not mentioned in Hylton's essay, but it makes clear how deep the differences are that underlie Putnam's rejection of Quine's indeterminacy thesis.

[36] See for example "Realism without Absolutes," pp. 281–284.

[37] *The Threefold Cord,* pp. 7–8.

[38] It appears that the issue might be put as one about their individuation. They may challenge any account of the notion of object. Interestingly, Quine mentions puzzles about quantum particles as a reason for adopting an ontology that dis-

sense talked of in intentional contexts, so that the problems they present are those of intentional contexts. Mirror images seem to me to be objects in a perfectly straightforward sense: they are phenomena as objective as is required by reference in everyday contexts, and spoken of with the same general devices of reference as other everyday "objects," even though they are not bodies. One can say much the same about the sky and about Putnam's lamp; the latter, in fact, is much closer to being a body.

Leaving aside the case of quantum particles, then, I don't see these examples as counterexamples to the formal concept of object.[39]

In this essay I have surveyed some things Putnam has said about ontology, especially as Quine conceives it. I have not always been successful in making the issues clear. I hope my attempt will stimulate the author and others to help me out. In some cases, where there is disagreement, I have tried to make out that a view either Quine's or close to it can be defended.[40]

---

penses with matter in favor of space-time regions and ultimately sets; see "Whither Physical Objects?," especially pp. 498–500.

[39] It must be admitted, however, that as usually described (in particular by me) this conception does involve some idealization. For example, the application of logic usually does not take account of vagueness, which can affect even identity statements. In the above I have taken this as going without saying.

[40] Sections I–V of this essay are a revision of my lecture at the Dublin conference, March 11, 2007 [commemorating Putnam's 80th birthday the previous summer]. Section VI has been added. [My debt to exchanges with Putnam over many years should be evident.]

Review Essay on Tait, *The Provenance of
Pure Reason: Essays on the Philosophy
of Mathematics and Its History*

1. Introduction

William Tait's standing in the philosophy of mathematics hardly needs to be argued for; for this reason the appearance of this collection will be welcomed. As noted in his preface, the essays in this book "span the years 1981–2002." The years given are evidently those of publication, although one essay (no. 6) was not previously published in its present form. It is, however, a reworking of papers published during that period. The introduction, one appendix, and some notes are new. Many of the essays will be familiar to the readers of *Philosophia Mathematica;* indeed two (nos. 4 and 12) first appeared there.

It should be no surprise to those who know Tait's work that this is a very rich collection, with contributions on a wide variety of issues, both systematic and historical. That poses a problem for a reviewer, because a serious discussion of even all the major issues would be beyond the scope of a review.

Before the essays in this collection, Tait's publications were almost all in mathematical logic, especially proof theory. He was a leading actor in the work on the revised Hilbert program stimulated after the war by Georg Kreisel, Kurt Schütte, and Gaisi Takeuti. Tait thus has an experience in mathematical research that is rare among those whose careers have been institutionally in philosophy. In the 1970s Tait became disillusioned with the proof-theoretic program on which he had worked. However, he draws on that background in several of these essays, and he has continued to work on mathematical questions. In particular, the set-theoretic project reflected in essay 6 is as much mathematical as philosophical.

It is not easy to characterize Tait's philosophical viewpoint briefly. Both the title of the book and some of its content mark him as a rationalist. His hero in the earlier history of philosophy is Plato, although he writes little about other canonical figures of the rationalist tradition.[1] But his view differs significantly from those of the canonical rationalists of modern foundations of mathematics, Frege and Gödel. Tait's philosophical method embodies the "linguistic turn," and he writes sympathetically about, and is clearly influenced by, the later Wittgenstein. Tait's rationalism is tempered by the aspiration to anchor reason and objectivity in human practices. And he is not sympathetic to any notion of rational or a priori intuition. His view of axioms in mathematics seems clearly at odds with that of Gödel and in important respects with that of Frege.

In what follows, I will discuss this outlook in more detail. The essays in the collection deal with more specific questions in foundations, as well as with historical figures, especially Plato, Dedekind, Cantor, Frege, Hilbert, and Wittgenstein. His point of view is not much influenced by the debates in philosophy of mathematics in the last generation or so, and not much more by what has come to be thought of as the central history of analytical philosophy from Frege, Moore and the early Russell, Wittgenstein's *Tractatus,* the Vienna Circle, to the later Wittgenstein and Quine (with the noted exception of the later Wittgenstein).

What is central for him and the object of some of his historical discussions is the history of the foundations of mathematics as it grew out of the nineteenth-century revolution in mathematics and continued into the twentieth century. In fact, these essays might be viewed as a plea for the philosophical importance of this history and especially for the contributions to it of Cantor, Dedekind, and Hilbert. Although he does not deny his importance, Tait undoubtedly deplores the virtual obsession of many philosophers of mathematics with Frege or ideas derived from Frege.[2] Thus his most extensive discussion

---

[1] Tait's "Reflections on the Concept of a priori Truth," not included here, discusses Leibniz sympathetically. His comments on Aristotle and Kant, generally critical, do not emphasize their rationalistic side. That is also true of most of his comment on Frege, but that is a more complicated case, which I will not go into.

[2] Curiously, he writes little about Russell.

of Frege is in essay 9, devoted to defending Cantor and Dedekind against him and criticizing his powerful advocate in our own time, Michael Dummett. Another figure about whom Tait has written at length is Gödel (see §5 below). There are also comments about Zermelo and Brouwer.

The plan of this essay is as follows: §2 attempts to explain and comment on Tait's general view of mathematical knowledge, what might with some violence be called his epistemology. The central places where that is presented are the introduction, essays 3 (the well-known "Truth and Proof") and 4, and, surprisingly to many readers, essays 7 and 8 on Plato. Section 3 concerns the issues about realism and "platonism" that Tait discusses in this connection. Section 4 discusses in some detail his analysis of Hilbertian finitism in essays 1 and 2. In §5 I will comment on a selection of issues arising from Tait's discussions of historical and contemporary figures.

2. The Axiomatic Conception

In his introduction, Tait writes that the axiomatic conception of mathematics is the only viable one (p. 4). What he calls the "axiomatic conception" certainly includes the axiomatic method. What is most distinctive about mathematics is that it is exact and that the criterion of truth is proof. These features are fully realized only when a branch of mathematics has been axiomatized and its primitive concepts identified, and proofs from axioms are made rigorous. Because of the exactness involved, mathematics is idealized when set against the world as represented in perception and in most common sense knowledge. Here he identifies his view with that of Plato, of whom he says that he recognized "the fact that the exact empirical sciences of his day—geometry, arithmetic, astronomy, and music theory, for example—were not literally true of the sensible world in the semantical sense and, indeed, did not literally apply to it" (p. 68; cf. p. 182). His emphasis on this point is one aspect of his kinship with Plato.

One element of the axiomatic conception as Tait understands it is that he rejects the demand that there be some prior standard that axioms have to meet in order to be treated as true or as known. One might say that they are true because they are axioms, in contrast to the view advanced by Frege in his correspondence with Hilbert that to be a candidate for axiomhood a statement must be true.[3] Problems that have troubled many philosophers lose a lot of their force on this position. Other well-known views, such as that axioms are evident to our reason (classical rationalists, apparently Gödel) or may become so from enough experience in working with them (possibly Bernays), or they may have an evidence that rests on an intuition that is not purely rational (Kant and other broadly Kantian views), or they may be accepted because a theory based on them is essential to scientific knowledge (Quine), lose a lot of their motivation. However, Tait seems to embrace the first view about the principles by which conclusions are drawn from axioms, that is principles of logic (which for him include a principle of choice that, in classical logic, implies the axiom of choice).

It might seem that the choice of axioms will then be quite arbitrary or at least depend on judgments about whether the resulting theory is useful or fruitful. Such a view calls in question the claim of mathematics to be a science or at least more than an auxiliary to empirical natural science. However, that is not Tait's view; he belongs among those who defend the autonomy of mathematics. He appropriates Plato's term "dialectic" for the process by which axioms are arrived at. Unfortunately he has rather little to say about it, even in the two essays about Plato. A passage on p. 97, although it mentions Plato, emphasizes grounds for accepting axioms that are familiar from modern discussions: the naturalness or fruitfulness of the resulting theory or its being "more in keeping with our intuitions than its alternative," or what appears to follow from "our conception" of the domain (e.g., of sets or ordinals). Something he does not mention, although I think it is implicit in his discussions of Plato, is that axiom systems have historically often been introduced to codify and make rigorous proofs carried out informally without making their assumptions explicit.

In essay 7, "Plato's Second Best Method," Tait represents Plato as concerned with a number of sciences and thus with the explanation of phenomena. But the context is the view that such theories never fit the

---

[3] Frege, Letter to Hilbert of December 27, 1899 (*Wissenschaftlicher Briefwechsel*, p. 63). Frege uses the word "axiom" so that axioms are not principles of logic. Although that usage does not agree with the most common usage of contemporary logicians, it happens to fit the way Tait uses the word in setting forth the axiomatic conception.

phenomena exactly. And whatever the process is by which first principles are arrived at, their development then proceeds by purely rational means. Tait sees the discussion of geometrical concepts as aiming at something like Euclid's axiomatization, although the *Elements* date from at least a couple of generations later than the time of Plato.

It would be natural to infer from Tait's statements that the phenomenological observation about how mathematics proceeds from axioms and the absence of an exact fit with sensible phenomena shows that mathematical knowledge is a priori.[4] One might object to that conclusion on roughly the following grounds: Physical theories are also idealized, but conclusions are also derived from them (with auxiliary mathematics) that are tested by observation, typically involving quantitative measurement. There is likely to be an unavoidable margin of error in the measurement, but that does not prevent observed results from calling the physical theories in question. Since physical theories are not a priori, the idealized character of a theory is not enough to show that it is a priori.

In this picture it is generally assumed that the mathematics involved will not be called in question. That would be a consequence of Tait's view that proof is the criterion of truth in mathematics. But one can question whether this really lays the issue to rest. It might be replied that the reason why proof works as a criterion of truth is that it doesn't yield results that conflict with experience, and in the scenario just envisaged, one will normally modify something other than the mathematics involved. Tait does not address the issues that Quine and others raise about the significance of this fact. He does cite (p. 8) a criticism of Quine made by me,[5] to the effect that if experience shows that some aspect of the physical world fails to instantiate a certain mathematical structure (e.g., space-time failing to be Euclidean), one will modify the theory by substituting a different structure, while the original structure does not lose its status as part of mathematics. Someone holding Quine's view could question whether that is the general case. Quine himself replied that we are disinclined to tamper with logic and mathematics because of the pervasiveness throughout science of their vo-

---

[4] I mean this in the sense standard today, somewhat distantly descended from Kant's usage. Tait himself prefers a different usage, which in the paper cited in note 2 he derives from Plato and Leibniz.

[5] *Mathematics in Philosophy*, pp. 195–196.

cabulary and of the application of their principles and results.[6] I believe that Tait is unmoved by these doubts because the separation of mathematical theory from empirical application is something like a starting point of his axiomatic conception; in particular, the physical theories have a significant mathematical component.

Before going further, we should say something about the propositions-as-types language that Tait uses in several essays. Such a language has an earlier history, going back at least to the Brouwer-Heyting-Kolmogorov (BHK) interpretation of the intuitionistic logical connectives. A standard source is William Howard's "Formulae-as-Types Notion of Construction," actually written in 1969. A simple language of this kind will have basic types such as the type N of natural numbers, for given types A, B a type $A \wedge B$ of ordered pairs of an object of type A and an object of type B, and a type $A \to B$ of functions that map any object of type A to an object of type B. An extension would admit also types $\forall x: A.B(x)$ and $\exists x: A.B(x)$, where, as indicated, B may depend on x. An object of the former type is a function mapping objects of type A to objects of type B(x); one of the latter type is a pair of an object of type A and an object of type B(x). So far the type symbols have the closure properties of formulae of first-order logic without negation or identity. Negation would be provided for with a null or empty type 0, so that $\neg A$ is $A \to 0$. Tait also introduces a two-element type 2 of truth-values (p. 77).[7] He views a proof as a construction of an object of the type represented by a formula. The type then takes on the role of a proposition, which is true if the type is nonempty.

For most authors the resulting logic is intuitionistic. Tait, however, uses this framework also for classical logic, which is obtained by adding for each formula A a function of type $\neg\neg A \to A$ (pp. 12, 120).[8] He refers to the objects thus introduced as "ideal," evidently echoing a usage of Hilbert, but he does not give much explanation of what this means. However, he notes that his logical principles other than this one admit only uniquely characterizable objects, but in general there is no

---

[6] "Reply to Charles Parsons," pp. 399–400.

[7] Then he shows how to define conjunction and disjunction. The finitist language of essay 1, discussed in §4 below, is much less expressive.

[8] Avigad (Review of Tait, p. 610) argues (necessarily briefly) that the use of this type-theoretic framework for classical mathematics has some awkward features. Avigad's review manages to present the content of Tait's book in an exemplary way, in much briefer compass than the present essay.

uniquely characterizable function from $\neg\neg A$ to A (p. 120). This remark echoes the early complaints about the axiom of choice that it introduces sets that are not definable, and it occurs in the context of the argument of essay 5 to the effect that the difficulties historically attributed to the axiom of choice are better ascribed to the law of excluded middle.

Tait is not as explicit as one would wish about the rules of proof for this logic, although what is said about the finitist fragment in essay 1 is helpful, and more is set forth in his "Against Intuitionism," e.g., p. 187.[9] And one can turn to other writers such as Per Martin-Löf. At all events, even with more types than are indicated above, one will arrive at a formal system that, assuming it has the basic rules for the type N of natural numbers, will be subject to the incompleteness theorem. Tait emphasizes that any particular such system can be expanded by introducing additional types; he clearly regards that as the natural way to expand resources for mathematical proof. Furthermore, he points out that what we can construct even of low types such as $N \to N$ can depend on what further types we have at our disposal.

Tait's view that proof is the criterion of truth has to make room for the fact that the incompleteness theorems are normally interpreted to imply that truth outruns provability in any particular formal system. Tait has quite a bit to say about the problems in this domain. I have still found it difficult to pin down his view. In particular, what he says about truth was somewhat confusing to this reader.

Tait does explicitly disclaim the idea that truth of a mathematical statement might be defined as provability from certain axioms. In the context of discussing the relation of excluded middle and bivalence, he writes:

> Of course, when I say that provability is the criterion for truth, I don't mean provability from some particular axioms. Rather, I am saying that, in mathematics, the assertion of truth is justified only by a proof, whatever we take the basis of the proof, the axioms, to be. (p. 96)

This is allowed for by the fact that the possibilities of constructing new types are open-ended. Introducing new types is, in effect, admitting new axioms. This raises three questions: (1) What can be said

---

[9] Tait chose not to include that paper in this collection; see p. vii.

about the introduction of new types? (2) Is it clear that this adequately captures the possibilities of introducing new axioms? (3) Where does this leave the idea that mathematical statements are either true or false? On Tait's assumption of classical logic we should be able to say this prior to the introduction of axioms that can settle them.

Concerning (1), Tait says little. He does, however, give one example: viewing the ordinals as an "incomplete type" Ω and, for a given ordinal β, the ordinals $\alpha < \beta$ as a type Ω_β (pp. 123–124). This enables him to explain how the Gödel sentence of a formal theory becomes provable by admitting a new type of the latter form. However, his development in essay 6 of a method of introducing large cardinals in set theory is carried out in terms of the usual language of set theory. Tait does not directly offer an answer to (2), and it does not seem to me that he is committed to an affirmative answer. It is possible that the propositions-as-types language in the end plays a less central role in Tait's thought than first appears. It is central in only three essays: essay 1, "Finitism," essay 3, "Truth and Proof," and essay 5 on the law of excluded middle and the axiom of choice. The latest of these, essay 5, was published in 1994. The theory of types reappears in the introduction, the last part of the book to be written. After describing its attractive features, Tait remarks (p. 14):

> However attractive the type-theoretic conception of mathematics is, it is not essential to the meaningfulness of mathematical propositions: the meaning of an arithmetic or set-theoretical proposition derives from the axioms that govern the basic concepts involved in it. The axiomatic conception need not be founded on the type-theoretic conception. (Nevertheless, it would be of interest to find out what implications the type-theoretic conception has for set theory.)

Concerning question (3), Tait argues that if A is understood as a mathematical proposition, then 'A is true' amounts essentially to A, so that bivalence collapses to the law of excluded middle (p. 95). It would follow that at any point we are able to say that A is either true or false.

It is not clear to me that that is the end of the story, even by Tait's lights. Some of what Tait writes about truth is confusing. Against Paul Benacerraf, he maintains that "it is difficult to understand how Tarski's 'account of truth' can have any significant bearing on any issue in the philosophy of mathematics" (p. 66).[10] The reason, as I understand it, is that a meaning given to 'S is true' based on Tarski's procedure would be another mathematical statement. But in the standard classical case at least, it would be in a theory with greater expressive power than the one it refers to. In the type-theoretic framework that would be one with additional types. In the case of a formal language where we can define 'S is true' and prove its equivalence to S, we are already presupposing greater expressive resources than those of the formal language in question. Tait relies on this fact in using the language of first-order arithmetic in the argument of pp. 95–96 to the effect that bivalence collapses to excluded middle. For the general case, incorporating for example the language of set theory, he would have to rely on the fact that the expressive resources of a formal language can always be expanded, or perhaps more accurately seen as partial.[11]

Tait does not address the manner in which the concept of truth is used in the philosophy of language, following especially the ideas of Davidson, who at a certain point made clear that truth had to be a primitive for him, in contrast to Tarski's emphasis on giving a definition. On this understanding, reference cannot so easily be avoided as Tait claims. Given the amount of discussion there has been about truth and its relation to meaning, Tait seems to me rather too quick with the concept of truth, although Davidson's procedure would not change the essential point of what was commented on above. Furthermore, as regards reference Tait has good arguments against regarding "the existence of mathematical objects" as a major problem.

The more serious issue does not depend on technicalities about truth and reference. We want to say, of statements that are not decidable from axioms that we now accept, that they are either true or false. How does Tait's explanation comport with the view of proof as the criterion of truth? At one point he remarks that there may be state-

---

[10] The immediately following statement seems to confuse truth with truth in a model. I don't think Tait's argument really turns on this, but the exposition is confusing because he seems to try to combine arguing against the relevance of a Tarskian conception of truth with arguing against the idea that mathematical statements are to be understood as relating to a "Model-in-the-Sky."

[11] In a communication of January 2009 commenting on an earlier version of this essay, Tait wrote, "Truth definitions are given in a metalanguage in which 'A is true' unfolds into a translation of A, and mathematics has no metalanguage." These remarks are my attempt to capture what he meant.

ments that have a proof from axioms that we *would* accept, even if we do not now accept them (p. 79). Although Tait does not claim otherwise, it is not evident that given a mathematical proposition A, either A or ¬A must satisfy this condition.

Tait considers the problematic example for this issue, the continuum hypothesis (CH). He notes that new axioms adopted in the future could lead to a particular decision about CH, say that it is true. Then he writes:

> The problem is that, on the axiomatic conception, the axioms define the subject matter and so there is no criterion of truth beyond what can be proved from the axioms. So, on this view, there is no issue of truth concerning the proposed new axiom prior to its acceptance. (p. 96)

I am not sure what he means by there being "no issue of truth" concerning such an axiom. Clearly we would not adopt it as a new axiom if it were not independent of the existing ones, so that the criterion of truth could not be antecedently satisfied for it. Perhaps that is all he means, since he does go on to mention possible grounds for accepting it, which he puts under the heading of "dialectic" (as noted above). But in a case like CH, as Tait notes (p. 98), the possibility is not easily ruled out that set theory might bifurcate, i.e., that present set theory might be extended by axioms that are equally "natural" or recommended by other considerations but yield opposite verdicts regarding CH. So long as we haven't ruled out that possibility, we can't say that either CH or ¬CH can be proved from axioms that we would accept, or even that both satisfy this condition.[12] As of now it is tempting to say, echoing discussions of vagueness, that CH is either true or false, but it may not be either determinately true or determinately false.[13]

In practice Tait relies on the idea that some axioms, and then some extensions of an existing axiom system, are implicit in the underlying concepts. In the context of set theory, he would accept Gödel's idea

---

[12] In the last case, if the axioms define the subject matter, it would be natural to conclude that CH is ambiguous in some way.

[13] In an unpublished manuscript, "Logic and Dialectic: The Truth of Axioms," Tait says that "CH ∨ ¬CH does not imply that the truth of one of the disjuncts is determined" (p. 10). I do not know what weight to put on this statement, in particular whether Tait means to embrace a notion of determinate truth.

that axioms may be "implied by the concept of set."[14] However, Tait's conception of the role of such considerations is quite different from Gödel's. For Gödel, making out that this is the case for a proposed axiom would be a direct argument for its truth, and the "perception" of the concepts involved would be an exercise of mathematical intuition. In Tait's case, such considerations evidently belong to dialectic. Furthermore, he views them as aspects of our understanding and use of the relevant language.

It seems to me that Tait's axiomatic conception is in an awkward position with respect to the consistency of theories. Obviously, to have a claim to truth and even to be of any use, a theory must be consistent. But although the consistency of a theory can be formulated as a mathematical proposition, Tait is very skeptical of the possibility of giving an informative proof of it in interesting cases.

> But we know that there is no relevant sense in which we can be sure that this requirement [of consistency] is met. No matter how familiar we become with set theory, or even arithmetic, as at least partially expressed by axioms, and no matter how intuitive these axioms are or become for us, consistency is just something that, ultimately, we must take on faith. (p. 90)

Of course for some theories, say first-order arithmetic PA, we can give a proof of consistency. Tait evidently holds that such a proof does little to add to such conviction of its consistency that we already have, in particular since it will use some principle (such as transfinite induction on a primitive recursive well-ordering of type ϵ_0) that is not derivable in PA. Furthermore, he would observe that for strong theories such as set theory we have no idea how to give the sort of consistency proof that proof theorists have sought and reason to doubt that one is possible that would be constructive in the sense that post-Hilbertian proof theory has worked with.

Still, I am not sure what Tait's claim is. Does he mean to say, for example, that we just "take on faith" that ZFC is consistent? The re-

---

[14] See for example the 1966 note to "What Is Cantor's Continuum Problem?," *Collected Works* (hereafter CW), vol. II, p. 260 n.20. Gödel apparently held that this is true for the axioms of ZFC and some large cardinal axioms, at least those implying the existence of inaccessible and Mahlo cardinals. But the note cited says that it has not yet been shown that this is true of axioms providing for much larger cardinals, e.g., measurables. I don't believe he thought the situation changed later.

mark just quoted seems to aim to deflect an answer that many would give, an appeal to mathematical experience. The resulting confidence in its consistency is at least in part empirical: Set theory has been working with these concepts since the time of Cantor and there is an understanding available of its basic ideas that deflects the arguments that led to paradoxes. It has been shown how to formalize many proofs in ZFC, and no contradiction has arisen in long experience. Among the consequences of the axioms, given constructions of the integers and real numbers in set theory, are theorems of arithmetic and analysis that have been accepted for a very long time, in some cases for centuries.

It would be natural to read Tait's talk of the axioms as "intuitive for us" as meaning that at least after this long experience of working with set theory, the axioms come to seem evident.[15] If one trusts this, one will say that they have "acquired evidence," to use a term due to Paul Bernays. Tait would surely emphasize that such evidence is fallible. But that is also the view of many modern rationalists who hold that mathematical axioms can be evident in a more direct way than Tait admits. Our beliefs about our immediate environment based on perception are also fallible in this way. Would Tait say that ultimately I must take on faith that I am now sitting in my study at home writing on my computer?

The more empirical side of this argument may well make Tait uncomfortable. But the most decisive part of it is not empirical in the usual sense but rather what Hilary Putnam once called quasi-empirical: the axioms yield a theory in which no contradiction has been derived, in which theorems that had been proved by more elementary means can be derived, which gives a satisfying picture of the mathematical phenomena in many domains. That we cannot be entirely certain of its consistency all would agree. It does not follow that our belief is without foundation.

That brings me to another point of tension in Tait's axiomatic conception. Many writers on set theory, beginning at least with Gödel, have argued that axioms of set theory can obtain warrant through their consequences, through the kind of theory they yield. The idea could hardly be expressed better than it was by Gödel:

---

[15] In the communication cited in note 10, Tait writes, "I mean rather that the notion of set which the axioms define becomes intuitive. We learn the language, so to speak." It seems to me that the reading I propose would follow from this.

There might exist axioms so abundant in their verifiable consequences, shedding so much light upon a whole field, and yielding
such powerful methods for solving problems . . . that, no matter
whether or not they are intrinsically necessary, they would have
to be accepted at least in the same sense as any well-established
physical theory.[16]

About another remark in about the same place and to much the same
effect, Tait writes:

> It is difficult to reconcile this with the iterative conception of the
> universe of sets we are discussing here. On the latter conception,
> the "intrinsic necessity" of an axiom arises from the fact that it
> expresses that some property possessed by the totality of ordinals
> is possessed by some ordinal. . . . A "probable decision" about
> the truth of a proposition from the point of view of the iterative
> conception can only be a probable decision about its derivability
> from that conception. (pp. 283–284)

Tait seems to be demanding of an axiom of set theory on the iterative
conception that it be intrinsically necessary, possibly not in the precise
sense he mentions, but at least in the looser sense of being "implied by
the concept of set," as Gödel would put it.[17] This is apparently a consequence of the view that the axioms define the subject matter. (Here
he clearly views the axioms as open-ended and not limited to a definite
formal system, but extensions of the formal system should be further
articulations of an underlying conception.) Later Tait suggests that
Gödel's comparison of mathematics with physical theory does make
sense on the more robustly realistic view that Gödel at least sometimes
expresses (p. 295).

Tait's conception of dialectic seems to make room for the considerations cited in the above quotation from Gödel. Possibly he means to
say no more than that a theory incorporating such axioms is a different
theory from a theory whose axioms are intrinsically necessary in the
sense that Gödel aimed at. However that may be, the demand for the
latter is an additional demand, not obviously part of the axiomatic
conception as Tait has outlined it.

---

[16] "What Is Cantor's Continuum Problem?," 1964 version, CW II, p. 261.
[17] On this point, Tait has changed his view since the publication of the book under
review.

3. Realism

These questions lead into another set of questions Tait discusses at some length, concerning realism or "platonism." That is the main theme of essay 3, "Truth and Proof," and it is also prominent in the following essay 4, "Beyond the Axioms." In essay 12 on Gödel, he discusses the question of Gödel's avowed platonism.[18]

Tait makes a very useful point in singling out what he calls "default realism":

> I want to save this term ["realism"] for the view that we can truthfully assert the existence of numbers and the like without explaining the assertion away as something else. Realism in this sense is the default position: when one believes mathematics is meaningful and has, as one inevitably must, finally become convinced that mathematical propositions cannot be reduced to propositions about something else or about nothing at all, then one is a realist. (p. 91)

This amounts to taking the language of mathematics at face value, with its reference to such objects as numbers, functions, and sets, and accepting what is proved in mathematics as true. Strictly, default realism does not involve holding that a paraphrase of language eliminating reference to mathematical objects cannot be carried out; the latter view is rather a strong motivating factor. But it does, I think, involve holding that no such paraphrase is necessary for mathematics to be sound.

In discussing Gödel, Tait says that "in a sense, realism in the default sense is not a substantive philosophical position" (p. 290). There he invokes something like Carnap's distinction of internal and external questions. An internal denial of the existence of numbers would have trivial counterexamples; an external denial

> ... depends on a univocal notion of existence which is simply not forthcoming. It is only as a meta-theoretical stance that realism is substantive, defending the legitimacy of ordinary mathematics against skepticism. (Ibid.)

---

[18] Note that I do not capitalize this word, although Tait consistently does. There is an issue here, namely what connection a view has or should have with the philosophy of Plato, about which I say a little later in this section.

Earlier, in essay 3, he writes that freed of confusion "Platonism will appear, not as a substantive philosophy or foundation of mathematics, but as a *truism*" (p. 62). I will say more about this shortly.

It is important to note that default realism as Tait understands it does not privilege classical over constructive mathematics (p. 291). That conflicts with the use of the terms "realism" and "anti-realism" by Michael Dummett and with the earlier use of the term "platonism" by Paul Bernays. It is not obvious that a finitist as Tait understands that view could not count as a realist. In contrast to the discussion of platonism by Bernays in "Sur le platonisme," Tait does not emphasize that there are different levels of mathematics that one can be realist or platonist *about*, but his view clearly leaves room for that. Tait's allowing that a constructivist mathematician can be a realist in the default sense coheres with his view that constructive mathematics is part of classical mathematics (introduction §4, parts of essay 5, and "Against Intuitionism").

A large part of essay 3 is devoted to criticizing alternative understandings of Platonism. Essay 4 introduces a contrast between default realism and what Tait calls superrealism (p. 91). This would be realism that goes beyond default realism, even in the context of accepting classical logic and set theory. This is a term belonging to a family of terms adopted by philosophers to describe versions of realism that they do not accept or even find coherent. Kant's "transcendental realism" and Hilary Putnam's "metaphysical realism" are well-known examples.

Of the different terms on offer, Kant's term "transcendental realism" is in my view preferable, because it reminds us that versions of realism that major philosophers find unacceptable have a genuine historical pedigree. Of course Kant wished to argue that the difficulties of transcendental realism leave no alternative but to accept transcendental idealism. Few would accept that conclusion, and Tait himself belongs to those philosophers in our time who have sought to arrive at a version of realism that does not carry heavy metaphysical or even epistemological baggage. In this he shows kinship with Putnam.

A difficulty with all these stances is that the philosopher describing the rejected version of realism attempts to argue that his own version, by some lights more attenuated, is the only alternative. The difficulty of this strategy is shown by the example of Putnam, who has tried different alternatives in some years of searching for an appropriately metaphysically modest version of realism. Another difficulty is that the alternative to what I will call transcendental realism may turn out to be just as counterintuitive or contrary to common sense, as is Kant's transcendental idealism on the most prevalent interpretation.

Part of the force of denying that default realism is a philosophical position is, I think, the view that superrealism and perhaps also intermediate positions that might be proposed are incoherent and not really meaningful. That would be a natural conclusion from the generally Wittgensteinian take on the issues that Tait adopts. It is less clear that the arguments he offers justify such a strong conclusion. The critical focus of essay 3 is a picture of Platonism he derives from writings of Benacerraf and Dummett. According to this picture, mathematical practice takes place in an object language that has to be interpreted.

> The Platonist's way is to interpret it by Tarski's truth definition, which interprets it as being about a model—a Model-in-the-Sky— which somehow exists independently of our mathematical practice and serves to adjudicate its correctness. So there are two layers of mathematics: the layer of mathematical practice in which we prove propositions such as (1) [There is a prime number greater than 10] and the layer of the Model at which (1) asserts the "real existence" of a number. (p. 67)

Tait's arguments against this picture are convincing, but it is only one way of articulating what a platonist view is supposed to be.

However, probably more central to this essay are various considerations favoring the view that default realism is sufficient, so that the attempt to articulate a stronger realism is not fruitful. Again, however, such a claim would not justify the further claim that any such attempt will lead to statements that are incoherent or lacking in meaning. However, the issue arises again in essay 4, where after introducing the notion of superrealism, Tait comments on some statements by Reuben Hersh (pp. 92–96). There he argues that the statements involved are either truisms, or grammatical remarks, or confused in some way, or expressions of superrealism. Except for the last point, this again is an argument to the effect that default realism is sufficient.

The relevant point about superrealism is that Tait objects that it makes mathematics "speculative"

> ... in the sense that even the most elementary computations, deductions, and propositions must answer to a reality which we, at best, can only partly comprehend and about which we could

be wrong. If there are grounds for truth and existence and they are not the axioms, then the axioms could be false. (p. 91)

I see this as a serious reason for taking superrealism to be false, but it is not clear that it licenses a stronger conclusion.

Two features of classical mathematics as we know them suggest a stronger realism: the acceptance of the law of excluded middle and the admission of impredicative specification of sets (and very likely of other concepts); the latter arises even in intuitionistic mathematics. Tait has a developed view of the former issue, already canvassed in §2. He is silent about the latter. The former was, as noted above, a criterion of platonism for Bernays and then the starting idea of the elaborate analysis of the notion of realism by Dummett. Apparently Tait attributes to Dummett the view that a realist by the latter's criterion is unavoidably a superrealist.[19] I don't find that quite evident, but that could well be due to my not being sufficiently clear about Dummett's conception. I think Tait's quarrel with Bernays's methodological platonism would merely be with his calling it platonism.

The role Tait gives to human practices and agreement raise a question whether his view gives mathematics the objectivity that we expect of it. In §5 of essay 4, Tait addresses this issue in the form of the "postmodern" challenge to the very idea of objectivity. He acknowledges that this affects also his view of logical inference (p. 99). I do not find his reply to a "social constructionist" view satisfying. One difficulty is in getting any handle on what an alternative form of life in Wittgenstein's sense might be. The issues are delicate, however, and I am not confident in advancing a view of my own, so that I have to leave the matter there.

Concerning the word "Platonism," Tait was fighting a battle that was already lost when "Truth and Proof" was published in 1986. Many contemporary writers on the philosophy of mathematics who use the term explicitly deny that they are attributing the view they discuss to Plato. That is one reason for not capitalizing the term. Tait, who knows far more about Plato than most logicians and philosophers of mathematics, is certainly entitled to point out that what has been

---

[19] I am guessing that the unnamed philosopher referred to in the first full sentence on p. 92 is Dummett.

called platonism in our time is fundamentally different from what Plato argued. Most of us would happily admit that. Bernays did make himself vulnerable when he wrote of the tendency that concerned him:

> This example [a contrast of Hilbert's geometry and Euclid's] shows that the tendency of which we are speaking consists in viewing the objects as cut off from all links with the reflecting subject. Since this tendency asserted itself especially in the philosophy of Plato, allow me to call it "platonism."[20]

Bernays seems to attribute to Plato a position in characteristically modern idealism/realism debates. Still, his usage does set up a useful contrast between what he calls platonism and different versions of constructivism. The dominant use of the term in post-war American philosophy of mathematics has been to apply it to views or theories that involve any commitment at all to abstract entities. From its origin in Goodman and Quine's "Steps toward a Constructive Nominalism" to its widespread use today, it has been connected with Plato by nothing more than the use of the word.

Although it is not the main issue for him, Tait's conscience concerning the word may be a cause of some injustice toward other writers. One of these is Dummett, especially in his paper "Platonism." Tait twice quotes the characterization with which the paper begins:

> Platonism, as a philosophy of mathematics, is founded on a simile: the comparison between the apprehension of mathematical truth to the perception of physical objects, and thus of mathematical reality to the physical universe. For the platonist, mathematical statements are true or false independently of our knowledge of their truth-values: they are rendered true or false by how things are in the mathematical realm. And this can be so only because, in turn, their meanings are not given by reference to our knowledge of mathematical truth, but to how things are in the realm of mathematical entities. (p. 202)

It is clear from the paper as a whole that Dummett's target is specifically Gödel's views, and more generally the conception of meaning as

---

[20] "Sur le platonisme," p. 53, translation from Benacerraf and Putnam, 2nd ed., p. 259.

given by truth-conditions. (I think it is the second that motivates his mention of Frege on the same page.) As regards the second, it is ancillary to an elaborate argument given mostly in other writings. Plato is not in view at all.

Tait first quotes the passage in essay 3 (p. 62), arguing against the view mainly attributed to Benacerraf's "Mathematical Truth" that cognition of the objects of perception is well understood, while that of mathematical objects is not. Benacerraf's argument was invoked in the revival of nominalism about mathematics that began several years later. Dummett, however, has never sympathized with nominalism, nor has he tried to argue that perceptual knowledge, or knowledge of the physical, is unproblematic in some important way in which knowledge of mathematics is not.

The second place where the passage is quoted is in a footnote to the discussion of Hersh commented on above. In note 10 to the end of this passage he writes that the carelessness he attributes to Hersh is not "the province only of amateur philosophers" (p. 103). He then quotes the beginning of the above passage and a passage from Tyler Burge.[21] I am puzzled by this note, because I don't find Dummett's statement careless, and Burge's seems to me quite carefully formulated.

However, what the carelessness is supposed to consist in might be not distinguishing a deflationary realism such as Tait's own from transcendental realism (cf. p. 95). In Burge's paper, there is much to show that he does make such a distinction, at least between the platonism he attributes to Frege and a Wittgensteinian gloss on realism such as Tait gives. I think Dummett is concerned to *argue* that the alternative to some version of transcendental realism is what he calls anti-realism, which characteristically requires the rejection of classical logic. Tait does not agree, and most of us are on his side. But Dummett's position does not involve missing a basic distinction.

However, Tait has indicated that the primary issue for him is the incoherence or even meaninglessness of stronger forms of realism. Evidently he thinks there is not a middle ground between default realism and what he calls superrealism. That is a philosophical disagreement, which does not show much sign of being resolved to this day.

---

[21] "Frege on Knowing the Third Realm," p. 304.

4. Finitism

Essay 1, "Finitism," is the earliest and one of the best known of the essays. It proposes an analysis of finitism in a sense that could reasonably be attributed to Hilbert and defends the thesis that the finitistically provable propositions of arithmetic are the theorems of primitive recursive arithmetic.[22] It is quite clear that Hilbert and Bernays intended that proofs in primitive recursive arithmetic (PRA) should be finitist, so that what is potentially controversial is the thesis that finitistically provable statements of arithmetic are theorems of PRA. I shall call this thesis Tait's thesis, as I have elsewhere.[23] I will make only a few comments on the historical issues this poses; Tait addresses many of them in the later essay 2.

An important respect in which Tait's analysis differs from that of Hilbert and Bernays (to the extent that they offer one) is that for the latter the finitary method should yield results that are intuitively evident where this implies an anchoring in intuition, of which the conception is of broadly Kantian provenance.[24] Tait sees the essential of finitism in avoiding infinite totalities. I think Hilbert and Bernays very likely thought that this requirement and the intuitiveness requirement went together.

Conceptually, Tait's analysis has been widely found convincing and has largely displaced the earlier efforts of Kreisel, which yielded

---

[22] An earlier analysis of finitism and defense of this thesis, based on somewhat different ideas, is given in Tait, "Constructive Reasoning."

[23] *Mathematical Thought and Its Objects* (cited hereafter *as* MTO), p. 239, following "Finitism and Intuitive Knowledge."

[24] I don't entirely agree with what Tait writes about the relevance of this sort of intuition to the issues about finitism. I will not pursue the matter here, because I have written on the subject elsewhere; see my "Intuition in Constructive Mathematics" and "Finitism and Intuitive Knowledge," the latter incorporated into chapter 7 of MTO. I conclude in "Finitism" and *MTO*, §44, that Hilbert and Bernays do not have a non-question-begging argument for the conclusion that intuitively evident arithmetic encompasses all of PRA. Thus I agree with Tait's conclusion that they do not give a satisfactory account of finitism based on their conception of intuition. This point of agreement may be more significant than the disagreements. More needs to be said about how Bernays in particular understood the concept of intuition. Tait criticizes my interpretation in "Gödel on Intuition," p. 101 n.14. I think the issue may concern more the notion of intuitive knowledge than that of intuition (i.e., intuition *of* in the sense of *MTO* §24]). The matter remains to be explored.

characterizations of finitism by systems proof-theoretically equivalent to first-order arithmetic PA.[25] I think one can make a little clearer what is convincing about Tait's analysis if some aspects of it are stated in other terms.

In 1958 Gödel inferred from the presumed intuitively evident character of the finitary method that it should exclude "abstract concepts . . . those that are essentially of second or higher order."[26] I am not sure exactly what Gödel had in mind, but Tait's analysis can be represented as based on a version of that idea.[27] Tait's view is that the finitist has no concept of function, because functions are "transfinite objects." The view implicitly attributed to finitism by Gödel, and to Hilbert by Tait, is that functions are not objects of intuition or are in some other relevant way not intuitive. However, Tait's language includes what are usually called function symbols. What is denied, then, is that the finitist needs to recognize them as designating functions.

We might say the same about a first-order language with function symbols, since (unless we have set-theoretic apparatus or the equivalent to do the work), we are not able to draw inferences from a statement involving a function symbol to the conclusion that there is a function satisfying the condition expressed by replacing the symbol by a suitable variable. That is a common point between Tait's finitism and nominalism.

To see how this works out, we need to consider how the propositions-as-types language that Tait uses (sketched in §2 above) is applied in this case. The decisive point is that Tait's finitist cannot admit function types, if admitting a type implies admitting objects of that type. Of course the type N of natural numbers is admitted; in fact Tait defines "finitist types of the first kind" as those resulting from starting with N and closing under $\wedge$, so that they are simply the types of, for each n, ordered n-tuples of natural numbers. It is the objects of these types that are serious objects. Tait's finitist language does contain expressions $A \rightarrow B$ for A, B such types, but evidently $A \rightarrow B$ is not a finitist type. Expressions of the form $f: A \rightarrow B$ are allowed, but they cannot be decomposed into a term designating a function and a type symbol. In

---

[25] See Kreisel, "Ordinal Logics" and "Mathematical Logic."

[26] "Über eine bisher noch nicht benützte Erweiterung," p. 240, trans. p. 241.

[27] Evidently without the inference from the presumed intuitively evident character of finitist results.

practice *f* must be a term that gives a "function" by an explicit expression, whose variables must be of the finitist types. This is compatible with holding that neither function symbols nor other expressions have to be taken as designating functions.[28]

The question then arises why, if that is what is essential, the language of first-order arithmetic is not finitist, so that there would be no relevant difference between finitist arithmetic and at least intuitionistic. Tait poses the issue about functions in ontological terms (although I doubt that he would be comfortable with that word). A similar view would hold that the predicates of a first-order language do not designate either classes or intensional entities (attributes). Does the language of first-order arithmetic eschew commitments incompatible with finitism?

It is clearly not Tait's view or Hilbert's that it does. Hilbert explicitly put severe restrictions on arithmetical quantification. He views a general statement about numbers (conceived for his purpose as strings of 1s) as a hypothetical: if one is presented with such a string, then it satisfies the relevant condition (which is to be computationally verifiable). An existential statement is an "incomplete assertion": it can be made only when one has an explicit construction of an instance. Hilbert explicitly states that a general statement cannot be negated, and evidently only the explicit form of an existential statement can be negated.[29]

Tait's view of finitist universal statements is not exactly Hilbert's but is similar: the explanation of the admitted types (including that of numbers, of which more shortly) licenses inferences about arbitrary elements of that type. Tait asserts that reasoning about an arbitrary object of a finitist type *A* is independent of the idea of the totality of objects of type *A* (p. 44, from essay 2). Does this express more than the general idea, not particularly connected with finitism, that understanding or using a predicate *F* does not commit one to a class or totality of *F*s? Apparently Tait thinks it does depend on the case at hand, even if it generalizes beyond finitism. In the cases that concern him, to be of a type *A* is to be constructed in a certain way (p. 25). So that reasoning about an arbitrary *a* of type *A* involves exercising understanding of the procedure of construction. About existential statements, Tait's version of finitism is equivalent to Hilbert's, since there is no existential

---

[28] Peter Koellner suggests speaking of virtual function types, on the model of Quine's virtual classes.

[29] "Über das Unendliche," p. 173.

quantification or counterpart of it in the finitist language, with the effect that an existential statement can be represented formally only by giving an explicit witness.

The key case for Tait's view of finitist general reasoning is clearly the type N of natural numbers. Tait gives a somewhat obscure description of Number as the "form of finite sequences," of which individual numbers are "subforms" (p. 25). 'x is a number' is apparently not to be quite an ordinary predicate; Tait writes:

> Thus, the relation of Number to the numbers is not that of a concept to the objects falling under it but that of a form or structure to its least specific subforms. (Ibid.)

A finitist might balk at talk of concepts, which Gödel, for example, might have counted among the higher-order objects that the remark quoted above was meant to exclude. But what is gained by substituting the relation of a "form or structure" to its "subforms"?

I don't know how Tait would answer this question. I think he may have come to think that the view of Number just alluded to was either insufficient or unnecessary for his purpose, since it drops out of his later writings and is not mentioned in his brief recapitulation and defense of his analysis in essay 2. I don't think his view requires that 'x is a number' should be fundamentally different from other predicates. What is important is that what counts as a number is determined by a construction, i.e., iterated application of the successor operation beginning with 0 or 1. In his view a statement of the form $n: N$, where n is a canonical numeral, is one for which it does not make sense to ask for a proof. But how can we draw conclusions about an arbitrary number? The common answer would be that any method would come down to mathematical induction. Tait treats as more fundamental a principle of iteration of functions: Given a type B, $a: B$ and $g: B \to B$, we can introduce a $f: N \to B$ such that $f0 = a$ and $f(Sx) = g(fx)$. This yields primitive recursion (pp. 27–28). He remarks:

> Suppose we have $k: B$, $g: B \to B$, and an arbitrary $n: N$. n is built up from 0 by iterating $m \to m'$:
> $$0, 1, 2, \ldots n = 0'' \ldots'$$
> By using exactly the same iteration, the same "$\ldots$', we may build up $fn: B$:
> $$k, gk, ggk, \ldots fn = g \ldots ggk. \text{ (p. 27)}$$

From the finitist point of view, cases of this have to be recognized one at a time; the finitist does not even regard the general principle as intelligible. It is not so clear that someone proceeding in this way could not get off at some point, for example admit addition and multiplication and perhaps other "functions" that do not grow much faster, but refuse to accept exponentiation. Or he might stop at iterated exponentiation. Such points of view have been supported by arguments, for example from computational feasibility or from predicativity. Tait does not consider these views, but he would probably have replies to these arguments. However, it is not clear to me that they would preserve the indubitability he claims for finitism. He would have to elaborate the claim that there is no "preferred or even equally preferable standpoint from which to criticize finitism" (p. 41). The holder of one of these views would claim that there is a limited but still workable mathematical practice that admits less than finitism does.

Thus far I have neglected a central part of Tait's analysis, his conception of finitist proofs. Proofs were mentioned by Gödel among the "abstract objects" that finitism should eschew. However, Tait's typed language requires proofs as objects. His solution to this problem is to use *formal* proofs and explicitly constructed operations on them. Proofs of general propositions, in particular the important hypothetical proofs $f: \forall x[F(x) \rightarrow G(x)]$, are of the latter sort. For particular values of x (of finitist types of the first kind), F and G are conjunctions of equations; this is what Tait calls finitist types of the second kind.

The result is a formalism equivalent to PRA, in its primitive means of expression more generous than an equation calculus but more limited than a version of PRA based on full classical propositional logic.[30] Tait obtains Hilbert's claim that a universal statement cannot be negated by the nature of the calculus: A proof of $\neg\forall x[F(x) \rightarrow G(x)]$ would have to be, as in the BHK interpretation of intuitionistic logic, a function g such that, for any $f: \forall x[F(x) \rightarrow G(x)]$, $g(f)$ is a "proof" of an absurdity. Thus functions would have to play the role of arguments. Independently of his calculus, however, Tait would probably question what the finitist could possibly understand by there not being an

---

[30] Tait's finitist language is evidently without negation or disjunction. Since PRA can be formulated as an equation calculus, this is not an essential limitation. But it seems there are some equivalences that a finitist should be able to state that cannot be stated in his language. However, this does not affect the thesis that the finitistically provable theorems of arithmetic are the theorems of PRA.

f: $\forall x[F(x) \rightarrow G(x)]$ other than there being a counterexample, which would have to be constructed explicitly. As to the admission to his language of existential quantification, $\forall\exists$ statements would also introduce commitment to functions.

These considerations still do not make it quite obvious why the language of first-order arithmetic, taken at face value, is not finitist in the sense of not presupposing or allowing infinite totalities. The doubt arises from the obvious distinction between a language quantifying only over natural numbers, and one admitting such objects as the set of natural numbers and functions on it. Hilbert, in the passage we have referred to above, appeals to an analogy between the usual quantification over numbers and infinite conjunction or disjunction. But why do we need to accept this analogy? If it is classical first-order arithmetic that we have in mind, then one might argue that the application of the law of excluded middle to generalizations about numbers presupposes at least *facts* about all numbers. I am not sure where this argument would lead, and I think the consideration is not especially relevant to Tait's analysis. His standpoint is constructivist from the start, and the understanding we have of a constructive reading of the language of first-order arithmetic is one in which proofs, and functions whose arguments are proofs and other functions, are in the ontology. We have seen how Tait solves the problem this poses for the analysis of finitism, so that functions are not in the ontology of finitism and only formal proofs are needed. But functions are in the ontology of the analysis. Tait states that he is analyzing finitism from a nonfinitist point of view. This was already true of Kreisel's rather different analysis of finitism.[31]

The obvious way in which, contrary to Tait, finitist arithmetic might go beyond PRA would be if one could see finitistically that some nonprimitive recursive function is total. Here is a very simple such function:

$$f(0, n) = 2n$$

$$f(m + 1, 0) = 1$$

$$f(m + 1, n + 1) = f[m, f(m + 1, n)].^{32}$$

[31] Both have the consequence that finitist arithmetic is captured by a single formal system.

[32] This is a slight variation of Rósza Péter's simplification of the original example of Ackermann, "Zum Hilbertschen Aufbau." See Péter, *Rekursive Funktionen*, §9 (any ed.).

$\lambda n.f(0, n)$ is total, and it is easy to prove that if $\lambda n.f(m, n)$ is total, so is $\lambda n.f(m + 1, n)$. But the argument either admits functions as entities or reasons by Π_2-induction, which on Tait's understanding would also introduce functions. Tait argues that some other ways of generating functions that might be accepted as finitist would not lead out of the primitive recursive functions (pp. 29–30).

Tait thus has a very convincing case that his thesis accords with the explanations of finitist generality given by Hilbert in the 1920s and in any event marks off a foundationally significant boundary within mathematics. It follows that Tait's criticism of Kreisel's analysis is just (pp. 39–40).[33] The question whether it accurately captures what the Hilbert school intended is somewhat tangled. It is clear that in some cases they admitted as finitist steps that are not formalizable in PRA; for example, Richard Zach has pointed out that Ackermann, "Begründung des *tertium non datur*," uses induction up to $\omega^{\omega^{\omega}}$. Tait points out (p. 55) that it is not evident that Hilbert or his colleagues knew that this went beyond PRA. The tendency to keep metamathematics on an intuitive level and therefore not to pose the question what formal system (if any) might capture it very likely inhibited them from focusing on questions like this. Still, it is harder to believe that, after Ackermann, "Zum Hilbertschen Aufbau der reellen Zahlen," it was not understood that $\omega^{\omega^{\omega}}$-induction would go beyond PRA. There has been quite a bit of controversy about such questions. Tait goes into these issues in essay 2 and then further in the appendix to essays 1 and 2.

5. Tait on Others

Like many commentators, Tait is at his best when writing about thinkers toward whose views he is sympathetic. Essays 7 and 8 on Plato are excellent pieces of philosophy; I'm not qualified to say how they stand as Plato scholarship, but they contain serious textual argument and certainly succeed in deriving from Plato views of real philosophical interest. Essay 11 on Cantor's *Grundlagen einer allgemeinen Mannigfaltigkeitslehre* (1883) brings to life a little read treatise. Others had

---

[33] That Kreisel's analysis might still capture something importantly related to finitism is suggested by Gödel in note 4 of "Über eine bisher noch nicht benützte Erweiterung" (also in the 1972 English version). I do not know of definite results that cash in these suggestions.

earlier defended Cantor against the charge that set theory as he developed it was "naive set theory" in the sense of being implicitly based on the inconsistent universal comprehension schema. However, Tait develops that observation (which might have been obvious but for the hold the paradoxes had on a whole generation of researchers in the foundations of mathematics) in a way that adds clarity to the matter, particularly, given the essay's focus, with respect to Cantor's early thought. Although he is less systematic in treating them, he has much of interest to say about Dedekind, Hilbert, and Zermelo.

Tait can often make effective use of Wittgenstein and offers in essay 9 a defensible reading of his discussion of rule-following; however, since this matter has so many subtleties and has been discussed so much by people better versed than I in Wittgenstein's thought, I will not venture to assess it further. Some interpreters of Wittgenstein see either the earlier or (more often) the later Wittgenstein as compatible with positive or constructive efforts in philosophy. Others see his discussions as deconstructing *all* positive constructions concerning the philosophical problems he addresses. Tait clearly belongs to the former camp, although he does invoke the later Wittgenstein in favor of relatively minimalist versions of, for example, realism and the objectivity of mathematics, as we saw in §3.[34]

Tait can be dismissive of views with which he does not sympathize and careless about their context. I have given one instance at the end of §3 and will not pursue the matter further.

The most recent writer to whom Tait devotes an essay is Kurt Gödel. Essay 12, a review article on volume III of the *Collected Works*, is a very rich essay. The long §1 is a deep and surprising analysis of the remarks about the construction of set theory in Gödel's 1933 lecture to the Mathematical Association of America (MAA) (CW III, 45–53). What is surprising is that by assuming that Gödel is working with second-order theories with the axiom of choice in the form of the assumption that the domain is well-ordered, Tait is able to explain how Gödel could arrive at a theory as strong as ZFC and then envisage going beyond it.

Section 3 concerns the vexed question of Gödel's platonism and is chiefly concerned to argue that very often Gödel's statements and argu-

---

[34] In his discussion of stronger forms of realism, Tait comes close to the deconstructive stance. (I have not found mention of the *Tractatus* in Tait's writings.)

ments would lead to default realism (see §3 above), but some remarks show that Gödel embraces a more far-reaching realism, for example in the well-known remark in the 1964 supplement to "What Is Cantor's Continuum Problem?" that conceptual "data" in mathematics "may represent an aspect of objective reality, but, as opposed to the sensations, their presence in us may be due to another kind of relationship between ourselves and reality" (CW II 268). It might be worth remarking that Gödel may come closest to Tait's default realism when he argues that mathematics has a real content (in opposition especially to conventionalism), since that claim applies to intuitionist as well as classical mathematics. I do not agree with all of Tait's comments on these issues, but there is no disagreeing with his general point.

One disagreement is worth mentioning. Late in the section Tait turns to Gödel's statements about mathematical intuition. It has long been noted that Gödel's realism extends to concepts and thus not only to the usual mathematical objects and statements referring to them. Tait quotes a remark of Hao Wang on "perception" of the concept of computability and then remarks:

> In particular, applied to the present case [i.e., set theory—CP], the concept of ordinal number is already determined in the sense that it is already determined what totalities of ordinals are bounded. It follows that the new axioms that we introduce to give upper bounds to such totalities are already contained in that concept. So realism about concepts, so understood, claims that mathematical truth is prior to the adoption of axioms rather than founded on them. (p. 294)

The latter is probably a consequence of the stronger realism to be found in some of Gödel's statements, transcendental realism in the sense of §3. However, Tait attempts to identify the line between realism about concepts and realism merely about mathematical objects with the line between default realism and transcendental realism. There is a tangle of issues here that I will not attempt to sort out. But there is clear evidence that Gödel thought that reference to concepts is involved in the use of predicates,[35] and that view opens the way for something that

---

[35] There are several passages in "Is Mathematics Syntax of Language?" to illustrate this point; see for example CW III 359 and the discussion of the issue in Essay 7 of this volume, §4.

might be called default realism about concepts. On the other hand, transcendental realism about set theory, in particular truth in set theory, is not obviously incompatible with a more "nominalist" view of predication.[36] So I would claim that the questions of transcendental realism and of realism about concepts are independent.

Among the concepts that Gödel remarks on is what he calls the concept of set. In his "Gödel's Conceptual Realism," D. A. Martin argues persuasively that Gödel has in mind much of the time what would more accurately be called the concept of the universe of sets. It is Gödel's realistic stance about that concept (or perhaps a parallel concept of the sequence of ordinals) that would lead to the conclusion Tait draws in the first sentence of the above quotation, so that it is specific to a particular concept. It is also "perception" of such a concept that might give insight into stronger axioms of set theory, although Gödel does not claim such insight for axioms strong enough to be incompatible with $V = L$.

Sections 2 and 4–7 of the essay largely concern issues about proof theory and constructive mathematics. I will not go into detail about them. They contain exposition and analysis of the discussion of proof theory and its foundations in some essays in the volume, particularly the MAA lecture, the remarkable lecture at Zilsel's of 1938,[37] the Yale lecture of 1941 where Gödel presented his primitive recursive functionals of finite type and their use for a consistency proof of arithmetic, and finally the published paper on that subject, the *Dialectica* paper "Über eine bisher noch nicht benützte Erweiterung."

Concerning the *Dialectica* interpretation, Tait makes an important claim: that its use to justify mathematical induction results in a circle (p. 301). This amounts to the claim that the consistency proof by means of the interpretation does not have simpler or more evident conceptual foundations than what one would obtain by reflecting directly on intu-

---

[36] Such a view is elaborated in the second part of Davidson, *Truth and Predication;* see also my "Quine's Nominalism," Essay 8 in this volume. It was that possibility that underlay the remark in §5 of Essay 2 of this volume (p. 57) that Tait quotes (p. 295). (It was p. 385 of the original publication; the reference to p. 185 is surely a typographical error.)

[37] Concerning this lecture, Tait notes the difficulty of "piecing together a coherent text," referring to Wilfried Sieg and me as "the editors" (p. 298). It is true that we wrote the introductory note jointly, but Cheryl Dawson is explicitly mentioned as also an editor of the text (see CW III 85).

itionistic first-order arithmetic. A fuller version of this discussion is in his "Gödel's Interpretation of Intuitionism."

In §7 Tait sketches a spelling out of Gödel's analysis in the Zilsel lecture of Gentzen's consistency proof for first-order arithmetic, from which he derived the no-counter-example interpretation, first published in 1951 in Kreisel, "On the Interpretation of Non-Finitist Proofs." Tait's "Gödel's Reformulation" elaborates this treatment.

In addition Tait has two further papers on Gödel, "Gödel's Correspondence" and "Gödel on Intuition." All the later papers were written either when the work on the present collection was far advanced or after its publication, so that they could not be included. It is still a matter of regret, though no reproach to the author or publisher, that the collection gives an inadequate idea of Tait's scholarship on Gödel, which in particular is as thorough a treatment of Gödel's thought about proof theory and constructive mathematics as there is.

My admiration for what these essays achieve should be evident to the reader. I will add that there are important essays and themes in this book that I have not gone into, such as the foundations of constructive mathematics, the relations of intuitionist to classical mathematics, and the construction of large cardinals in set theory.

The book contains a fair number of typographical and editorial errors. One worth mentioning is that the name "Skolem" is consistently misspelled "Skølem." Something more distressing is that the binding of my copy already shows some signs of cracking. I haven't worked *that* hard on this review.[38]

---

[38] I am grateful to Peter Koellner and the author for very helpful comments on an earlier draft.

Particular conventions are observed for citations of two authors:

Paul Bernays. Essays by Bernays originally written in German and reprinted in *Abhandlungen zur Philosophie der Mathematik* are cited according to its pagination. Other essays are cited in the original pagination. Both will be given in the margins of the reprints in *Essays on the Philosophy of Mathematics.* Note that "Sur le platonisme" and "Quelques points de vue" occur in *Abhandlungen* in German translation, which we ignore here. But cf. the translator's note in *Essays* to the latter essay. Previously published English translations reprinted in *Essays* are generally subjected to some revision.

Kurt Gödel. His writings are cited by volume and page number of his *Collected Works,* generally abbreviated CW.

Ackermann, Wilhelm. "Begründung des *tertium non datur* mittels der Hilbertschen Theorie der Widerspruchsfreiheit." *Mathematische Annalen* 93 (1924), 1–36.

———. "Zum Hilbertschen Aufbau der reellen Zahlen." *Mathematische Annalen* 99 (1928), 118–133.

———. "Zur Widerspruchsfreiheit der Zahlentheorie." *Mathematische Annalen* 117 (1940), 162–194.

Avigad, Jeremy. Review of Tait, *The Provenance of Pure Reason. The Bulletin of Symbolic Logic* 12 (2006), 608–611.

Awodey, Steve, and A. W. Carus. "Carnap, Completeness, and Categoricity: The *Gabelbarkeitssatz* of 1928." *Erkenntnis* 54 (2001), 145–172.

———. "Gödel and Carnap." In Feferman, Parsons, and Simpson, pp. 252–274.

———. "How Carnap Could Have Replied to Gödel." In Steve Awodey and Carsten Klein (eds.), *Carnap Brought Home: The View from Jena,* pp. 199–220. Chicago: Open Court, 2004.

Baire, René, Émile Borel, Jacques Hadamard, and Henri Lebesgue. "Cinq lettres sur la théorie des ensembles." *Bulletin de la Société Mathématique de France* 33 (1905), 261–273. Translated as appendix 1 to Moore, *Zermelo's Axiom of Choice.*

Barwise, Jon (ed.). *Handbook of Mathematical Logic.* Amsterdam: North-Holland, 1977.

Benacerraf, Paul. "Mathematical Truth." *The Journal of Philosophy* 70 (1973), 661–679.

Benacerraf, Paul, and Hilary Putnam (eds.). *Philosophy of Mathematics: Selected Readings.* Englewood Cliffs, N.J.: Prentice-Hall, 1964. 2nd ed., Cambridge: Cambridge University Press, 1983.

Bernays, Paul. *Abhandlungen zur Philosophie der Mathematik.* Darmstadt: Wissenschaftlicher Buchgesellschaft, 1976.

———. "Die Bedeutung Hilberts für die Philosophie der Mathematik." *Die Naturwissenschaften* 10, no. 4 (1922), 93–99. Translation in Mancosu, *From Brouwer to Hilbert.* Reprinted with the translation in *Essays.*

———. "Bemerkungen zu Lorenzens Stellungnahme in der Philosophie der Mathematik." In Kuno Lorenz (ed.), *Konstruktionen versus Positionen: Beiträge zur Diskussion um die konstruktive Wissenschaftstheorie (Paul Lorenzen zum 60. Geburtstag),* Band I: *Spezielle Wissenschaftstheorie,* pp. 3–16. Berlin: de Gruyter, 1979.

———. "Bemerkungen zur Grundlagenfrage." Appendix IV to Ferdinand Gonseth, *Philosophie mathématique.* Actualités scientifiques et industrielles, 837. Paris: Hermann, 1939.

———. "Bemerkungen zur Philosophie der Mathematik." *In Akten des XIV. Internationalen Kongresses für Philosophie (Wien, 2.–9. September 1968),* vol. 6, pp. 192–198. Vienna: Herder, 1969. Reprinted in *Abhandlungen.* Reprinted with translation in *Essays.*

———. "Betrachtungen zu Ludwig Wittgensteins *Bemerkungen über die Grundlagen der Mathematik.*" *Ratio* 3, no. 1 (1959), 1–18. Reprinted in *Abhandlungen.* Translation in the simultaneous English edition of *Ratio.* Reprinted with the translation in *Essays.*

———. "Charakterzüge der Philosophie Gonseths." *Dialectica* 14 (1960), 151–156.

———. "Dritte Gespräche von Zürich." *Dialectica* 6 (1952), 130–140.

———. *Essays on the Philosophy of Mathematics.* Edited by Wilfried Sieg, W. W. Tait, Steve Awodey, and Dirk Schlimm. Chicago: Open Court, forthcoming. Cited as *Essays.*

———. "Die Grundbegriffe der reinen Geometrie in ihrem Verhältnis zur Anschauung." *Die Naturwissenschaften* 16, no. 12 (1928), 197–203. Reprinted with translation in *Essays.*

———. "Die Grundgedanken der Fries'schen Philosophie in ihrem Verhältnis zum heutigen Stand der Wissenschaften." *Abhandlungen der Fries'schen Schule* N. F. 5 (1930), 99–113. Reprinted with translation in *Essays.*

———. "Grundsätzliche Betrachtungen zur Erkenntnistheorie." *Abhandlungen der Fries'schen Schule* N. F. 6 (1937), 275–290. Reprinted with translation in *Essays.*

———. "Grundsätzliches zur 'philosophie ouverte'." *Dialectica* 2 (1948), 273–279.

———. "Kurze Biographie." In Müller, *Sets and Classes,* pp. xiv–xvi. Reprinted with new translation in *Essays.* (The Müller volume contains an English version, which is abbreviated and not always accurate.)

———. "Die Mathematik als ein zugleich Vertrautes und Unbekanntes." *Synthese* 9 (1955), 465–471. Reprinted in *Abhandlungen.* Reprinted with translation in *Essays.*

———. "Mathematische Existenz und Widerspruchsfreiheit." In *Études de philosophie des sciences, en hommage à Ferdinand Gonseth à l'occasion de son soixantième anniversaire,* pp. 11–25. Neuchâtel: Éditions du Griffon, 1950. Reprinted in *Abhandlungen.* Reprinted with translation in *Essays.*

———. "Die Philosophie der Mathematik und die Hilbertsche Beweistheorie." *Blätter für deutsche Philosophie* 4 (1930), 326–367. Reprinted with postscript in *Abhandlungen.* Translation in Mancosu, *From Brouwer to Hilbert.* Reprinted with the translation in *Essays.*

———. "Quelques points de vue concernant le problème de l'évidence." *Synthese 5* (1946), 321–326. Reprinted with translation in *Essays.*

———. Review of Gödel, "Russell's Mathematical Logic." *The Journal of Symbolic Logic* 11 (1946), 75–79.

———. "Die schematische Korrespondenz und die idealisierten Strukturen." *Dialectica* 24 (1970), 53–66. Reprinted in *Abhandlungen.* Reprinted with translation in *Essays.*

———. "Some Empirical Aspects of Mathematics." In S. Dockx and Bernays (eds.), *Information and Prediction in Science,* pp. 123–128. New York: Academic Press, 1965.

———. "Sur le platonisme dans les mathématiques." *L'enseignement mathématique* 34 (1935), 52–69. Translation in Benacerraf and Putnam. Reprinted with the translation in *Essays.*

———. "Über Hilberts Gedanken zur Grundlegung der Arithmetik." *Jahresbericht der deutschen Mathematiker-Vereinigung* 31 (1922), 10–19. Translation in Mancosu, *From Brouwer to Hilbert.* Reprinted with the translation in *Essays.*

———. "Über Nelsons Stellungnahme in der Philosophie der Mathematik." *Die Naturwissenschaften* 16, no. 9 (1928), 142–145. Reprinted with translation in *Essays.*

———. "Überlegungen zu Ferdinand Gonseths Philosophie." *Dialectica* 31 (1977), 119–128.

———. "Von der Syntax der Sprache zur Philosophie der Wissenschaften." *Dialectica* 11 (1957), 233–246.

———. "Zum Begriff der Dialektik." *Dialectica* 1 (1947), 172–175.

———. "Zur Frage der Anknüpfung an die kantische Erkenntnistheorie." *Dialectica* 9 (1955), 23–65, 195–221.

———. "Zur Rolle der Sprache in erkenntnistheoretischer Hinsicht." *Synthese* 13 (1961), 185–200. Reprinted in *Abhandlungen.* Reprinted with translation in *Essays.*

See also Hilbert and Bernays.

Blackwell, Kenneth. "A Non-Existent Revision to *Introduction to Mathematical Philosophy.*" *Russell: The Journal of the Bertrand Russell Archives* no. 20 (1976), 16–18.

Bolzano, Bernard. *Paradoxien des Unendlichen.* Leipzig, 1851.

———. *Wissenschaftslehre.* Sulzbach, 1837.

Boolos, George. "The Iterative Conception of Set." *The Journal of Philosophy* 68 (1971), 215–231. Reprinted in *Logic, Logic, and Logic.*

———. *Logic, Logic, and Logic.* Cambridge, Mass.: Harvard University Press, 1998.

———. "Nominalist Platonism." *Philosophical Review* 94 (1985), 327–344. Reprinted in *Logic, Logic, and Logic*.

———. "On Second-Order Logic." *The Journal of Philosophy* 72 (1975), 509–527. Reprinted in *Logic, Logic, and Logic*.

Borchert, Donald M. (General ed.). *The Encyclopedia of Philosophy*, 2nd ed. 10 vols. Detroit: Macmillan Reference USA, 2006.

Borel, Émile. "Sur les principes de la théorie des ensembles." In G. Castelnuovo (ed.), *Atti del IV Congresso Internationale dei Matematici* (Roma, 6–11 Aprile 1908), vol. 2: *Communicazioni delle sezione I e II*. Rome: Accademia dei Lincei, 1909. Cited according to reprint in Borel, *Leçons sur la théorie des fonctions*, 4th ed. (Paris: Gauthier-Villars, 1950).

See also Baire et al.

Brouwer, L. E. J. *Cambridge Lectures on Intuitionism*. Edited by Dirk van Dalen. Cambridge: Cambridge University Press, 1981.

———. *Collected Works*, vol. 1: *Philosophy and the Foundations of Mathematics*. Edited by Arend Heyting. Amsterdam: North-Holland, 1975. Cited as CW.

———. "Consciousness, Philosophy, and Mathematics." *Proceedings of the Tenth International Congress of Philosophy, Amsterdam 1948*, pp. 1235–1249. Amsterdam: North-Holland, 1949. Reprinted in CW.

———. "Historical Background, Principles, and Methods of Intuitionism." *South African Journal of Science* 49 (1952), 139–146. Reprinted in CW.

———. *Intuitionisme en formalisme*. Amsterdam 1912. Translated as "Intuitionism and formalism." *Bulletin of the American Mathematical Society* 20 (1913), 81–96. Translation reprinted in CW.

———. *Intuitionismus*. Edited by Dirk van Dalen. Mannheim: Bibliographisches Institut, 1992.

———. "Intuitionistische Betrachtungen über den Formalismus." *Koninklijke Akademie van Wetenschappen, Proceedings of the Section of Sciences* 31 (1928), 374–379. Reprinted in CW. Translation in Mancosu, *From Brouwer to Hilbert*.

———. *Leven, kunst, en mystiek*. Delft: J. Waltman Jr., 1905. Excerpts translated in CW. Full translation by W. P. van Stigt, *Notre Dame Journal of Formal Logic* 37 (1996), 389–429.

———. "Mathematik, Wissenschaft, und Sprache." *Monatshefte für Mathematik und Physik* 36 (1929), 153–164. Reprinted in CW. Translation in Mancosu, *From Brouwer to Hilbert*, pp. 45–53.

———. *Over de grondslagen der wiskunde*. Amsterdam: Maas and van Suchtelen, 1907. Reprinted in van Dalen, *L. E. J. Brouwer en de grondslagen van de wiskunde*. Translation in CW.

———. "Points and Spaces." *Canadian Journal of Mathematics* 6 (1954), 1–17. Reprinted in CW.

———. "Richtlijnen der intuitionistische wiskunde." *Koninklijke Akademie van Wetenschappen, Proceedings of the Section of Sciences* 50 (1947), 339. Translation in CW 477.

———. *Die Struktur des Kontinuums*. Vienna 1930. Reprinted in CW. Translation in Mancosu, *From Brouwer to Hilbert*, pp. 54–63.

————. "Willen, weten, spreken." *Euclides* 9 (1933), 177–193. Excerpts translated in CW. Full translation in W. P. van Stigt, *Brouwer's Intuitionism* (Amsterdam: North-Holland, 1990), pp. 418–431.

————. *Het wezen der meetkunde*. Amsterdam, 1909. Translation in CW.

See also van Dalen.

Burge, Tyler. "Frege on Knowing the Third Realm." *Mind* 101 (1992), 634–650. Reprinted in *Truth, Thought, Reason: Essays on Frege* (Oxford: Clarendon Press, 2005).

————. "Predication and Truth." *The Journal of Philosophy* 104 (2007), 590–608.

Burgess, John P. "Cats, Dogs, and So On." In Zimmerman, *Oxford Studies in Metaphysics*, vol. 4, pp. 56–77.

————. *Fixing Frege*. Princeton, N.J.: Princeton University Press, 2005.

Butts, Robert E., and Jaakko Hintikka (eds.). *Logic, Foundations of Mathematics, and Computability Theory*. Dordrecht: Reidel, 1977.

Cantor, Georg. *Gesammelte Abhandlungen mathematischen und philosophischen Inhalts*. Edited by Ernst Zermelo. Berlin: Springer, 1932.

Carnap, Rudolf. "Empiricism, Semantics, and Ontology." *Revue internationale de philosophie* 4 (1950), 20–40. Reprinted in 2nd ed. of *Meaning and Necessity*.

————. "Ein Gültigkeitskriterium für die Sätze der klassischen Mathematik." *Monatshefte für Mathematik und Physik* 42 (1935), 163–190. Translated as §34 of *Logical Syntax*.

————. "Intellectual Autobiography." In Schilpp, *The Philosophy of Rudolf Carnap*, pp. 3–86.

————. *The Logical Syntax of Language*. Translation, with additions, of *Logische Syntax*. London: Kegan Paul, 1937.

————. *Logische Syntax der Sprache*. Vienna: Springer, 1934.

————. "Die logizistische Grundlegung der Mathematik." *Erkenntnis* 2 (1931), 91–105. Translation in Benacerraf and Putnam.

————. *Meaning and Necessity*. Chicago: University of Chicago Press, 1947. 2nd ed., 1956.

————. "Replies and Systematic Expositions." In Schilpp, *The Philosophy of Rudolf Carnap*, pp. 859–1013.

See also Quine and Carnap.

Cassirer, Ernst. *Gesammelte Werke. Hamburger Ausgabe*. Edited by Birgit Recki. Hamburg: Meiner, 1995–.

————. "Kant und die moderne Mathematik." *Kant-Studien* 12 (1907), 1–49.

————. *Substance and Function and Einstein's Theory of Relativity*. Translated by William Curtis Swabey and Marie Collins Swabey. Chicago: Open Court, 1923.

————. *Substanzbegriff und Funktionsbegriff. Untersuchungen über die Grundfragen der Erkenntniskritk*. Berlin: Bruno Cassirer, 1910. Reprinted as *Werke*, vol. 6, 2000. Translation in *Substance and Function*. Cited according to original.

Chihara, Charles S. "A Gödelian Thesis Regarding Mathematical Objects: Do They Exist? And Can We Perceive Them?" *Philosophical Review* 91 (1982), 211–227.

———. *Ontology and the Vicious Circle Principle*. Ithaca, N.Y.: Cornell University Press, 1973.

Church, Alonzo. "Carnap's *Introduction to Semantics*." *The Philosophical Review* 52 (1943), 298–304.

———. "A Set of Postulates for the Foundation of Logic." *Annals of Mathematics* (2) 33 (1932), 346–366, 34 (1933), 839–864.

Chwistek, Leon. "Die nominalistische Grundlegung der Mathematik." *Erkenntnis* 3 (1932–1933), 367–389.

Cori, René, Alexander Razborov, Stevo Todorcevic, and Carol Wood (eds.). *Logic Colloquium 2000. Proceedings of the Annual European Summer Meeting of the Association for Symbolic Logic, Held in Paris, France, July 23–31, 2000.* Lecture Notes in Logic 19. Association for Symbolic Logic and Wellesley, Mass.: A. K. Peters, 2005.

Davidson, Donald. *Truth and Predication*. Cambridge, Mass.: Harvard University Press, 2005.

Davis, Martin. Review of Dawson, *Logical Dilemmas*. *Philosophia Mathematica* (III) 6 (1998), 116–128.

———. "What Did Gödel Believe and When Did He Believe It?" *The Bulletin of Symbolic Logic* 11 (2005), 194–206.

——— (ed.). *The Undecidable*. Hewlett, N.Y.: The Raven Press, 1965.

Dawson, John W., Jr. *Logical Dilemmas: The Life and Work of Kurt Gödel*. Wellesley, Mass.: A. K. Peters, 1997.

See also Gödel.

Dedekind, Richard. *Was sind und was sollen die Zahlen?* Braunschweig: Vieweg, 1888. 2nd ed., 1893. 3rd ed., 1911. Translation in Ewald, *From Kant to Hilbert*, vol. 2, 787–833.

Demopoulos, William. "The 1910 *Principia's* Theory of Functions and Classes and the Theory of Descriptions." *European Journal of Analytic Philosophy* 3 (2007), 159–177.

Dreben, Burton. "Quine." In Robert B. Barrett and Roger Gibson (eds.), *Perspectives on Quine*, pp. 82–95. Oxford: Blackwell, 1990.

Dreben, Burton, and John Denton. "Herbrand-Style Consistency Proofs." In John Myhill, Akiko Kino, and Richard E. Vesley (eds.), *Intuitionism and Proof Theory*, pp. 419–433. Amsterdam: North-Holland, 1970.

Dummett, Michael. Critical Notice of Brouwer, *Collected Works*. *Mind* 89 (1980), 605–616.

———. "Platonism." In *Truth and Other Enigmas*, pp. 202–214. (Written and presented in 1967.)

———. *Truth and Other Enigmas*. London: Duckworth, 1978.

Edelman, Gerald M. *Bright Air, Brilliant Fire*. New York: Basic Books, 1992.

Edwards, Paul (ed.). *The Encyclopedia of Philosophy*. 8 vols. New York: Macmillan, 1967. For 2nd ed. see Borchert.

Evans, Gareth. "Identity and Predication." *The Journal of Philosophy* 72 (1975), 343–363.

Ewald, William. "Hilbert's Wide Program." In Cori et al., pp. 228–251.

——— (ed.). *From Kant to Hilbert: A Source Book in the Foundations of Mathematics*. 2 vols. Oxford: Clarendon Press, 1996.

Ewald, William, and Wilfried Sieg (eds.). *David Hilbert's Lectures on the Foundations of Logic and Arithmetic, 1917–1933*. Berlin: Springer-Verlag, 2013.

Feferman, Solomon. "Gödel's Life and Work." In Gödel, CW I, 1–36.

———. *In the Light of Logic*. New York: Oxford University Press, 1998.

———. "Kurt Gödel: Conviction and Caution." *Philosophia Naturalis* 21 (1984), 546–562. Reprinted in *In the Light of Logic*.

———. "Lieber Herr Bernays! Lieber Herr Gödel! Gödel on Finitism, Constructivity, and Hilbert's Program." *Dialectica* 62 (2008), 179–203. Also in Matthias Baaz, Christos H. Papadimitriou, Hilary Putnam, Dana Scott, and Charles L. Harper Jr. (eds.), *Kurt Gödel and the Foundations of Mathematics: Horizons of Truth*, pp. 111–133. Cambridge: Cambridge University Press, 2011.

———. "Predicative Provability in Set Theory." *Bulletin of the American Mathematical Society* 72 (1966), 486–489.

———. "Predicativity." In Stewart Shapiro (ed.), *The Oxford Handbook of Philosophy of Mathematics and Logic*, pp. 590–621. New York: Oxford University Press, 2005.

———. "Systems of Predicative Analysis." *The Journal of Symbolic Logic* 29 (1964), 1–30.

———. "Theories of Finite Type Related to Mathematical Practice." In Barwise, *Handbook*, pp. 913–971.

———. "Toward Useful Type-Free Theories, I." *The Journal of Symbolic Logic* 49 (1984), 75–111.

———. "Transfinite Recursive Progressions of Axiomatic Theories." *The Journal of Symbolic Logic* 27 (1962), 259–316.

———. "Weyl Vindicated: *Das Kontinuum* 70 Years Later." In Carlo Cellucci and Giovanni Sambin (eds.), *Temi e prospettive della logica e della scienza contemporanee*, vol. 1, pp. 59–93. Bologna: CLUEB, 1988. Reprinted with corrections and additions in *In the Light of Logic*.

See also Gödel.

Feferman, Solomon, and Geoffrey Hellman. "Challenges to Predicative Foundations of Arithmetic." In Sher and Tieszen, pp. 317–335.

———. "Predicative Foundations of Arithmetic." *Journal of Philosophical Logic* 22 (1995), 1–17.

Feferman, Solomon, Charles Parsons, and Stephen G. Simpson (eds.). *Kurt Gödel: Essays for His Centennial*. Lecture Notes in Logic 33. Association for Symbolic Logic and Cambridge University Press, 2010.

Field, Hartry. *Realism, Mathematics, and Modality*. Oxford: Blackwell, 1989.

Floyd, Juliet. "Wang and Wittgenstein." In Parsons and Link, *Hao Wang*, pp. 145–193.

Føllesdal, Dagfinn. "Developments in Quine's Behaviorism." *American Philosophical Quarterly* 48 (2011), 273–282.

Fraenkel, Adolf (A. A.). *Einleitung in die Mengenlehre*. Berlin: Springer, 1919. 2nd ed., 1923. 3rd ed., 1928. Cited according to 2nd ed. (After his emigration to Palestine [eventually Israel], Fraenkel published under the name A. A. Fraenkel or Abraham A. Fraenkel.)

Franchella, Miriam. "Paul Bernays' Philosophical Way." *Grazer philosophische Studien* 70 (2005), 47–66.

Frege, Gottlob. *Grundgesetze der Arithmetik, begriffschriftlich abgeleitet.* 2 vols. Jena: Pohle, 1893, 1903.

———. *Wissenschaftlicher Briefwechsel.* Edited by Gottfried Gabriel, Hans Hermes, Friedrich Kambartel, Friedrich Kaulbach, Christian Thiel, and Albert Veraart. Hamburg: Meiner, 1976.

Friedman, Michael. *Kant and the Exact Sciences.* Cambridge, Mass.: Harvard University Press, 1992.

———. "Logical Truth and Analyticity in Carnap's *Logical Syntax of Language.*" In William Aspray and Philip Kitcher (eds.), *History and Philosophy of Modern Mathematics,* pp. 82–95. Minneapolis: University of Minnesota Press, 1988. Reprinted as "Analytic Truth in Carnap's Logical Syntax of Language," in *Reconsidering Logical Positivism.*

———. *Reconsidering Logical Positivism.* Cambridge: Cambridge University Press, 1999.

———. "Tolerance and Analyticity in Carnap's Philosophy of Mathematics." In *Reconsidering Logical Positivism,* pp. 198–233.

Frost-Arnold, Gregory. "Carnap, Tarski, and Quine's Year Together: Conversations on Logic, Science, and Mathematics." Ph.D. dissertation, University of Pittsburgh, 2006.

Glanzberg, Michael. "Quantification and Realism." *Philosophy and Phenomenological Research* 69 (2004), 541–572.

Gödel, Kurt. *Collected Works, Volume I: Publications 1929–36,* and *Volume II: Publications 1938–1974.* Solomon Feferman (Editor-in-chief), John W. Dawson Jr., Stephen C. Kleene, Gregory H. Moore, Robert M. Solovay, and Jean van Heijenoort (eds.). New York: Oxford University Press, 1986, 1990.

———. *Collected Works, Volume III: Unpublished Essays and Lectures.* Solomon Feferman (Editor-in-chief), John W. Dawson Jr., Warren Goldfarb, Charles Parsons, and Robert M. Solovay (eds.). New York: Oxford University Press, 1995.

———. *Collected Works, Volume IV: Correspondence A–G,* and *Volume V: Correspondence H–Z.* Solomon Feferman and John W. Dawson Jr. (Editors-in-chief), Warren Goldfarb, Charles Parsons, and Wilfried Sieg (eds.). Oxford: Clarendon Press, 2003.
These volumes are cited as CW.

———. "The Consistency of the Axiom of Choice and of the Generalized Continuum Hypothesis." *Proceedings of the National Academy of Sciences, U.S.A.* 24 (1938), 556–557. Reprinted in CW II.

———. *The Consistency of the Axiom of Choice and of the Generalized Continuum Hypothesis with the Axioms of Set Theory.* Princeton, N.J.: Princeton University Press, 1940. Reprinted with additional notes, 1951, and with further notes, 1966. All reprinted in CW II.

———. "Is Mathematics Syntax of Language?" Unfinished paper dating from the 1950s, surviving in six drafts. Versions III and V published in CW III, 334–362.

———. "The Modern Development of the Foundations of Mathematics in the Light of Philosophy." Shorthand draft of an undelivered lecture, 1961 or later.

Transcription of the German shorthand with English translation in CW III, 374–387.

———. "On an Extension of Finitary Mathematics Which Has Not Yet Been Used." Expanded English version of "Über eine bisher noch nicht benützte Erweiterung." In CW II, 271–280.

———. "Ontological Proof." In CW III, 403–404.

———. "The Present Situation in the Foundations of Mathematics." Address to the Mathematical Association of America, 1933. In CW III, 45–53.

———. "A Remark about the Relationship between Relativity Theory and Idealistic Philosophy." In Paul Arthur Schilpp (ed.), *Albert Einstein: Philosopher-Scientist*, pp. 555–562. Evanston, Ill.: The Library of Living Philosophers, 1949. Reprinted in CW II.

———, "Remarks before the Princeton Bicentennial Conference on Problems in Mathematics." Remarks delivered in 1946. In Davis, *The Undecidable*, pp. 84–88. Reprinted in CW II.

———. "Russell's Mathematical Logic." In P. A. Schilpp (ed.), *The Philosophy of Bertrand Russell*, pp. 125–153. Evanston, Ill.: Northwestern University, 1944. Reprinted in CW II.

———. "Some Basic Theorems of the Foundations of Mathematics and Their Philosophical Implications." Josiah Willard Gibbs Lecture, American Mathematical Society, 1951. In CW III, 304–323. (The word "philosophical" was omitted from the title in CW III.)

———. "Some Observations about the Relationship between Relativity Theory and Kantian Philosophy." Unfinished paper from late 1940s, surviving in five drafts. Versions B2 and C1 in CW III, 247–259.

———. "Some Remarks on the Undecidability Results." In CW II, 305–306.

———. "Über eine bisher noch nicht benützte Erweiterung des finiten Standpunktes." *Dialectica* 12 (1958), 280–287. Reprinted with English translation in CW II.

———. "Über formal unentscheidbare Sätze der *Principia Mathematica* und verwandter Systeme." *Monatshefte für Mathematik und Physik* 38 (1931), 173–198. Reprinted with English translation in CW I.

———. "Undecidable Diophantine Propositions." Notes for an undelivered lecture, c. 1939–1940. In CW III, 164–175.

———. "Die Vollständigkeit der Axiome des logischen Funktionenkalküls." *Monatshefte für Mathematik und Physik* 37 (1930), 349–360. Reprinted with translation in CW I.

———. "Vortrag bei Zilsel." Lecture to an informal circle, Vienna, 1938. Transcription of shorthand notes and English translation in CW III, 86–113.

———. "What Is Cantor's Continuum Problem?" *American Mathematical Monthly* 54 (1947), 515–525. Revised and expanded version in Benacerraf and Putnam, pp. 258–273 of 1st ed. Both versions reprinted in CW II.

———. "Zur intuitionistischen Arithmetik und Zahlentheorie." *Ergebnisse eines mathematischen Kolloquiums* 4 (1933), 34–38. Reprinted with translation in CW I.

Goldfarb, Warren. "On Gödel's Way In: The Influence of Rudolf Carnap." *The Bulletin of Symbolic Logic* 11 (2005), 185–193.

Goldfarb, Warren, and Thomas Ricketts. "Carnap and the Philosophy of Mathematics." In David Bell and Wilhelm Vossenkühl (eds.), *Science and Subjectivity*, pp. 61–78. Berlin: Akademie-Verlag, 1992.

Gonseth, Ferdinand (ed.). *Les entretiens de Zurich sur les fondements et la méthode des sciences mathématiques, 6–9 décembre 1938*. Zurich: Leeman, 1941.

Goodman, Nelson. "Nominalisms." In Hahn and Schilpp, pp. 159–161.

———. *Problems and Projects*. Indianapolis, Ind.: Bobbs-Merrill, 1972.

———. *The Structure of Appearance*. Cambridge, Mass.: Harvard University Press, 1951.

———. "A World of Individuals." In I. M. Bochenski, Alonzo Church, and Nelson Goodman, *The Problem of Universals*, pp. 13–31. Notre Dame, Ind.: University of Notre Dame Press, 1956. Reprinted in *Problems and Projects*.

Goodman, Nelson, and W. V. Quine. "Steps toward a Constructive Nominalism." *The Journal of Symbolic Logic* 12 (1947), 105–122. Reprinted in Goodman, *Problems and Projects*. Cited according to original.

Hahn, Lewis Edwin, and Paul Arthur Schilpp (eds.). *The Philosophy of W. V. Quine*. The Library of Living Philosophers 18. La Salle, Ill.: Open Court, 1986. 2nd ed., 1998.

Hallett, Michael. *Cantorian Set Theory and Limitation of Size*. Oxford: Clarendon Press, 1984.

Harman, Gilbert. "Immanent and Transcendent Approaches to the Theory of Meaning." In Robert B. Barrett and Roger Gibson (eds.), *Perspectives on Quine*. Oxford: Blackwell, 1990.

———. Review of Quine, *The Roots of Reference*. *The Journal of Philosophy* 72 (1975), 388–396.

Hauser, Kai. "Gödel's Program Revisited I: The Turn to Phenomenology." *The Bulletin of Symbolic Logic* 12 (2006), 529–590.

Heinzmann, Gerhard. "Paul Bernays et la philosophie ouverte." In James Gasser and Henri Volken (eds.), *Logic and Set Theory in 20th Century Switzerland*. Bern, 2001. Online at www.philosophie.ch.

Heis, Jeremy. "'Critical Philosophy Begins at the Very Point Where Logistic Leaves Off': Cassirer's Response to Frege and Russell." *Perspectives on Science* 18 (2010), 383–408.

Hellman, Geoffrey. *Mathematics without Numbers: Toward a Modal-Structural Interpretation*. Oxford: Clarendon Press, 1989.

See also Feferman and Hellman.

Herbrand, Jacques. *Écrits logiques*. Edited by Jean van Heijenoort. Paris: Presses Universitaires de France, 1968. Translation by Warren Goldfarb. Dordrecht: Reidel, 1971.

———. *Recherches sur la théorie de la démonstration*. (Dissertation, University of Paris, 1930.) Prace Towartzystwa Naukowego Warszawskiego, Wydzial III, no. 33. Reprinted in *Écrits logiques*.

———. "Sur la non-contradiction de l'arithmétique." *Journal für die reine und angewandte Mathematik* 166 (1931), 1–8. Reprinted in *Écrits logiques*.

Hilbert, David. "Axiomatisches Denken." *Mathematische Annalen* 78 (1918), 405–415.

————. *Grundlagen der Geometrie.* Leipzig: Teubner, 1899. Many subsequent editions. Translation of the first edition (with some additional material) by E. J. Townsend. La Salle, Ill.: Open Court, 1902. Translation of the 10th ed. by Leo Unger, 1971.

————. "Neubegründung der Mathematik. Erste Mitteilung." *Abhandlungen aus dem mathematischen Seminar der Universität Hamburg* 1, 157–177. Translation in Mancosu, *From Brouwer to Hilbert.*

————. "Über das Unendliche." *Mathematische Annalen* 95 (1926), 161–190. Translation in van Heijenoort.

Hilbert, David, and Paul Bernays. *Grundlagen der Mathematik.* Berlin: Springer. Vol. 1, 1934, vol. 2, 1939. 2nd ed., vol. 1, 1968, vol. 2, 1970.

Howard, W. A. "The Formulae-as-Types Notion of Construction." In J. R. Hindley and J. P. Seldin (eds.), *To H. B. Curry: Essays on Combinatory Logic, Lambda Calculus, and Formalism,* pp. 479–490. London: Academic Press, 1980.

Husserl, Edmund. "Vom Ursprung der Geometrie." Beilage III of *Die Krisis der europaischen Wissenschaften und die transzendentale Phänomenologie.* Husserliana, vol. 6. The Hague: Martinus Nijhoff, 1954.

Hylton, Peter. "Quine on Reference and Ontology." In Roger F. Gibson Jr. (ed.), *The Cambridge Companion to Quine,* pp. 115–150. Cambridge: Cambridge University Press, 2004.

Isaacson, Daniel. "Mathematical Intuition and Objectivity." In Alexander George (ed.), *Mathematics and Mind.* New York: Oxford University Press, 1994.

Kennedy, Juliette, and Mark van Atten. "Gödel's Modernism: On Set-Theoretic Incompleteness." *Graduate Faculty Philosophy Journal* 25 (2004), 289–349. *See also* van Atten and Kennedy.

Kleene, Stephen C. *Introduction to Metamathematics.* New York: Van Nostrand, 1952. *See also* Gödel.

Kleene, Stephen C., and J. Barkley Rosser. "The Inconsistency of Certain Formal Logics." *Annals of Mathematics* (2) 36 (1935), 630–636.

Koellner, Peter. "On Reflection Principles." *Annals of Pure and Applied Logic* 157 (2009), 206–219.

————. "On the Question of Absolute Undecidability." *Philosophia Mathematica* (III) 14 (2006), 153–188. Reprinted with revisions and Postscript in Feferman, Parsons, and Simpson, pp. 189–215. Cited according to reprint.

————. "Strong Logics of First and Second Order." *The Bulletin of Symbolic Logic* 16 (2010), 1–36.

Köhler, Eckehart. "Gödel und der Wiener Kreis." In Paul Kruntorad (ed.), *Jour fixe der Vernunft,* pp. 127–158. Vienna: Hölder-Pichler-Tempsky, 1991.

Kreisel, G. "The Axiom of Choice and the Class of Hyperarithmetic Functions." *Indagationes Mathematicae* 24 (1962), 307–319.

————. "Mathematical Logic." In T. L. Saaty (ed.), *Lectures on Modern Mathematics,* vol. 3, pp. 95–195. New York: Wiley, 1965.

————. "On the Interpretation of Non-Finitist Proofs—Part I." *The Journal of Symbolic Logic* 16 (1951), 241–267.

————. "Ordinal Logics and the Characterization of Informal Concepts of Proof." In *Proceedings of the International Congress of Mathematicians,*

Edinburgh 1958, pp. 289–299. Cambridge: Cambridge University Press, 1960.

———. "La prédicativité." *Bulletin de la Societé mathématique de France* 88 (1960), 371–391.

Kuiper, John. "Ideas and Explorations: Brouwer's Road to Intuitionism." Ph.D. dissertation, University of Utrecht, 2004.

Langford, C. H. "On Inductive Relations." *Bulletin of the American Mathematical Society* 33 (1927), 599–607.

Leibniz, Gottfried Wilhelm. "Meditations on Knowledge, Truth, and Ideas." In Leibniz, *Philosophical Papers* (Roger Ariew and Daniel Garber [trans. and eds.]), pp. 23–27. Indianapolis, Ind.: Hackett, 1989. Latin original in C. J. Gerhardt (ed.), *Die philosophischen Schriften von G. W. Leibniz,* IV, 422–426. Berlin, 1875–1890.

Link, Montgomery. *See* Parsons and Link.

Lorenzen, Paul. *Einführung in die operative Logik und Mathematik.* Berlin: Springer, 1955. 2nd ed., 1969.

———. "Maß und Inegral in der konstruktiven Analysis." *Mathematische Zeitschrift* 54 (1951), 275–290.

———. "Die Widerspruchsfreiheit der klassischen Analysis." *Mathematische Zeitschrift* 54 (1951), 1–24.

Maddy, Penelope. "Believing the Axioms." *The Journal of Symbolic Logic* 53 (1988), 481–511, 736–764.

———. "The Roots of Contemporary Platonism." *The Journal of Symbolic Logic* 54 (1989), 1121–1144.

Mancosu, Paolo (ed.). *From Brouwer to Hilbert: The Debate on Foundations of Mathematics in the 1920s.* New York: Oxford University Press, 1998.

———. "Harvard 1940–41: Tarski, Carnap, and Quine on a Finitistic Language of Mathematics and Science." *History and Philosophy of Logic* 26 (2005), 327–357.

———. "Quine and Tarski on Nominalism." In Zimmerman, *Oxford Studies in Metaphysics,* vol. 4, pp. 22–55.

Martin, Donald A. "Gödel's Conceptual Realism." *The Bulletin of Symbolic Logic* 11 (2005), 207–224. Reprinted in Feferman, Parsons, and Simpson.

Mates, Benson. *The Philosophy of Leibniz.* New York: Oxford University Press, 1986.

McCarty, David Charles. Review of Sieg, Sommer, and Talcott. *The Bulletin of Symbolic Logic* 11 (2005), 239–241.

Menzler-Trott, Eckart. *Logic's Lost Genius: The Life of Gerhard Gentzen.* London Mathematical Society and American Mathematical Society, 2007.

Mirimanoff, Dimitri. "Les antinomies de Russell et de Burali-Forti et le problème fondamental de la théorie des ensembles." *L'enseignement mathématique* 19 (1917), 37–52.

Moore, Gregory H. *Zermelo's Axiom of Choice: Its Origin, Development, and Influence.* New York: Springer-Verlag, 1982.

See also Gödel.

Müller, G. H. (ed.). *Sets and Classes: On the Work by Paul Bernays.* Amsterdam: North-Holland, 1976.

Myhill, John. "The Undefinability of the Set of Natural Numbers in the Ramified *Principia*." In Geoge Nakhnikian (ed.), *Bertrand Russell's Philosophy*, pp. 19–27. London: Duckworth, 1974.

Myrvold, Wayne C. "Computability in Quantum Mechanics." In Werner DePauli-Schimanovich et al. (eds.), *The Foundational Debate*, pp. 33–46. Dordrecht: Kluwer, 1995.

Natorp, Paul. *Die logischen Grundlagen der exakten Wissenschaften*. Leipzig: Teubner, 1910.

Neale, Stephen. "The Philosophical Significance of Gödel's Slingshot." *Mind* 104 (1995), 761–825.

Nelson, Leonard. "Kritische Philosophie und mathematische Axiomatik." *Unterrichtsblätter für Mathematik und Naturwissenschaften* 34, 108–115, 136–142. With preface by David Hilbert and discussion remarks by Wilhelm Ackermann and Paul Bernays. Reprinted with further preface and notes by Wilhelm Ackermann in *Beiträge zur Philosophie der Logik und Mathematik*. Frankfurt a. M.: Verlag Öffentliches Leben, 1959. Also reprinted in *Gesammelte Schriften* (Paul Bernays et al., eds.), vol. 3. Hamburg: Meiner, 1974.

Parsons, Charles. "Analyticity for Realists." In Juliette Kennedy (ed.), *Interpreting Gödel*. Cambridge: Cambridge University Press, forthcoming.

———. "Brouwer, Luitzen Egbertus Jan." In Edwards, *Encyclopedia*, vol. 1, pp. 399–401. In 2nd ed. with supplement and bibliographical updating by Mark van Atten, vol. 1, pp. 700–703.

———. "A Distinction in Platonism." In *Under Tall Trees: A Tribute to Dirk van de Kaa*, pp. 9–10. Wassenaar: Netherlands Institute for Advanced Study, 1995.

———. "Finitism and Intuitive Knowledge." In Matthias Schirn (ed.), *The Philosophy of Mathematics Today*, pp. 249–270. Oxford: Clarendon Press, 1998.

———. *From Kant to Husserl: Selected Essays*. Cambridge, Mass.: Harvard University Press, 2012.

———. "Gödel and Philosophical Idealism." *Philosophia Mathematica* (III) 18 (2010), 166–192.

———. "Intuition in Constructive Mathematics." In Jeremy Butterfield (ed.), *Language, Mind, and Logic*, pp. 211–229. Cambridge: Cambridge University Press, 1986.

———. *Mathematical Thought and Its Objects*. New York: Cambridge University Press, 2008. Cited as MTO.

———. *Mathematics in Philosophy: Selected Essays*. Ithaca, N.Y.: Cornell University Press, 1983. Paperback ed. with some corrections, 2005.

———. "Objects and Logic." *The Monist* 65 (1982), 491–516. Reprinted in expanded form as chapter 1, §§1–6, of MTO.

———. "The Problem of Absolute Universality." In Agustín Rayo and Gabriel Uzquiano (eds.), *Absolute Generality*, pp. 203–219. Oxford: Clarendon Press, 2006.

———. "Putnam on 'Realism' and 'Empiricism' in Mathematics." In Randall E. Auxier and Lewis Edwin Hahn (eds.), *The Philosophy of Hilary Putnam*. The Library of Living Philosophers 34. Chicago: Open Court, forthcoming.

———. "Quine on the Philosophy of Mathematics." Essay 7 of *Mathematics in Philosophy*. Also in Hahn and Schilpp, pp. 369–395.

————. Review Essay: John P. Burgess, *Fixing Frege*. *The Journal of Philosophy* 106 (2009), 404–417.

————. Review of Leonard Nelson, *Beiträge zur Philosophie der Logik und Mathematik*. *The Journal of Philosophy* 59 (1962), 242–246.

————. "Sets and Classes." *Noûs* 8 (1974), 1–12. Reprinted as Essay 8 of *Mathematics in Philosophy*.

————. "Some Consequences of the Entanglement of Logic and Mathematics." In Michael Frauchiger (ed.), *Reference, Rationality, and Phenomenology: Themes from Føllesdal*, pp. 153–178. Lauener Library of Aenalytical Philosophy 2. Frankfurt: Ontos Verlag, 2013.

————. "Some Remarks on Frege's Conception of Extension." In Matthias Schirn (ed.), *Studien zu Frege I: Logik und Philosophie der Mathematik*, pp. 265–277. Stuttgart: Frommann-Holzboog, 1976. Reprinted with Postscript as Essay 5 of *From Kant to Husserl*.

————. "Structuralism and the Concept of Set." In Walter Sinnott-Armstrong (ed. in collaboration with others), *Modality, Morality, and Belief: Essays in Honor of Ruth Barcan Marcus*, pp. 74–92. Cambridge: Cambridge University Press, 1995.

————. "The Structuralist View of Mathematical Objects." *Synthese* 84 (1990), 303–346.

————. "What Is the Iterative Conception of Set?" In Butts and Hintikka, pp. 365–377. Reprinted as Essay 10 of *Mathematics in Philosophy*.

See also Feferman, Parsons, and Simpson.

See also Gödel.

See also Wang.

Parsons, Charles, and Montgomery Link (eds.). *Hao Wang, Logician and Philosopher*. Texts in Philosophy 16. London: College Publications, 2011.

Pears, David, ed. *Bertand Russell: A Collection of Critical Essays*. Garden City, N.Y.: Doubleday Anchor Books, 1972.

Peckhaus, Volker. *Hilbertprogramm und kritische Philosophie*. Göttingen: Vandenhoeck & Rupprecht, 1990.

Péter, Rósza. *Rekursive Funktionen*. Budapest: Akadémiai Kiadó, 1951. 2nd ed., 1957. Translation (described as 3rd ed.), *Recursive Functions*. New York: Academic Press, 1967.

Potter, Michael. *Reason's Nearest Kin: Philosophies of Arithmetic from Kant to Carnap*. Oxford University Press, 2000.

Putnam, Hilary. *Ethics without Ontology*. Cambridge, Mass.: Harvard University Press, 2004.

————. *The Many Faces of Realism*. La Salle, Ill.: Open Court, 1987.

————. *Mathematics, Matter, and Method: Philosophical Papers*, vol. 2. Cambridge: Cambridge University Press, 1975. 2nd ed., 1979.

————. "Mathematics without Foundations." *The Journal of Philosophy* 64 (1967), 5–22. Reprinted in *Mathematics, Matter, and Method*, pp. 43–59.

————. *Philosophy in an Age of Science: Physics, Mathematics, and Skepticism*. Edited by Mario de Caro and David Macarthur. Cambridge, Mass.: Harvard University Press, 2012.

————. *Philosophy of Logic*. New York: Harper and Row, 1971. Reprinted in *Mathematics, Matter, and Method*, 2nd ed., pp. 323–357.

———. "The Question of Realism." In *Words and Life*, pp. 295–312.

———. "Realism and Reason." In *Meaning and the Moral Sciences*. London: Routledge and Kegan Paul, 1978.

———. *Realism and Reason: Philosophical Papers*, vol. 3. Cambridge: Cambridge University Press, 1983.

———. *Realism with a Human Face*. Edited by James Conant. Cambridge, Mass.: Harvard University Press, 1990.

———. "Realism without Absolutes." *International Journal of Philosophical Studies* 1 (1993), 179–192. Reprinted in *Words and Life*.

———. *The Threefold Cord: Mind, Body, and World*. New York: Columbia University Press, 1999.

———. "Truth and Convention." *Dialectica* 41 (1987), 69–77. Reprinted in *Realism with a Human Face*.

———. "Was Wittgenstein *Really* an Anti-realist about Mathematics?" In Timothy McCarthy and Sean C. Stidd (eds.), *Wittgenstein in America*, pp. 140–194. Oxford: Clarendon Press, 2001. Reprinted in *Philosophy in an Age of Science*.

———. *Words and Life*. Edited by James Conant. Cambridge, Mass.: Harvard University Press, 1994.

Quine, W. V. "Autobiography of W. V. Quine." In Hahn and Schilpp, pp. 3–46.

———. "Carnap and Logical Truth." *Synthese* 12 (1960), 350–374. Reprinted in Schilpp, *The Philosophy of Rudolf Carnap*, and in *The Ways of Paradox*.

———. *Confessions of a Confirmed Extensionalist and Other Essays*. Edited by Douglas B. Quine and Dagfinn Føllesdal. Cambridge, Mass.: Harvard University Press, 2008.

———. "Designation and Existence." *The Journal of Philosophy* 36 (1939), 701–709.

———. *Elementary Logic*. Boston: Ginn and Company, 1941.

———. *From a Logical Point of View: 9 Logico-Philosophical Essays*. Cambridge, Mass.: Harvard University Press, 1953. 2nd ed., 1961. Cited as FLPV.

———. "Homage to Rudolf Carnap." Boston Sudies in the Philosophy of Science 8 (1971), xxii–xxv. Reprinted in *The Ways of Paradox*, 2nd ed., pp. 40–44.

———. "Logic and the Reification of Universals." In FLPV.

———. "A Logistical Approach to the Ontological Problem." In *The Ways of Paradox*, 1st ed., pp. 64–69. Also in 2nd ed., pp. 197–202.

———. *Methods of Logic*. New York: Holt, 1950. 2nd ed., 1959. 3rd ed., New York: Holt, Rinehart, and Winston, 1972. 4th ed., Cambridge, Mass.: Harvard University Press, 1982.

———. "Nominalism." Lecture to the Harvard Philosophy Colloquium, March 11, 1946. In Zimmerman, *Oxford Studies in Metaphysics*, vol. 4, pp. 3–21. Reprinted in *Confessions of a Confirmed Extensionalist*, pp. 7–23.

———. "Notes on Existence and Necessity." *The Journal of Philosophy* 40 (1943), 113–127.

———. "On Carnap's Views on Ontology." *Philosophical Studies* 2 (1951), 65–72. Reprinted in *The Ways of Paradox*.

———. "On Frege's Way Out." *Mind* 64 (1955), 145–169. Reprinted in SLP.

———. "On the Logic of Quantification." *The Journal of Symbolic Logic* 10 (1945), 1–12. Reprinted in SLP, pp. 181–195.

————. "On Universals." *The Journal of Symbolic Logic* 12 (1947), 74–84.

————. "On What There Is." *Review of Metaphysics* 2 (1948), 21–38. Reprinted in FLPV.

————. *Ontological Relativity and Other Essays*. New York: Columbia University Press, 1969.

————. "Ontological Remarks on the Propositional Calculus." *Mind* 43 (1934), 472–476. Reprinted in *The Ways of Paradox*.

————. *Philosophy of Logic*. Englewood Cliffs, N.J.: Prentice-Hall, 1970. 2nd ed., Cambridge, Mass.: Harvard University Press, 1986.

————. "The Problem of Interpreting Modal Logic." *The Journal of Symbolic Logic* 12 (1947), 43–48.

————. "Reactions." In Paolo Leonardo and Marco Santambrogio (eds.), *On Quine: New Essays,* pp. 347–361. Cambridge: Cambridge University Press, 1995.

————. "Reply to Charles Parsons." In Hahn and Schilpp, pp. 397–403.

————. "Reply to Joseph S. Ullian." In Hahn and Schilpp, pp. 590–593.

————. "Reply to Nelson Goodman." In Hahn and Schilpp, pp. 162–163.

————. Review of Parsons, *Mathematics in Philosophy. The Journal of Philosophy* 81 (1984), 783–794. Reprinted in *Quine in Dialogue,* pp. 194–205. Edited by Dagfinn Føllesdal and Douglas B. Quine. Cambridge, Mass.: Harvard University Press, 2008.

————. *The Roots of Reference*. La Salle, Ill.: Open Court, 1974.

————. *Selected Logic Papers*. New York: Random House, 1966. Enlarged edition, Cambridge, Mass.: Harvard University Press, 1995. Cited as SLP. Note that all papers first published before 1966 have the same pagination in both editions.

————. "Semantics and Abstract Objects." *Proceedings of the American Academy of Arts and Sciences* 80 (1951), 90–96.

————. *O sentido da nova lógica*. São Paulo: Martins, 1944.

————. *Set Theory and Its Logic*. Cambridge, Mass.: Harvard University Press, 1963. Revised ed., 1969.

————. *Theories and Things*. Cambridge, Mass.: The Belknap Press of Harvard University Press, 1981.

————. "Three Grades of Modal Involvement." In *Actes du XIème Congrès international de philosophie, Bruxelles, 20–26 août 1953*, XIV, 65–81. Amsterdam: North-Holland, 1953. Reprinted in *The Ways of Paradox,* pp. 158–164.

————. *The Time of My Life: An Autobiography*. Cambridge, Mass.: MIT Press, 1985.

————. "Truth by Convention." In O. H. Lee (ed.), *Philosophical Essays for A. N. Whitehead,* pp. 90–124. New York: Longmans, 1936. Reprinted in *The Ways of Paradox*.

————. "Two Dogmas of Empiricism." *Philosophical Review* 60 (1951), 20–43. Reprinted in FLPV.

————. "The Variable." In Rohit Parikh (ed.), *Logic Colloquium: Symposium on Logic Held in Boston, 1972–73*. Lecture Notes in Mathematics 453. Springer-Verlag, 1975. Reprinted in *The Ways of Paradox*, 2nd ed., pp. 272–282.

————. *The Ways of Paradox and Other Essays*. New York: Random House, 1966. 2nd ed., enlarged, Cambridge, Mass.: Harvard University Press, 1976.

———. "Whitehead and the Rise of Modern Logic." In Paul Arthur Schilpp (ed.), *The Philosophy of Alfred North Whitehead,* pp. 125–163. Evanston, Ill.: Northwestern University, 1941. Reprinted in *Selected Logic Papers,* pp. 3–36.

———. "Whither Physical Objects?" In Robert S. Cohen, Paul K. Feyerabend, and Marx W. Wartofsky (eds.), *Essays in Memory of Imre Lakatos,* pp. 497–504. Boston Studies in the Philosophy of Science 39. Dordrecht: Reidel, 1976.

———. *Word and Object.* Cambridge, Mass.: Technology Press (later MIT Press), 1960.

Quine, W. V., and Rudolf Carnap. *Dear Carnap, Dear Van: The Quine-Carnap Correspondence and Related Work.* Edited by Richard Creath. Berkeley: University of California Press, 1990.

See also Goodman and Quine.

Ramsey, F. P. "The Foundations of Mathematics." *Proceedings of the London Mathematical Society* 25, no. 2 (1926), 338–384. Cited according to reprint in *The Foundations of Mathematics and Other Logical Essays.*

———. *The Foundations of Mathematics and Other Logical Essays.* Edited by R. B. Braithwaite, with a preface by G. E. Moore. London: Routledge and Kegan Paul, 1931.

———. "Mathematical Logic." *The Mathematical Gazette* 13, no. 184, 185–193. Cited according to reprint in *The Foundations of Mathematics.*

Rang, Bernhard, and Wolfgang Thomas. "Zermelo's Discovery of the Russell Paradox." *Historia Mathematica* 8 (1981), 15–22.

Rawls, John. "The Independence of Moral Theory." *Proceedings and Addresses of the American Philosophical Association* 49 (1974–75), 1–22.

———. *Political Liberalism.* New York: Columbia University Press, 1993.

———. *A Theory of Justice.* Cambridge, Mass.: The Belknap Press of Harvard University Press, 1971. 2nd ed., 1999. Cited according to 1st ed.

Reck, Erich H. "Carnap and Modern Logic." In Michael Friedman and Richard Creath (eds.), *The Cambridge Companion to Carnap,* pp. 176–199. Cambridge: Cambridge University Press, 2007.

Reid, Constance. *Hilbert.* Berlin: Springer, 1970.

Resnik, Michael. "Second-Order Logic Still Wild." *The Journal of Philosophy* 85 (1988), 75–87.

Ricketts, Thomas. "Carnap's Principle of Tolerance, Empiricism, and Conventionalism." In Peter Clark and Bob Hale (eds.), *Reading Putnam,* pp. 176–200. Oxford: Blackwell, 1994.

See also Goldfarb and Ricketts.

Rosser, J. Barkley. *See* Kleene and Rosser.

Rouilhan, Philippe de. *Russell et le cercle des paradoxes.* Paris: Presses Universitaires de France, 1996.

———. "Russell's Logics." In Cori et al., pp. 335–349.

Russell, Bertrand. *The Autobiography of Bertrand Russell, 1914–1944.* London: Allen and Unwin, 1968.

———. *An Essay on the Foundations of Geometry.* Cambridge: Cambridge University Press, 1897.

———. *Introduction to Mathematical Philosophy.* London: Allen and Unwin, 1919. 2nd ed., 1920. Many reprints.

———. "Mathematical Logic as Based on the Theory of Types." *American Journal of Mathematics* 30 (1908), 222–262. Reprinted in van Heijenoort.

———. "On Some Difficulties in the Theory of Transfinite Numbers and Order Types." *Proceedings of the London Mathematical Society* (2) 4 (1906), 29–46. Reprinted in Russell, *Essays in Analysis*. Edited by Douglas Lackey. New York: Braziller, 1973.

———. *Principles of Mathematics*. Cambridge: Cambridge University Press, 1903. 2nd ed., London: Allen and Unwin, 1937.

Scanlon, T. M. "The Consistency of Number Theory via Herbrand's Theorem." *The Journal of Symbolic Logic* 38 (1973), 29–58.

Schilpp, Paul Arthur (ed.). *Albert Einstein: Philosopher-Scientist*. The Library of Living Philosophers 6. Evanston, Ill.: The Library of Living Philosophers, 1949.

———. *The Philosophy of Bertrand Russell*. The Library of Living Philosophers 5. Evanston, Ill.: Northwestern University, 1944.

———. *The Philosophy of Rudolf Carnap*. The Library of Living Philosophers 10. La Salle, Ill.: Open Court, 1963.

See also Hahn and Schilpp.

Schütte, Kurt. "Eine Grenze für die Beweisbarkeit der transfiniten Induktion in der verzweigten Typenlogik." *Archiv für mathematische Logik und Grundlagenforschung* 7 (1965), 45–60.

———. "Predicative Well-Orderings." In J. N. Crossley and Michael Dummett (eds.), *Formal Systems and Recursive Functions*, pp. 280–303. Amsterdam: North-Holland, 1965.

———. *Proof Theory*. Berlin: Springer-Verlag, 1977.

Sher, Gila, and Richard Tieszen (eds.). *Between Logic and Intuition: Essays in Honor of Charles Parsons*. Cambridge: Cambridge University Press, 2000.

Shoenfield, Joseph R. "Axioms of Set Theory." In Barwise, *Handbook*, pp. 321–344.

Sieg, Wilfried. "Beyond Hilbert's Reach?" In David B. Malament (ed.), *Reading Natural Philosophy: Essays in the History and Philosophy of Science and Mathematics*, pp. 363–405. Chicago: Open Court, 2002.

Sieg, Wilfried, Richard Sommer, and Carolyn Talcott (eds.). *Reflections on the Foundations of Mathematics: Essays in Honor of Solomon Feferman*. Lecture Notes in Logic 15. Association for Symbolic Logic and Natick, Mass.: A. K. Peters, 2002.

See also Bernays.

See also Ewald and Sieg.

See also Gödel.

Simpson, Stephen G. "The Gödel Hierarchy and Reverse Mathematics." In Feferman, Parsons, and Simpson, pp. 109–127.

———. *Subsystems of Second Order Arithmetic*. Springer-Verlag, 1999. 2nd ed., Association for Symbolic Logic and Cambridge University Press, 2009.

See also Feferman, Parsons, and Simpson.

Smullyan, Arthur F. "Modality and Description." *The Journal of Symbolic Logic* 13 (1948), 31–37.

Spector, Clifford. "Recursive Ordinals and Predicative Set Theory." In *Summaries of Talks Presented at the Summer Institute for Symbolic Logic, Cornell University, 1957*, pp. 377–382.

Strawson, P. F. "Reference and Its Roots." In Hahn and Schilpp, pp. 518–532.

Stroud, Barry. *The Significance of Philosophical Scepticism*. Oxford: Clarendon Press, 1984.

Tait, William (W. W.). "Against Intuitionism: Constructive Mathematics Is Part of Classical Mathematics." *Journal of Philosophical Logic* 12 (1983), 173–196.

———. "Constructive Reasoning." In B. van Rootselaar and J. F. Staal (eds.), *Logic, Methodology, and Philosophy of Science III*, pp. 185–198. Amsterdam: North-Holland, 1968.

———. "Finitism." *The Journal of Philosophy* 78 (1981), 524–546. Reprinted in *The Provenance*.

———. "Gödel on Intuition and on Hilbert's Finitism." In Feferman, Parsons, and Simpson, pp. 88–108.

———. "Gödel's Correspondence on Proof Theory and Constructive Mathematics." *Philosophia Mathematica* (III) 14 (2006), 76–111.

———. "Gödel's Interpretation of Intuitionism." *Philosophia Mathematica* (III) 14 (2006), 208–228.

———. "Gödel's Reformulation of Gentzen's First Consistency Proof for Arithmetic: The No-Counter-Example Interpretation." *The Bulletin of Symbolic Logic* 11 (2005), 225–238. Reprinted in Feferman, Parsons, and Simpson.

———. "Gödel's Unpublished Papers on Foundations of Mathematics." *Philosophia Mathematica* (III) 9, 87–126. Reprinted in *The Provenance*.

———. *The Provenance of Pure Reason: Essays on the Philosophy of Mathematics and Its History*. New York: Oxford University Press, 2005.

———. "Reflections on the Concept of *a priori* Truth and Its Corruption by Kant." In Michael Detlefsen (ed.), *Proof and Knowledge in Mathematics*, pp. 33–64. London: Routledge, 1992.

———. "Truth and Proof." *Synthese* 69 (1986), 341–370. Reprinted with some additional notes in *The Provenance*, pp. 61–88.

See also Bernays.

Takeuti, Gaisi. *Memoirs of a Proof Theorist: Gödel and Other Logicians*. Edited and translated by Mariko Yasugi and Nicholas Passell. World Scientific, 2003.

Tieszen, Richard. *After Gödel: Platonism and Rationalism in Mathematics and Logic*. Oxford University Press, 2011.

———. "Kurt Gödel and Phenomenology." *Philosophy of Science* 59 (1992), 176–194.

———. "Kurt Gödel's Path from the Incompleteness Theorems (1931) to Phenomenology (1961)." *The Bulletin of Symbolic Logic* 4 (1998), 181–203.

Toledo, Sue. *Tableau Systems for First Order Number Theory and Certain Higher Order Systems*. Lecture Notes in Mathematics 447. Berlin: Springer-Verlag, 1975.

Väänänen, Jouko. "Second-Order Logic and the Foundations of Mathematics." *The Bulletin of Symbolic Logic* 7 (2001), 504–520.

van Atten, Mark. *Brouwer Meets Husserl: On the Phenomenology of Choice Sequences*. Dordrecht: Springer, 2007.

———. *On Brouwer*. Belmont, Calif.: Thomson-Wadsworth, 2004.

See also Parsons, "Brouwer."

van Atten, Mark, and Juliette Kennedy. "On the Philosophical Development of Kurt Gödel." *The Bulletin of Symbolic Logic* 9 (2003), 425–476. Reprinted in Feferman, Parsons, and Simpson.

See also Kennedy and van Atten.

van Dalen, Dirk. "Hermann Weyl's Intuitionistic Mathematics." *The Bulletin of Symbolic Logic* 1 (1995), 145–169.

———— (ed.). *L. E. J. Brouwer en de grondslagen van de wiskunde.* Utrecht: Epsilon Uitgaven, 2001. 2nd ed., 2005. Contains a reprint of Brouwer, *Over de grondslagen der wiskunde,* with additional material.

————. *Mystic, Geometer, and Intuitionist: The Life of L. E. J. Brouwer,* vol. 1: *The Dawning Revolution.* Oxford University Press, 1999.

See also Brouwer.

van Heijenoort, Jean (ed.). *From Frege to Gödel: A Source Book in Mathematical Logic, 1879–1931.* Cambridge, Mass.: Harvard University Press, 1967.

von Neumann, J. "Über eine Widerspruchsfreiheitsfrage in der axiomatischen Mengenlehre." *Journal für die reine und angewandte Mathematik* 160 (1929), 494–508.

Wang, Hao. "The Axiomatization of Arithmetic." *The Journal of Symbolic Logic* 22 (1957), 145–158. Reprinted in *A Survey.*

————. *Beyond Analytic Philosophy: Doing Justice to What We Know.* Cambridge, Mass.: MIT Press, 1985.

————. *Computation, Logic, Philosophy: A Collection of Essays.* Beijing: Science Press, 1990.

————. "Eighty Years of Foundational Studies." *Dialectica* 12 (1958), 466–497. Reprinted in *A Survey.*

————. "The Formalization of Mathematics." *The Journal of Symbolic Logic* 19 (1954), 241–266. Reprinted in *A Survey.*

————. *From Mathematics to Philosophy.* London: Routledge and Kegan Paul, 1974. Cited as FMP.

————. "Large Sets." In Butts and Hintikka, pp. 309–333.

————. *A Logical Journey: From Gödel to Philosophy.* Cambridge, Mass.: MIT Press, 1996.

————. "On Physicalism and Algorithmism: Can Machines Think?" *Philosophia Mathematica* (III) 1 (1993), 97–138.

————. "Ordinal Numbers and Predicative Set Theory." *Zeitschrift für mathematische Logik und Grundlagen der Mathematik* 5 (1959), 624–651. Reprinted in *A Survey.*

————. "Process and Existence in Mathematics." In Yehoshua Bar-Hillel, E. I. J. Ponznanski, M. O. Rabin, and Abraham Robinson (eds.), *Essays on the Foundations of Mathematics, Dedicated to Professor A. A. Fraenkel on His Seventieth Anniversary,* pp. 328–351. Jerusalem: Magnes Press, The Hebrew University of Jerusalem, 1961. Partly incorporated into chapter 7 of FMP. Reprinted in *Computation, Logic, Philosophy.*

————. "Quine's Logical Ideas in Historical Perspective." In Hahn and Schilpp, pp. 623–643.

————. *Reflections on Kurt Gödel.* Cambridge, Mass.: MIT Press, 1987.

————. "Sets and Concepts, on the Basis of Discussions with Gödel." Edited with introduction and notes by Charles Parsons. In Parsons and Link, *Hao Wang,* pp. 77–118.

———. "Some Facts about Kurt Gödel." *The Journal of Symbolic Logic* 46 (1981), 653–659. Reprinted with some revisions in *Reflections,* chapter 2.

———. *A Survey of Mathematical Logic.* Beijing: Science Press, 1962; Amsterdam: North-Holland, 1963.

———. "To and from Philosophy: Discussions with Gödel and Wittgenstein." *Synthese* 88 (1991), 229–277.

———. "Two Commandments of Analytic Empiricism." *The Journal of Philosophy* 82 (1985), 449–462.

———. "What Is an Individual?" *Philosophical Review* 62 (1953), 413–420. Reprinted in FMP, appendix.

Weyl, Hermann. "Der *circulus vitiosus* in der heutigen Begründung der Analysis." *Jahresbericht der deutschen Mathematiker-Vereinigung* 28 (1919), 85–92.

———. *Das Kontinuum. Kritische Untersuchungen über die Grundlagen der Analysis.* Leipzig: Veit, 1918. Translated by Stephen Pollard and Thomas Bole as *The Continuum: A Critical Examination of the Foundations of Analysis.* Kirksville, Mo.: Thomas Jefferson University Press, 1987.

———. Review of Schilpp, *The Philosophy of Bertrand Russell. American Mathematical Monthly* 53 (1946), 208–214.

———. Über die neue Grundlagenkrise der Mathematik. *Mathematische Zeitschrift* 10 (1921), 39–79. Translation in Mancosu, *From Brouwer to Hilbert.*

Wittgenstein, Ludwig. *Remarks on the Foundations of Mathematics.* Edited by G. E. M. Anscombe, Rush Rhees, and G. H. von Wright. With a translation by Anscombe. Oxford: Blackwell, 1956.

Yourgrau, Palle. Review of Wang, *Reflections on Kurt Gödel. Philosophy and Phenomenological Research* 50 (1989), 391–408.

Zach, Richard. "The Practice of Finitism: Epsilon Calculus and Consistency Proofs in Hilbert's Program." *Synthese* 137 (2003), 211–259.

Zermelo, Ernst. *Collected Works/Gesammelte Werke.* Edited by Heinz-Dieter Ebbinghaus, Craig Fraser, and Akihiro Kanamori. Vol. I, *Set Theory, Miscellanea/ Mengenlehre, Varia.* Edited by Ebbinghaus and Kanamori. Berlin: Springer-Verlag, 2010.

———. "Über Grenzzahlen und Mengenbereiche." *Fundamenta Mathematicae* 16 (1930), 29–47. Reprinted with a translation in *Collected Works,* vol. I.

———. "Untersuchungen über die Grundlagen der Mengenlehre I." *Mathematische Annalen* 65 (1908), 261–281. Translation in van Heijenoort, *From Frege to Gödel.* Original and translation reprinted in *Collected Works,* vol. I.

See also Cantor.

Zimmerman, Dean W. (ed.). *Oxford Studies in Metaphysics,* vol. 4: *Nominalism.* Oxford University Press, 2008.

COPYRIGHT ACKNOWLEDGMENTS

Essay 2 appeared in Wilfried Sieg, Richard Sommer, and Carolyn Talcott (eds.), *Reflections on the Foundations of Mathematics: Essays in Honor of Solomon Feferman*, pp. 372–389. Lecture Notes in Logic 15. Association for Symbolic Logic and Natick, Mass.: A. K. Peters, 2002. Copyright © 2002 by Association for Symbolic Logic. Reprinted by permission of the Association for Symbolic Logic and the editors.

Essay 3 appeared in Costas Dimitracopoulos, Ludomir Newelski, Dag Normann, and John R. Steel (eds.), *Logic Colloquium 2005: Proceedings of the Annual European Summer Meeting of the Association for Symbolic Logic, Held in Athens, Greece, July 28–August 3, 2005*, pp. 129–150. Lecture Notes in Logic 28. Association for Symbolic Logic and Cambridge University Press, 2008. Copyright © 2008 by Association for Symbolic Logic. Reprinted by permission of the Association for Symbolic Logic and the editors.

Essay 4 first appeared as "Gödel, Kurt Friedrich (1906–1978)," in John R. Shook (gen. ed.), *Dictionary of Modern American Philosophers*, vol. 2, pp. 940–946. Bristol: Thoemmes Press, 2005. Reprinted by permission of the General Editor and Bloomsbury Publishing, plc.

Essay 5 appeared as "Introductory Note to *1944*" in Kurt Gödel, *Collected Works, Volume II: Publications 1938–1974*, Solomon Feferman et al., eds., pp. 102–118. New York and Oxford: Oxford University Press, 1990. Used by permission of Oxford University Press, Inc., and the editors.

Essay 6 appeared in Paolo Leonardi and Marco Santambrogio (eds.), *On Quine: New Essays*, pp. 297–313. Copyright © Cambridge University Press, 1995. Reprinted with permission of Cambridge University Press and the editors.

Essay 7 appeared in *The Bulletin of Symbolic Logic*, vol. 1 (1995), pp. 44–74. Copyright © 1995 by Association for Symbolic Logic. Reprinted by permission of the Association for Symbolic Logic.

Essay 8 appeared in *American Philosophical Quarterly* 48 (2011): 213–228. Copyright © 2011 by the Board of Trustees of the University of Illinois. Reprinted by permission of University of Illinois Press.

Essay 9 appeared in Robert B. Barrett and Roger F. Gibson (eds.), *Perspectives on Quine*, pp. 273–290. Oxford: Blackwell, 1990. Reproduced by permission of Wiley-Blackwell, Ltd., and the editors.

Essay 10 first appeared in *Philosophia Mathematica*, series III, vol. 6 (1998): 3–24. Now published by Oxford University Press. Used by permission.

Essay 11 appears in Maria Baghramian (ed.), *Reading Putnam*, pp. 182–201. London: Routledge, 2012. Reprinted by permission of the editor and publisher. Copyright retained by Charles Parsons.

Essay 12 appeared in *Philosolphia Mathematica*, series III, vol. 17 (2009): 220–247. Copyright retained by Charles Parsons.